Springer Proceedings in Mathematics & Statistics

Volume 74

More information about this series at http://www.springer.com/series/10533

Springer Proceedings in Mathematics & Statistics

This book series features volumes composed of select contributions from workshops and conferences in all areas of current research in mathematics and statistics, including OR and optimization. In addition to an overall evaluation of the interest, scientific quality, and timeliness of each proposal at the hands of the publisher, individual contributions are all refereed to the high quality standards of leading journals in the field. Thus, this series provides the research community with well-edited, authoritative reports on developments in the most exciting areas of mathematical and statistical research today.

Michael G. Akritas • S.N. Lahiri
Dimitris N. Politis
Editors

Topics in Nonparametric Statistics

Proceedings of the First Conference of the
International Society for Nonparametric
Statistics

Springer

Editors
Michael G. Akritas
Department of Statistics
The Pennsylvania State University
University Park, PA, USA

S.N. Lahiri
Department of Statistics
North Carolina State University
Raleigh, NC, USA

Dimitris N. Politis
Department of Mathematics
University of California
San Diego, CA, USA

ISSN 2194-1009   ISSN 2194-1017 (electronic)
ISBN 978-1-4939-4581-8   ISBN 978-1-4939-0569-0 (eBook)
DOI 10.1007/978-1-4939-0569-0
Springer New York Heidelberg Dordrecht London

Preface

This volume presents a selection of papers developed from talks presented at the First Conference of the International Society of Nonparametric Statistics (ISNPS), held on June 15–19, 2012, at Chalkidiki, Greece. The papers cover a wide spectrum of topics of current interest and provide a glimpse of some of the most significant recent advances in theory, methodology, and applications of nonparametric statistics. Some of the topics covered in this volume include: Curve estimation and smoothing, Distribution of clusters of exceedances, Frequency domain and multiscale methods, Inference for extremes, Level set estimation, Model selection and variable selection in high dimensions, Multiple testing, Nonparametric methods for image analysis and telecommunication networks, Nonparametric signal filtration, Particle filters, Statistical learning, and Resampling methods (Bootstrap, Permutation tests, etc.). All papers in the volume are refereed.

ISNPS was formed in 2010 with the mission "to foster the research and practice of nonparametric statistics, and to promote the dissemination of new developments in the field via conferences, books and journal publications." The nature of ISNPS is uniquely global, and its international conferences are designed to facilitate the exchange of ideas and latest advances among researchers from all around the world in cooperation with established statistical societies such as the Institute of Mathematical Statistics (IMS) and the International Statistical Institute (ISI). ISNPS has a distinguished Advisory Committee that includes R. Beran, P. Bickel, R. Carroll, D. Cook, P. Hall, R. Johnson, B. Lindsay, E. Parzen, P. Robinson, M. Rosenblatt, G. Roussas, T. Subba Rao, and G. Wahba. The Charting Committee of ISNPS consists of over 50 prominent researchers from all over the world.

The First Conference of ISNPS included over 275 talks (keynote, special invited, invited and contributed) with presenters coming from all five continents. After the success of the First Conference, a second conference is currently being organized. The second ISNPS Conference is scheduled to take place in Cádiz, Spain, June 12–16, 2014, and is projected to include over 350 presentations. More information on ISNPS and the conferences can be found at http://www.isnpstat. org/.

We would like to thank Marc Strauss and Jon Gurstelle of Springer for their immediate support of this project. It has been a pleasure working with Hannah Bracken of Springer during the production of the volume. We would like to acknowledge her patience, support, and cheer at all stages of preparation of the manuscript. Finally, we are grateful to our referees who provided reports and feedback on the papers on a tight schedule for timely publication of the proceedings; this volume is much improved as a result of their efforts.

University Park, PA, USA   Michael G. Akritas
Raleigh, NC, USA   S.N. Lahiri
San Diego, CA, USA   Dimitris N. Politis

Editors' Biography

Michael G. Akritas is a Professor of Statistics at the Pennsylvania State University. His research interests include nonparametric modeling for factorial designs, rank-based inference, modeling and inference for high dimensional data, and censored data. He served as director of the National Statistical Consulting Center for Astronomy (1994–2001), co-editor of the *Journal of Nonparametric Statistics* (2007–2010), and guest editor and associate editor for a number of other journals. He is a fellow of ASA and IMS.

Soumendra N. Lahiri is a Distinguished Professor of Statistics at the North Carolina State University. His research interests include Nonparametric Statistics, Time Series, Spatial Statistics, and Statistical Inference for high dimensional data. He served as an editor of Sankhya, Series A (2007–2009) and currently he is on the editorial boards of the *Annals of Statistics* and the *Journal of Statistical Planning and Inference*. He is a fellow of the ASA and the IMS, and an elected member of the International Statistical Institute.

Dimitris N. Politis is a Professor of Mathematics and Adjunct Professor of Economics at the University of California, San Diego. His research interests include Time Series Analysis, Resampling and Subsampling, Nonparametric Function Estimation, and Modelfree Prediction. He has served as an editor of the *IMS Bulletin* (2010–2013), co-editor of the *Journal of Time Series Analysis* (2013–present), co-editor of the *Journal of Nonparametric Statistics* (2008–2011), and associate editor for several journals including *Bernoulli*, the *Journal of the American Statistical Association*, and the *Journal of the Royal Statistical Society, Series B*. He is also a founding member of the Editorial Board of the Springer book series Frontiers in Probability and the Statistical Sciences (2012–present). He is a fellow of the ASA and IMS, and former fellow of the Guggenheim Foundation.

Contents

Contributors

László Győrfi Department of Computer Science and Information Theory, Budapest University of Technology and Economics, Budapest, Hungary

Alessio Sancetta Department of Economics, Royal Holloway University of London, London, UK

Andreas Artemiou Department of Mathematical Sciences, Michigan Technological University, Houghton, MI, USA

D. Bagkavos Accenture, Athens, Greece

Patrice Bertail MODAL'X, University Paris-Ouest, Nanterre, France

CREST-LS, Malakoff, France

Stefano Bonnini Department of Economics and Management, University of Ferrara, Ferrara, Italy

Francesco Bravo Department of Economics, University of York, York, UK

Alessandro Cardinali University of Bristol, Bristol, UK

Livio Corain Department of Management and Engineering, University of Padova, Padova, Italy

P.-A. Cornillon Université Rennes 2, 35043 Rennes, France

Ariadni Papana Dagiasis Department of Mathematics, Cleveland State University, Cleveland, OH, USA

Justin Davis Tulane University, New Orleans, LA, USA

Alexander V. Dobrovidov Institute of Control Sciences of Russian Academy of Sciences, Moscow, Russia

Anna E. Dudek AGH University of Science and Technology, Krakow, Poland

David van Duin Department of Infectious Disease, Cleveland Clinic, Cleveland, OH, USA

Raed A. Dweik Department of Pulmonary, Allergy and Critical Care Medicine and Pathobiology/ Lerner Research Institute, Cleveland Clinic, Cleveland, OH, USA

Leif Ellingson Texas Tech University, Lubbock, TX, USA

Gang Feng Technische Universität Braunschweig, Institut für Mathematische Stochastik, Pockelsstrasse 14, 38106 Braunschweig, Germany

Frédéric Ferraty Institut de Mathématiques de Toulouse Équipe LSP, Université Paul Sabatier, 118, route de Narbonne, 31062 Toulouse Cedex, France

Emmanuelle Gautherat CREST-LS, Malakoff, France

REGARDS, University of Reims, Reims, France

Subhashis Ghoshal North Carolina State University, Raleigh, NC, USA

Francesco Giordano Department of Economics and Statistics, University of Salerno, Salerno, Italy

Aldo Goia Dipartimento di Studi per l'Economia e l'Impresa, Università del Piemonte, Orientale "A. Avogadro", Via Perrone, 18, 28100 Novara, Italy

M. Ivette Gomes Centro de Estatística e Aplicações, Faculdade de Ciências, Universidade de Lisboa, Lisboa, Portugal

W. González-Manteiga Universidad de Santiago de Compostela, La Coruña, Spain

Hugo Harari-Kermadec ENS-Cachan, Cachan, France

Harrie Hendriks Radboud University Nijmegen, Nijmegen, The Netherlands

N. Hengartner Los Alamos National Laboratory, Los Alamos, NM, USA

Javier Hidalgo London School of Economics, London, UK

Maarten Jansen Department of Mathematics and Computer Science, Université Libre de Bruxelles (ULB), Brussel, Belgium

Jens-Peter Kreiss Technische Universität Braunschweig, Institut für Mathematische Stochastik, Pockelsstrasse 14, 38106 Braunschweig, Germany

Dimitris Kugiumtzis Aristotle University of Thessaloniki, Thessaloniki, Greece

Catherine Kyrtsou University of Macedonia, Thessaloniki, Greece

BETA, University of Strasbourg, Strasbourg, France

Economix, University of Paris 10, Nanterre, France

ISC-Paris, Ile-de-France, Paris, France

Jacek Leśkow Institute of Mathematics, Cracow University of Technology, Krakow, Poland

Sofiane Maiz Universite de Saint Etienne, Jean Monnet, Saint Etienne, France

LASPI, IUT de Roanne, Roanne, France

Natalia Markovich Institute of Control Sciences, Russian Academy of Sciences, Moscow, Russia

E. Matzner-Løber Université Rennes 2, 35043 Rennes, France

Simos G. Meintanis Department of Economics, National and Kapodistrian University of Athens, Athens, Greece

Unit for Business Mathematics and Informatics, North–West University, Potchefstroom, South Africa

Angeliki Papana University of Macedonia, Thessaloniki, Greece

Maria Lucia Parrella Department of Economics and Statistics, University of Salerno, Salerno, Italy

P.N. Patil School of Mathematics, University of Birmingham, Birmingham, UK

Vic Patrangenaru Florida State University, Tallahassee, FL, USA

M. Pavlidou Aristotle University of Thessaloniki, Thessaloniki, Greece

Marianna Pensky University of Central Florida, Orlando, FL, USA

Dimitris N. Politis University of California at San Diego, La Jolla, CA, USA

Silvia Polettini Department of Methods and Models for Economics, Territory and Finance, Sapienza University of Rome, Rome, Italy

Stefanos Poulis University of California at San Diego, La Jolla, CA, USA

A. Rodríguez-Casal Universidad de Santiago de Compostela, La Coruña, Spain

P. Saavedra-Nieves Universidad de Santiago de Compostela, La Coruña, Spain

Ernesto Salinelli Dipartimento di Studi per l'Economia e l'Impresa, Unversità del Piemonte, Orientale "A. Avogadro", Via Perrone, 18, 28100 Novara, Italy

Luigi Salmaso Department of Management and Engineering, University of Padova, Padova, Italy

Min Shu Department of Applied Mathematics and Statistics, Stony Brook University, Stony Brook, NY, USA

Francesca Solmi Department of Statistical Sciences, University of Padova, Padova, Italy

Pedro C.L. Souza London School of Economics, London, UK

Milan Stehlík Department of Applied Statistics, Johannes Kepler University, Linz, Austria

Departamento de Matemática, Universidad Técnica Federico Santa Maríka, Casilla 110-V, Valparaíkso, Chile

B. Thieurmel Data Knowledge, 75008 Paris, France

Kostas Triantafyllopoulos School of Mathematics and Statistics, University of Sheffield, Sheffield, UK

Paul San Valentin Florida State University, Tallahassee, FL, USA

Vyacheslav Vasiliev Tomsk State University, Tomsk, Russia

Philippe Vieu Institut de Mathématiques, Université Paul Sabatier, Toulouse, France

John Thomas White SAS Institute Inc., Cary, NC, USA

Yuping Wu Department of Mathematics, Cleveland State University, Cleveland, OH, USA

G. Zioutas Aristotle University of Thessaloniki, Thessaloniki, Greece

Chapter 1
A Cost Based Reweighted Scheme of Principal Support Vector Machine

Andreas Artemiou and Min Shu

Abstract Principal Support Vector Machine (PSVM) is a recently proposed method that uses Support Vector Machines to achieve linear and nonlinear sufficient dimension reduction under a unified framework. In this work, a reweighted scheme is used to improve the performance of the algorithm. We present basic theoretical results and demonstrate the effectiveness of the reweighted algorithm through simulations and real data application.

Keywords Support vector machine • Sufficient dimension reduction • Inverse regression • Misclassification penalty • Imbalanced data

1.1 Introduction

The recent increased capability of computers in storing large datasets allows researchers to collect large amount of data. This creates the need to analyze them and a number of methods have been proposed to efficiently and accurately reduce the dimensionality of several problems, in order for their analysis to become feasible. A set of methods were developed in what is known as sufficient dimension reduction (SDR) for regression problems. See for example [2, 4, 9–12].

In SDR the objective is to estimate a $p \times d$ matrix $\boldsymbol{\beta}$ $(d \leq p)$, such that:

$$Y \perp\!\!\!\perp X | \boldsymbol{\beta}^{\mathsf{T}} X \qquad (1.1)$$

where Y is the response variable, X is a p dimensional predictor and $\boldsymbol{\beta}^{\mathsf{T}} X$ gives d linear combinations of the predictors. As long as d is less than p then dimension

A. Artemiou (✉)
Department of Mathematical Sciences, Michigan Technological University,
306 Fisher Hall, 1400 Townsend Drive, Houghton, MI 49931, USA
e-mail: aartemio@mtu.edu

M. Shu
Department of Applied Mathematics and Statistics, Stony Brook University,
Stony Brook, NY 11794-3600, USA
e-mail: min.shu@stonybrook.edu

© Springer Science+Business Media New York 2014
M.G. Akritas et al. (eds.), *Topics in Nonparametric Statistics*, Springer Proceedings
in Mathematics & Statistics 74, DOI 10.1007/978-1-4939-0569-0_1

reduction is achieved. There are many possible $\boldsymbol{\beta}$ that can satisfy model (1.1), which is also known as the conditional independence model. The space spanned by the column vectors of $\boldsymbol{\beta}$ spans a dimension reduction space, $\mathscr{S}(\boldsymbol{\beta})$. If the intersection of all dimension reduction subspaces is a dimension reduction subspace itself, then it is called the Central Dimension Reduction Subspace (CDRS), denoted as $\mathscr{S}_{Y|X}$. CDRS has the minimum dimension d among all dimension reduction subspaces. CDRS does not always exist, but if it exists it is unique. For more see [3].

Recently there has been interest towards nonlinear SDR, where nonlinear functions of the predictors are extracted without losing information for the conditional distribution of $Y|X$, that is:

$$Y \perp\!\!\!\perp X | \phi(X) \tag{1.2}$$

where $\phi : \mathbb{R}^p \to \mathbb{R}^d$. See for example [7, 17, 18]. Li et al. [13] used Support Vector Machine [5] to achieve linear and nonlinear SDR under a unified framework.

To estimate the CDRS, Li et al. [13] used the idea of slicing the response as was proposed by Li [9]. One of the algorithms they suggested to implement is the "left vs right" (LVR) approach.

In the LVR approach at each dividing point between slices $\hat{y}_r, r = 1, \ldots, H - 1$, where H the number of slices, a binary response variable is defined as follows:

$$\tilde{Y}_i^r = I(Y_i \geq \hat{y}_r) - I(Y_i < \hat{y}_r) \tag{1.3}$$

where Y_i the ith response in the dataset and $I(\cdot)$ is the indicator function. Using soft margin SVM approach at each dividing point the optimal hyperplane $\boldsymbol{\psi}^\top x + t$ which separates the two classes is found, where $(\boldsymbol{\psi}, t) \in \mathbb{R}^p \times \mathbb{R}$ and it was shown that $\boldsymbol{\psi} \in \mathscr{S}_{Y|X}$. As it will be shown in the next section, to use the soft margin SVM approach, one needs to minimize an objective function where a tuning parameter exists. This is known as the misclassification penalty, or the "cost" and in the classical setting is the same for both classes. In this work our focus is to estimate the CDRS by a scheme that changes the misclassification penalty.

Our proposal involves using a separate cost for each class and it was inspired by the fact that there are different numbers of observations in each class at each dividing point. This will have an effect on the performance of the algorithm since for the first few comparisons there is a much smaller number of observations on the left of the dividing points than the number of points on the right of it. Similarly, on the last few comparisons the number of observations on the right of the dividing points is much smaller than those on the left. For example if the sample size $n = 100$ and there are $H = 20$ slices then if there are equal number of observations in each slice, at the first and the last (19th) comparisons one class will have 5 points and the other class will have 95 points. Similarly at the second and second to last comparisons one class will have 10 points and the other 90 points and so on. A similar idea was presented by Veropoulos et al. [15] in the classic SVM literature.

The notation and the main idea of Li et al. [13] is reviewed in Sect. 1.2. In Sect. 1.3 the new method which is called Cost Reweighted PSVM (CRPSVM) is introduced for the linear case. In Sect. 1.4 we discuss the estimation procedure, and in Sect. 1.5 we have some simulations. Finally a small discussion follows at the end. Due to space constraints results for the nonlinear case are omitted, but the results are discussed. Finally asymptotic theory and properties of the method are omitted but the interested reader is referred to Li et al. [13] as the results are similar.

1.2   Principal Support Vector Machine

In this section the main idea behind PSVM is outlined. Let $(Y_i, X_i), i = 1, \ldots, n$ be iid observations, and let $\tilde{Y}_i$ be the value of Y_i based on Eq. (1.3) for any cutoff point $\hat{y}_r$. Let $\Sigma = \text{var}(X)$ and $(\psi, t) \in \mathbb{R}^p \times \mathbb{R}$ be a vector and a constant characterizing the optimal hyperplane.

In the linear case it was suggested to minimize the following objective function in the sample level

$$
\begin{aligned}
&\text{minimize} \quad \psi^\mathsf{T} \Sigma_n \psi + \lambda n^{-1} \sum_{i=1}^n \xi_i \text{ among } (\psi, t, \xi) \in \mathbb{R}^p \times \mathbb{R} \times \mathbb{R}^n \\
&\text{subject to } \tilde{Y}_i [\psi^\mathsf{T}(X_i - \bar{X}) - t] \geq 1 - \xi_i, \ \xi_i \geq 0 \ i = 1, \ldots, n,
\end{aligned}
\tag{1.4}
$$

where Σ_n is the sample estimator of the population covariance matrix Σ, $\lambda > 0$ is the misclassification penalty and $\xi = (\xi_1, \ldots, \xi_n)^\mathsf{T}$, where $\xi_i, i = 1, \ldots, n$ is the misclassification distance for the ith observation, which is 0 if the point is correctly classified. Σ_n in the first term of the objective function is not present in the traditional SVM literature, but it was introduced by Li et al. [13] as it was essential in the dimension reduction framework for two reasons. First it gives a unified framework for linear and nonlinear dimension reduction and in the linear case it allows for dimension reduction without matrix inversion thus allowing researchers to use the method in large p small n problems. Fixing (ψ, t) and minimizing over the ξ_i's one can show that the above constraint problem takes the form

$$
\psi^\mathsf{T} \Sigma_n \psi + \frac{\lambda}{n} \sum_{i=1}^n \{1 - Y_i [\psi^\mathsf{T}(X_i - \bar{X}) - t]\}^+
\tag{1.5}
$$

where $a^+ = \max\{0, a\}$. At the population level this is written as

$$
L(\psi, t) = \psi^\mathsf{T} \Sigma \psi + \lambda E\{1 - \tilde{Y}[\psi^\mathsf{T}(X - EX) - t]\}^+.
\tag{1.6}
$$

It was shown that if (ψ^*, t^*) minimizes the objective function (1.6) among all $(\psi, t) \in \mathbb{R}^p \times \mathbb{R}$ then $\psi^* \in \mathscr{S}_{Y|X}$, under the assumption that $E(X|\beta^\mathsf{T} X)$ is a linear function of $\beta^\mathsf{T} X$, where β is as defined in (1.1). This assumption is known as the linear conditional mean (LCM) assumption and it is very common in the SDR framework. It is equivalent to the assumption of X being elliptically distributed.

1.3   Cost Reweighted PSVM

To reweight based on our description in the Sect. 1.1 two separate costs $\lambda_i, i = -1, 1$ are introduced in the objective function. Each point will have a cost based on which class it belongs. The idea is that it is much more costly to get a misclassification from the class that has a smaller portion of the data (i.e. 10 %) than the class that has the biggest portion of the data (i.e. 90 %).

As in classic linear PSVM at the sample level of the problem one minimizes the following objective function:

$$\text{minimize} \quad \boldsymbol{\psi}^{\mathsf{T}} \boldsymbol{\Sigma}_n \boldsymbol{\psi} + \frac{1}{n} \sum_{i=1}^{n} \lambda_{\tilde{y}_i} \xi_i \text{ among } (\boldsymbol{\psi}, t, \boldsymbol{\xi}) \in \mathbb{R}^p \times \mathbb{R} \times \mathbb{R}^n \tag{1.7}$$
$$\text{subject to } \tilde{y}_i [\boldsymbol{\psi}^{\mathsf{T}}(\boldsymbol{x}_i - \bar{\boldsymbol{x}}) - t] \geq 1 - \xi_i, \ \xi_i \geq 0 \ i = 1, \ldots, n,$$

where we clearly denote the dependence of the misclassification penalty λ on the value of $\tilde{y}_i$. A similar derivation as in the PSVM case can lead to the following population level objective function:

$$L_R(\boldsymbol{\psi}, t) = \boldsymbol{\psi}^{\mathsf{T}} \boldsymbol{\Sigma} \boldsymbol{\psi} + E \left(\lambda_{\tilde{Y}} \{ 1 - \tilde{Y} [\boldsymbol{\psi}^{\mathsf{T}}(\boldsymbol{X} - E\boldsymbol{X}) - t] \}^+ \right) \tag{1.8}$$

The only difference of this with (1.6) is the definition of the misclassification penalty (cost) and its dependence on the value of $\tilde{Y}$.

The following theorem shows that indeed the minimizer $\boldsymbol{\psi}^*$ of $L_R(\boldsymbol{\psi}, t)$ is in the CDRS. The proof is left for the Appendix.

Theorem 1. *Suppose $E(\boldsymbol{X} | \boldsymbol{\beta}^{\mathsf{T}} \boldsymbol{X})$ is a linear function of $\boldsymbol{\beta}^{\mathsf{T}} \boldsymbol{X}$, where $\boldsymbol{\beta}$ is as defined in (1.1). If $(\boldsymbol{\psi}^*, t^*)$ minimizes the objective function (1.8) among all $(\boldsymbol{\psi}, t) \in \mathbb{R}^p \times \mathbb{R}$, then $\boldsymbol{\psi}^* \in \mathscr{S}_{Y|X}$.*

1.4   Estimation

In order to estimate the vectors that span the CDRS a quadratic problem programming needs to be solved. It is a similar procedure as the one that appears in Cortes and Vapnik [5] for estimating the hyperplane for standard SVM algorithms and in Veropoulos et al. [15] for the different costs in each class problem. We emphasize that the problem we are solving is essentially different as we are multiplying the first term with $\boldsymbol{\Sigma}_n$. That is, to solve the sample version of the objective function (1.8) which is:

$$\hat{L}_R(\boldsymbol{\psi}, t) = \boldsymbol{\psi}^{\mathsf{T}} \boldsymbol{\Sigma}_n \boldsymbol{\psi} + \frac{1}{n} \sum_{i=1}^{n} \lambda_{\tilde{y}_i} \{ 1 - \tilde{Y}_i [\boldsymbol{\psi}^{\mathsf{T}}(\boldsymbol{X}_i - \bar{\boldsymbol{X}}) - t] \}^+$$

one needs to standardize the predictors using $Z_i = \Sigma_n^{-1/2}(X_i - \bar{X})$ and $\zeta = \Sigma_n^{1/2}\psi$ and the objective function becomes:

$$\hat{L}_R(\zeta, t) = \zeta^{\mathsf{T}}\zeta + \frac{1}{n}\sum_{i=1}^{n}\lambda_{\tilde{y}_i}\{1 - \tilde{Y}_i[\zeta^{\mathsf{T}}Z_i - t]\}^{+} \tag{1.9}$$

This looks closer to the classic SVM objective function that was used in the classification setting. Then the following theorem holds:

Theorem 2. *If ζ^* minimizes the objective function in (1.9) over $\mathbb{R}^p$, then $\zeta^* = \frac{1}{2}Z^{\mathsf{T}}(\alpha \odot \tilde{y})$ where α is found by solving the quadratic programming problem:*

$$\text{maximize}\quad \alpha^{\mathsf{T}}\mathbf{1} - \frac{1}{4}(\alpha \odot \tilde{y})^{\mathsf{T}}ZZ^{\mathsf{T}}(\alpha \odot \tilde{y})$$
$$\text{subject to}\quad 0 < \alpha < \frac{1}{n}\lambda^*,\quad (\alpha \odot \tilde{y})^{\mathsf{T}}\mathbf{1} = 0, \tag{1.10}$$

where $\mathbf{1} = (1, \ldots, 1)^{\mathsf{T}} \in \mathbb{R}^n$ and $\lambda^* \in \mathbb{R}^n$ has entries the value of the cost corresponding to each data point.

Thus to find the dividing hyperplane for the original data we have that $\psi^* = \Sigma_n^{-1/2}\zeta^*$. The proof is shown in the Appendix.

The following algorithm is proposed for estimation:

1. Compute the sample mean $\bar{X}$ and sample variance matrix $\hat{\Sigma} = n^{-1}\sum_{i=1}^{n}(X_i - \bar{X})(X_i - \bar{X})^{\mathsf{T}}$ and use them to standardize the data.
2. Let $q_r, r = 1, \ldots, H - 1$, be $H - 1$ dividing points. In the simulation section we choose them to be, the $(100 \times r/H)$th sample percentile of $\{Y_1, \ldots, Y_n\}$. For each r, let $\tilde{Y}_i^r = I(Y_i > q_r) - I(Y_i \leq q_r)$ and λ^{*r} be the n dimensional vector of costs for the respective cutoff point and let $(\hat{\zeta}_r, \hat{t}_r)$ be the minimizer of $\zeta^{\mathsf{T}}\zeta + \lambda^{*r}E_n\{1 - \tilde{Y}^r[\zeta^{\mathsf{T}}Z - t]\}^{+}$, which are found using Theorem 2. This process yields $H - 1$ normal vectors $\hat{\zeta}_1, \ldots, \hat{\zeta}_{H-1}$.
3. Use the normal vectors to calculate $\hat{\psi}_i = \hat{\Sigma}^{-1/2}\hat{\zeta}_i$ to construct matrix $\hat{V}_n = \sum_{r=1}^{H-1}\hat{\psi}_r\hat{\psi}_r^{\mathsf{T}}$.
4. Let $\hat{v}_1, \ldots, \hat{v}_d$ be the eigenvectors of the matrix $\hat{V}_n$ corresponding to its d largest eigenvalues. We use subspace spanned by $\hat{v} = (\hat{v}_1, \ldots, \hat{v}_d)$ to estimate the CDRS, $\mathcal{S}_{Y|X}$.

1.5 Simulation Results

The following three models are used for the simulations:

$$\text{Model I:}\quad Y = X_1 + X_2 + \sigma\varepsilon,$$

$$\text{Model II:} \quad Y = X_1/[0.5 + (X_2 + 1)^2] + \sigma\varepsilon,$$

$$\text{Model III:} \quad Y = 4 + X_1 + (X_2 + X_3 + 2) * \sigma * \varepsilon,$$

where $X \sim N(0, I_{p \times p})$, $p = 10, 20, 30$, $\varepsilon \sim N(0, 1)$. There are 500 simulations with sample size $n = 100$, $\sigma = 0.2$ and $H = 5, 10, 20, 50$.

To evaluate the performance the distance measure proposed by Li et al. [12] is used where $\mathscr{S}_1$ and $\mathscr{S}_2$ are two subspaces of $\mathbb{R}^p$. The distance is defined by the matrix norm $\|P_{\mathscr{S}_1} - P_{\mathscr{S}_2}\|$, where $P_{\mathscr{S}_i}$ denotes orthogonal projection on $\mathscr{S}_i$. In this work the Frobenius norm is used.

To compare the performance between the two algorithms the ratio between the two costs is set equal to the inverse ratio between the number of observations between the two classes (a similar idea was used in the SVM literature by Lee et al. [8]). This implies that

$$\frac{\lambda_{-1}}{\lambda_1} = \frac{n_1}{n_{-1}}.$$

For CRPSVM, there are multiple iterations of the algorithm for different dividing points. At each dividing point, q_r, there are different number of observations on the right slice and the left slice, so the above ratio will be different each time. Thus, for the simulations the larger class always has misclassification penalty λ_{base} (which is defined) and then the above ratio is used to define the misclassification penalty for the smaller class.

We compare the performance of CRPSVM with the PSVM algorithm and the results are presented in Table 1.1. For models I and III the two methods have similar performance for small number of slices but we can see that the performance for the reweighted algorithm is improved as the number of slices increases. This is expected as the larger the number of slices, the larger the difference between the number of points in each slice at the first few and last few comparisons and therefore the higher the effect when using the reweighted method. For model II CRPSVM is better for all cases.

In this paper the nonlinear dimension reduction is not discussed for briefness. Our simulations show that reweighted method outperforms PSVM but to a lesser extend than the linear case. This is probably due to the fact that in the nonlinear setting we move into a higher dimensional space in which it is likely that the data are separable or at least there are fewer misclassifications. Thus, the effect of the few misclassifications is smaller than that of the linear case where it is clear that the data are not separable.

A BIC type criterion is used to estimate the dimension of the CDRS. This is a common approach in SDR literature (see [19]). We try to find the value k that maximizes:

$$G_n(k) = \sum_{i=1}^{k} \mu_i(V_n) - \alpha n^{-\frac{3}{8}} \mu_i(V_n)k \tag{1.11}$$

Table 1.1 Performance of PSVM and CRPSVM for different number of slices where $\lambda_{base} = \lambda_{PSVM} = 1$ for linear sufficient SDR

p	Model	Methods	Number of slices H			
			5	10	20	50
10	I	PSVM	0.23 (0.061)	0.22 (0.060)	0.22 (0.056)	0.22 (0.059)
		CRPSVM	0.22 (0.062)	0.17 (0.048)	0.15 (0.046)	0.13 (0.043)
	II	PSVM	1.00 (0.222)	0.95 (0.204)	0.92 (0.219)	0.94 (0.214)
		CRPSVM	0.78 (0.170)	0.75 (0.165)	0.73 (0.164)	0.75 (0.170)
	III	PSVM	1.35 (0.119)	1.35 (0.113)	1.33 (0.128)	1.34 (0.119)
		CRPSVM	1.33 (0.133)	1.28 (0.153)	1.26 (0.169)	1.26 (0.171)
20	I	PSVM	0.35 (0.069)	0.34 (0.064)	0.33 (0.065)	0.33 (0.068)
		CRPSVM	0.32 (0.069)	0.25 (0.056)	0.21 (0.056)	0.19 (0.049)
	II	PSVM	1.27 (0.153)	1.20 (0.157)	1.19 (0.154)	1.19 (0.156)
		CRPSVM	1.10 (0.140)	1.04 (0.149)	1.04 (0.142)	1.02 (0.154)
	III	PSVM	1.44 (0.061)	1.44 (0.057)	1.44 (0.059)	1.43 (0.064)
		CRPSVM	1.42 (0.068)	1.40 (0.080)	1.38 (0.086)	1.38 (0.088)
30	I	PSVM	0.45 (0.076)	0.44 (0.077)	0.43 (0.075)	0.42 (0.074)
		CRPSVM	0.40 (0.070)	0.32 (0.065)	0.27 (0.064)	0.25 (0.061)
	II	PSVM	1.40 (0.111)	1.36 (0.117)	1.34 (0.116)	1.34 (0.124)
		CRPSVM	1.27 (0.113)	1.23 (0.118)	1.21 (0.124)	1.20 (0.124)
	III	PSVM	1.50 (0.050)	1.50 (0.049)	1.50 (0.043)	1.49 (0.046)
		CRPSVM	1.48 (0.050)	1.45 (0.055)	1.44 (0.057)	1.44 (0.057)

where $\mu_i(V_n)$ the eigenvalues of V_n and α is a constant that needs to be determined and should depend on p, H, d and λ. Li et al. [13] used a slightly modified criterion and determined α using cross validation at the expense of computation time. Here, based on a small study of how each parameter affects the estimation of the dimension through the criterion above we choose a specific value which gives good estimation under various combinations of the parameters above.

To compare the performance between the cross validated BIC (CVBIC) criterion for PSVM with the performance of the BIC criterion for CRPSVM we use $\alpha = \frac{\log(\lambda_{base}+2)p^{1/4}}{H^{1/4}}$ for models I and II. Also $\sigma = 0.2$, $\lambda_{base} = \lambda_{PSVM} = 1$, $p = 10, 20, 30$, sample size $n = 100, 200, 300, 400, 500$, $H = 20, 50$ and the percentage of correct dimension estimation is summarized in Table 1.2 for 500 simulations. We can see that for model I which has $d = 1$ the CVBIC of PSVM works better for smaller sample sizes while for large sample sizes the performance is perfect for both the CVBIC of PSVM and the BIC of CRPSVM. On the other hand in almost all the simulation for model II where $d = 2$ the proposed BIC criterion of CRPSVM performs better especially for small sample sizes.

Table 1.2 Performance of CVBIC criterion for dimension determination of PSVM and the BIC criterion for CRPSVM with $\alpha = \frac{\log(\lambda_{\text{base}}+2)p^{1/4}}{H^{1/4}}$ for different number of slices where $\lambda_{\text{base}} = \lambda_{\text{PSVM}} = 1$

			Model I, $d = 1$					Model II, $d = 2$				
			Number of observations n					Number of observations n				
H	p	Method	100	200	300	400	500	100	200	300	400	500
20	10	PSVM	1	1	1	1	1	0.748	0.978	1	1	1
		CRPSVM	0.992	1	1	1	1	0.842	0.97	0.998	1	1
	20	PSVM	1	1	1	1	1	0.31	0.862	0.98	0.944	1
		CRPSVM	0.926	0.998	1	1	1	0.89	0.946	0.996	1	1
	30	PSVM	0.996	1	1	1	0.93	0.174	0.584	0.912	0.98	0.874
		CRPSVM	0.566	0.994	1	1	1	0.916	0.96	0.996	0.998	1
50	10	PSVM	1	1	1	1	1	0.732	0.992	1	1	1
		CRPSVM	0.974	1	1	1	1	0.964	0.994	1	1	1
	20	PSVM	1	1	1	1	1	0.358	0.866	0.984	1	1
		CRPSVM	0.752	0.994	1	1	1	0.97	1	1	1	1
	30	PSVM	0.996	1	1	1	1	0.162	0.624	0.914	0.992	1
		CRPSVM	0.336	0.952	1	1	1	0.924	1	1	1	1

1.6 Real Dataset Analysis

In this section we show through a real dataset the advantage one can gain by working with the reweighted method. We use the dataset from the UC Irvine machine learning repository [1]. The dataset was first used in Ein-Dor and Feldmesser [6] and the objective is to create a regression model that estimates relative performance of the Central Processing Unit (CPU) of a computer using some of its characteristics, including cache memory size, cycle time, minimum and maximum input/output channels, and minimum and maximum main memory. Relative performance was calculated using observations from users of different machines in the market. The dataset consists of 209 models where performance is not available.

We apply both PSVM and CRPSVM on this dataset. We use $\lambda = 1$ and we use the dr package in R [16] to separate the data into ten slices. Figure 1.1 shows clearly that in the first direction both methods capture the nonlinear relationship which [6] expect to see in this case. Although the correlation (ρ) between the first direction of the two methods is $\rho = -0.9$ which indicates the strong similarity, we can also see from the plots that the reweighted algorithm performs slightly better because the points with smaller performance are closer to the curve they form and also points with larger performance seem to be more aligned with the curve.

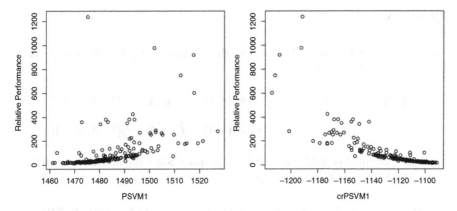

Fig. 1.1 First direction for the computer hardware dataset for the PSVM (*left*) and CRPSVM (*right*)

1.7 Discussion

In this work, it is shown that reweighting using different costs improve the performance of PSVM which was introduced by Li et al. [13].

PSVM is a method that achieves both linear and nonlinear dimension reduction in a unified framework. It brings together SDR and machine learning, two areas used separately to solve high dimensional problems. With this work we have shown that one of the ideas used in the machine learning literature to improve the performance of Support Vector Machines (SVM) can also be used in the SDR framework to improve the performance of PSVM.

The original algorithm uses slices of different size without correcting for this. This manuscript shows through simulation and real data examples that reweighting can improve the performance of the algorithm. To achieve this we apply a different misclassification penalty for each class. Different sample size between classes can lead the separating hyperplane to be more biased towards the bigger class. Since the way the algorithm works separates the data in two unequal sample size classes, reweighting reduces the bias towards the bigger class and therefore produces more accurate results. To demonstrate the effectiveness of this procedure we chose the ratio between the two penalties to be the inverse ratio between the two class sizes. The increase in performance reinforces previous results by Lee et al. [8] in the classification framework. We note that although only one approach to find the relationship between the two different penalties is used, there is still an open question on how to find an optimal relationship between the two penalties as well as applying other ideas that target samples with different sample sizes.

Appendix

Proof of Theorem 1. Without loss of generality, assume that $EX = 0$. First we note that for $i = 1, -1$

$$
\begin{aligned}
E\left(\lambda_{\tilde{Y}}[1 - \tilde{Y}(\boldsymbol{\psi}^\mathsf{T}X - t)]^+\right) &= E\{E[(\lambda_{\tilde{Y}}[1 - \tilde{Y}(\boldsymbol{\psi}^\mathsf{T}X - t)]^+) \mid Y, \boldsymbol{\beta}^\mathsf{T}X]\} \\
&= E\{E[(\lambda_{\tilde{Y}}[1 - \tilde{Y}(\boldsymbol{\psi}^\mathsf{T}X - t)])^+ \mid Y, \boldsymbol{\beta}^\mathsf{T}X]\}
\end{aligned}
$$

where the last equality holds because $\lambda_{\tilde{Y}}$ is positive. Since the function $a \mapsto a^+$ is convex, by Jensen's inequality we have

$$
\begin{aligned}
E[(\lambda_{\tilde{Y}}[1 - \tilde{Y}(\boldsymbol{\psi}^\mathsf{T}X - t)])^+ \mid Y, \boldsymbol{\beta}^\mathsf{T}X] &\geq \{E[\lambda_{\tilde{Y}}[1 - \tilde{Y}(\boldsymbol{\psi}^\mathsf{T}X - t)] \mid Y, \boldsymbol{\beta}^\mathsf{T}X]\}^+ \\
&= \{\lambda_{\tilde{Y}}[1 - \tilde{Y}(E(\boldsymbol{\psi}^\mathsf{T}X \mid \boldsymbol{\beta}^\mathsf{T}X) - t)]\}^+
\end{aligned}
$$

where the equality follows from the model assumption (1.1). Thus

$$
E\left(\lambda_{\tilde{Y}}[1 - \tilde{Y}(\boldsymbol{\psi}^\mathsf{T}X - t)]^+\right) \geq E\{\lambda_{\tilde{Y}}[1 - \tilde{Y}(E(\boldsymbol{\psi}^\mathsf{T}X \mid \boldsymbol{\beta}^\mathsf{T}X) - t)]\}^+. \qquad (1.12)
$$

On the first term now we have:

$$
\begin{aligned}
\boldsymbol{\psi}^\mathsf{T}\boldsymbol{\Sigma}\boldsymbol{\psi} = \mathrm{var}(\boldsymbol{\psi}^\mathsf{T}X) &= \mathrm{var}[E(\boldsymbol{\psi}^\mathsf{T}X \mid \boldsymbol{\beta}^\mathsf{T}X)] + E[\mathrm{var}(\boldsymbol{\psi}^\mathsf{T}X \mid \boldsymbol{\beta}^\mathsf{T}X)] \\
&\geq \mathrm{var}[E(\boldsymbol{\psi}^\mathsf{T}X \mid \boldsymbol{\beta}^\mathsf{T}X)]. \qquad (1.13)
\end{aligned}
$$

Combining (1.12) and (1.13),

$$
L_R(\boldsymbol{\psi}, t) \geq \mathrm{var}[E(\boldsymbol{\psi}^\mathsf{T}X \mid \boldsymbol{\beta}^\mathsf{T}X)] + E\{\lambda_{\tilde{Y}}[1 - \tilde{Y}(E(\boldsymbol{\psi}^\mathsf{T}X \mid \boldsymbol{\beta}^\mathsf{T}X) - t)]\}^+.
$$

By the definition of the linearity condition in the theorem $E(\boldsymbol{\psi}^\mathsf{T}X \mid \boldsymbol{\beta}^\mathsf{T}X) = \boldsymbol{\psi}^\mathsf{T}P_{\boldsymbol{\beta}}^\mathsf{T}(\boldsymbol{\Sigma})X$ and therefore the right-hand side of the inequality is equal to $L_R(P_{\boldsymbol{\beta}}(\boldsymbol{\Sigma})\boldsymbol{\psi}, t)$. If $\boldsymbol{\psi}$ is not in the CDRS then the inequality is strict which implies $\boldsymbol{\psi}$ is not the minimizer. $\square$

Proof of Theorem 2. Following the same argument as in Vapnik [14] it can be shown that minimizing (1.9) is equivalent tov

$$
\begin{aligned}
&\text{minimizing } \boldsymbol{\zeta}^\mathsf{T}\boldsymbol{\zeta} + \frac{1}{n}(\boldsymbol{\lambda}^*)^\mathsf{T}\boldsymbol{\xi} \text{ over } (\boldsymbol{\zeta}, t, \boldsymbol{\xi}) \\
&\text{subject to } \boldsymbol{\xi} \geq \mathbf{0}, \ \tilde{y} \odot (\boldsymbol{\zeta}^\mathsf{T}Z - t\mathbf{1}) \geq \mathbf{1} - \boldsymbol{\xi}.
\end{aligned} \qquad (1.14)
$$

The Lagrangian function of this problem is

$$
L(c, t, \boldsymbol{\xi}, \boldsymbol{\alpha}, \boldsymbol{\beta}) = \boldsymbol{\zeta}^\mathsf{T}\boldsymbol{\zeta} + \frac{1}{n}(\boldsymbol{\lambda}^*)^\mathsf{T}\boldsymbol{\xi} - \boldsymbol{\alpha}^\mathsf{T}[\tilde{y} \odot (\boldsymbol{\zeta}^\mathsf{T}Z - t\mathbf{1}) - \mathbf{1} + \boldsymbol{\xi}] - \boldsymbol{\beta}^\mathsf{T}\boldsymbol{\xi}. \qquad (1.15)
$$

where $\xi = (\xi_1, \ldots, \xi_n)$. Let (ζ^*, ξ^*, t^*) be a solution to problem (1.14). Using the Kuhn–Tucker Theorem, one can show that minimizing over (ζ, t, ξ) is similar as maximizing over (α, β). So, differentiating with respect to ζ, t, and ξ to obtain the system of equations:

$$\begin{cases} \partial L/\partial \zeta = 2\zeta - Z^{\mathsf{T}}(\alpha \odot \tilde{y}) = 0 \\ \partial L/\partial t = \alpha^{\mathsf{T}} \tilde{y} = 0 \\ \partial L/\partial \xi = \frac{1}{n}\lambda^* - \alpha - \beta = 0. \end{cases} \tag{1.16}$$

Substitute the last two equations above into (1.15) to obtain

$$\zeta^{\mathsf{T}}\zeta - \alpha^{\mathsf{T}}[\tilde{y} \odot (\zeta^{\mathsf{T}}Z) - \mathbf{1}] \tag{1.17}$$

Now substitute the first equation in (1.16) $(\zeta = \frac{1}{2}Z^{\mathsf{T}}(\alpha \odot \tilde{y}))$ in the above:

$$\mathbf{1}^{\mathsf{T}}\alpha - \frac{1}{4}(\alpha \odot \tilde{y})^{\mathsf{T}}ZZ^{\mathsf{T}}(\alpha \odot \tilde{y}). \tag{1.18}$$

Thus to minimize (1.15) we need to maximize (1.18) over the constraints

$$\begin{cases} \alpha^{\mathsf{T}}\tilde{y} = 0 \\ \frac{1}{n}\lambda^* - \alpha - \beta = 0. \end{cases} \tag{1.19}$$

which are equivalent to the constraints in (1.10).   □

Acknowledgements Andreas Artemiou is supported in part by NSF grant DMS-12-07651. The authors would like to help the editors and the referees for their valuable comments.

References

1. Bache, K., Lichman, M.: UCI Machine Learning Repository. University of California, School of Information and Computer Science, Irvine (2013). http://archive.ics.uci.edu/ml
2. Cook, R.D.: Principal Hessian directions revisited (with discussion). J. Am. Stat. Assoc. **93**, 84–100 (1998a)
3. Cook, R.D.: Regression Graphics: Ideas for Studying Regressions through Graphics. Wiley, New York (1998b)
4. Cook, R.D., Weisberg, S.: Discussion of "Sliced inverse regression for dimension reduction". J. Am. Stat. Assoc. **86**, 316–342 (1991)
5. Cortes, C., Vapnik, V.: Support vector networks. Mach. Learn. **20**, 1–25 (1995)
6. Ein-Dor, P., Feldmesser, J.: Attributes of the performance of central processing units: a relative performance prediction model. Commun. ACM **30**(4), 308–317 (1987)
7. Fukumizu, K., Bach, F.R., Jordan, M.I.: Kernel dimension reduction in regression. Ann. Stat. **4**, 1871–1905 (2009)

8. Lee, K.K., Gunn, S.R., Harris, C.J., Reed, P.A.S.: Classification of imbalanced data with transparent kernels. In: Proceedings of International Joint Conference on Neural Networks (IJCNN '01), vol. 4, pp. 2410–2415, Washington, D.C. (2001)

9. Li, K.-C.: Sliced inverse regression for dimension reduction (with discussion). J. Am. Stat. Assoc. **86**, 316–342 (1991)

10. Li, K.-C.: On principal Hessian directions for data visualization and dimension reduction: another application of Stein's lemma. J. Am. Stat. Assoc. **86**, 316–342 (1992)

11. Li, B., Wang, S.: On directional regression for dimension reduction. J. Am. Stat. Assoc. **102**, 997–1008 (2007)

12. Li, B., Zha, H., Chiaromonte, F.: Contour regression: a general approach to dimension reduction. Ann. Stat. **33**, 1580–1616 (2005)

13. Li, B., Artemiou, A., Li, L.: Principal support vector machine for linear and nonlinear sufficient dimension reduction. Ann. Stat. **39**, 3182–3210 (2011)

14. Vapnik, V.: Statistical Learning Theory. Wiley, New York (1998)

15. Veropoulos, K., Campbell, C., Cristianini, N.: Controlling the sensitivity of support vector machines. In: Proceedings of the Sixteenth International Joint Conference on Artificial Intelligence (IJCAI '99), Workshop ML3, Stockholm, pp. 55–60

16. Weisberg, S.: Dimension reduction regression in R. J. Stat. Softw. **7**(1) (2002) (Online)

17. Wu, H.M.: Kernel sliced inverse regression with applications on classification. J. Comput. Graph. Stat. **17**, 590–610 (2008)

18. Yeh, Y.-R., Huang, S.-Y., Lee, Y.-Y.: Nonlinear dimension reduction with Kernel sliced inverse regression. IEEE Trans. Knowl. Data Eng. **21**, 1590–1603 (2009)

19. Zhu, L.X., Miao, B., Peng, H.: On sliced inverse regression with large dimensional covariates. J. Am. Stat. Assoc. **101**, 630–643 (2006)

Chapter 2
Power Assessment of a New Test of Independence

P.N. Patil and D. Bagkavos

Abstract A new nonparametric test of independence between the components of bivariate random vectors (X, Y) is motivated and evaluated in practice. The test statistics is based on the fact that under independence, every quantile of Y given $X = x$ is constant. This is in contrast to the most commonly used basis that the joint probability density or distribution function of X and Y, equals to the product of their marginal probability densities or distributions, respectively. Emphasis is given on the small sample power properties of the test. Through numeric simulations with distributional data, the power of the test is benchmarked against standard independence tests, already existing in the literature.

Keywords Independence • Hypothesis test • Statistical power • Test size Quantile regression

2.1 Introduction

The present note is concerned with the general problem of testing the stochastic independence between the components of bivariate random vectors. Historically, the Cramer–Von Mises distance measure has provided the basis for several hypothesis tests on this topic, e.g., [2, 4–6, 11], and [12]. See also [14] for a broader view of the subject.

A test of independence which originates from a different basis is investigated here. The concept, first considered in [10] and further explored in [3] is that independence is implied if every regression quantile of Y versus $X = x$ is constant. Under the linear quantile regression framework, c.f. [1], this idea is put to action by applying density weighted conditional expectation on the first order condition for minimization of the least absolute deviation criterion and integrating across

P.N. Patil
School of Mathematics, University of Birmingham, Edgbaston, Birmingham, B152TT, UK

D. Bagkavos (✉)
Accenture, Rostoviou 39–41, 11526, Athens, Greece
e-mail: dimitrios.bagkavos@gmail.com

© Springer Science+Business Media New York 2014   13
M.G. Akritas et al. (eds.), *Topics in Nonparametric Statistics*, Springer Proceedings in Mathematics & Statistics 74, DOI 10.1007/978-1-4939-0569-0_2

all possible quantiles. This results in a new condition which under independence between X and Y equals to zero, while otherwise takes large positive values. Naturally, this provides a proper candidate for developing a consistent independence test with its sample, kernel based, analogue as the proposed test statistic.

The purpose of this note is to provide insight on the practical performance of the suggested hypothesis test. Specifically, the small sample distribution of the test statistic under the null is approximated and then utilized in calculating the power of the test as a function of the level of correlation between X and Y. Furthermore, the power functions of the tests developed in [2] and [6] are used as a benchmark in assessing the practical performance of the test presented here. We note here that the test's theoretical properties, including its asymptotic distribution under both the null and alternative hypotheses, establishment of its consistency against all dependence alternatives as well as a bandwidth choice rule which controls the trade-off between the test's power and size functions will be provided in future work.

The rest of the paper is organized as follows. Section 2.2 discusses the development of the test and provides the test statistic. Numerical evidence on the power of the test and comparison with the powers of the [2] and [6] tests is given in Sect. 2.3.

2.2   Motivation and Test Statistic

Let $(X, Y) \in \mathbb{R} \times \mathbb{R}$ be a random variable with cumulative distribution function $F(x, y)$ and probability density function $f(x, y)$. Denote with $F_Y(y|x)$ the marginal distribution of Y conditional on $X = x$ and with $F_Y^{-1}(y|x)$ its inverse. The marginal (unconditional) distribution of Y is denoted by $F_Y(y)$ and by $F_Y^{-1}(y)$ its inverse. Obviously under independence between X and Y, $F_Y(y|x) = F_Y(y)$ and $F_Y^{-1}(y|x) = F_Y^{-1}(y)$.

The basis of the proposed test is that under independence, for every quantile p, where $0 < p < 1$, we have that $F_Y^{-1}(p|x) = c_p$ where c_p does not vary with x.

For example, $F_Y^{-1}(p|x)$ can be modeled (see also [1]) by

$$F_Y^{-1}(p|x) = \beta x + F_Y^{-1}(p). \tag{2.1}$$

Under independence between X and Y, for any fixed $p \in (0, 1)$ the conditional quantile function of Y given $X = x$ does not depend on x and therefore

$$F_Y^{-1}(p|x) = F_Y^{-1}(p).$$

Now, let $\psi_p(u) = \mathrm{sign}(u) + 2p - 1$ and let (X_1, Y_1) and (X_2, Y_2) be two independent random vectors with common probability density function $f(x, y)$. Now set

$$J(p) = \mathbb{E}\left\{K_h(X_1, X_2)\psi_p\left(Y_2 - F_Y^{-1}(p)\right)\psi_p\left(Y_1 - F_Y^{-1}(p)\right)\right\}$$

where $K_h(X_1, X_2) = h^{-1}K((X_1 - X_2)h^{-1})$, K is a second order kernel and h denotes bandwidth, i.e., the spread of the kernel. Observe that for every fixed $p \in (0, 1)$, under the hypothesis of independence of X and Y,

$$g_p(x) = E\{\psi_p(Y_1 - F_Y^{-1}(p))|X_1 = x\} = 0.$$

Therefore, under independence of X and Y, for every $p \in (0, 1)$ we have,

$$J(p) = \mathbb{E}\left\{K_h(X_1, X_2)\psi_p(Y_2 - F_Y^{-1}(p))\mathbb{E}\{\psi_p(Y_1 - F_Y^{-1}(p))|X_1\}\right\} = 0$$

and consequently

$$J = \int_0^1 J(p)\, dp = 0.$$

Under dependence we have

$$g_p(X_1) = E\{\psi_p(Y_1 - F_Y^{-1}(p))|X_1\} \neq 0 \text{ a.s.}$$

for at least one $p \in (0, 1)$. Assuming that g_p is twice differentiable,

$$
\begin{aligned}
J(p) &= \mathbb{E}\left\{K_h(X_1, X_2)\psi_p\left(Y_2 - F_Y^{-1}(p)\right)\psi_p\left(Y_1 - F_Y^{-1}(p)\right)\right\} \\
&= \mathbb{E}\left\{K_h(X_1, X_2)\mathbb{E}\left\{\psi_p\left(Y_2 - F_Y^{-1}(p)\right)\psi_p\left(Y_1 - F_Y^{-1}(p)\right)|(X_1, X_2)\right\}\right\} \\
&= \int_0^1 g_p^2(x)f_X^2(x)dx + O(h^2).
\end{aligned}
$$

Thus as $h \to 0$, $J(p) > 0$ and since $J(p)$ is a continuous function of p,

$$J = \int_0^1 J(p)dp > 0.$$

Therefore, J can be used for the following hypothesis test

$$H_0 : \bigcap_{0<p<1} H_{0p}, \quad H_{0p} : F_Y^{-1}(p|x) = c_p$$

with the alternative specified by

$$H_1 : \bigcup_{0<p<1} H_{1p}, \quad H_{1p} : F_Y^{-1}(p|x) = c(x)$$

where now $c(x)$ varies with x for at least one $p \in (0, 1)$.

For deriving a test statistic, assume a sample $(X_i, Y_i), i = 1, \ldots, n$ from $F(x, y)$ and denote by $\hat{F}_Y^{-1}(y)$ the inverse of the empirical marginal distribution of Y, $\hat{F}_Y(y)$. The density weighted conditional expectation

$$\mathbb{E}\left\{\psi_p\left(Y - F_Y^{-1}(p)\right)|x\right\} f_X(x),$$

can be reasonably estimated (fixing $x = X_i$) by

$$\frac{1}{(n-1)h} \sum_{j=1,\, j\neq i}^n K\left(\frac{X_i - X_j}{h}\right)\psi_p(Y_j - \hat{F}^{-1}(p)) \tag{2.2}$$

where the real valued function K is called kernel and integrates to 1, while h is called bandwidth and controls the spread of the kernel and therefore the amount of smoothing applied. Based on (2.2), the sample version of J is

$$T_n(h) = \int_0^1 \frac{1}{n} \sum_{i=1}^n \psi_p(Y_i - \hat{F}_Y^{-1}(p)) \frac{1}{(n-1)h} \sum_{j \neq i} K\left(\frac{X_i - X_j}{h}\right) \psi_p(Y_j - \hat{F}^{-1}(p)) \, dp$$

$$= \frac{1}{n(n-1)h} \sum_{1 \leq i < j \leq n} \sum K\left(\frac{X_i - X_j}{h}\right) \int_0^1 \psi_p\left(Y_i - \hat{F}_Y^{-1}(p)\right) \psi_p(Y_j - \hat{F}^{-1}(p)) \, dp$$

which is also the proposed test statistic. By denoting with R_{ni} the rank of Y_i after ordering the random sample $(X_i, Y_i), i = 1, 2, \ldots, n$ with respect to Y_i and after slightly modifying the definition of the empirical distribution function from $\hat{F}_Y$ to $n(n+1)^{-1}\hat{F}_Y$, a suitable for computational purposes form of the statistic is

$$T_n(h) = \frac{2}{n(n-1)h} \sum_{1 \leq i < j \leq n} \sum K\left(\frac{X_i - X_j}{h}\right)$$

$$\times \left\{ \frac{\min(R_{ni}, R_{nj})^2}{(n+1)^2} + \frac{n - \max(R_{ni}, R_{nj})^2}{(n+1)^2} - \frac{1}{3} \right\}. \quad (2.3)$$

The next section discusses the test's operational characteristics and provides numerical evidence on its power.

2.3   Numerical Evaluation of the Test's Power

In this section, the implementation details of the suggested test are discussed and then distributional data is used to exhibit the performance of the proposed test's power properties and asses its practical performance.

Throughout this section $T_n(h)$ is calculated on bivariate samples of size 50 by (2.3) with K being the uniform kernel. The bandwidth, h, is chosen so as to maximize the test's power under the null, subject to keeping the significance level constant. Specifically, in each $T_n(h)$ implementation, the bandwidth is given by

$$h = ch^*. \quad (2.4)$$

In (2.4), for each given sample, h^* is the optimal MISE regression bandwidth of [13], implemented in package lokern, R. The factor c is determined by a grid search in a probe analysis and applies to all bandwidth calculations with samples from the same distribution.

The probe analysis is designed to find c so that the test's size under the null matches the user supplied level confidence level a. Specifically, for each of the 30 equidistant c points in $(0, 3)$, $10{,}000$ test statistic values, $T_n(ch^*)$, under the null are calculated. The probability $a(ch^*) = P(T_n(ch^*) > l_a)$ is then calculated where l_a is the $(1 - a)100\%$ quantile of the $10{,}000$ T_n values, calculated as described in the next paragraph. The c that corresponds to the highest among the 30 $a(ch^*)$'s is selected and used in (2.4).

The bandwidth choice rule employed here implicitly works as a selection rule based on bootstrap/edgeworth expansion ideas considered in the past by [8]. The benefit of such an approach is that it offers the means to control both power and size. An additional advantage of this bandwidth procedure is that it offers the best bandwidth both under the null and the alternative hypothesis.

As a cut-off point in the test's power function approximation, we use the $(1 - a)100\%$ quantile of the numerical distribution of $T_n(h)$. For this purpose, definition 7 of [7], which is readily implemented in R by the quantile() function, is applied on $100{,}000$ $T_n(h)$ values. The $T_n(h)$ values result by applying (2.3), implemented as described in the previous paragraph, on $100{,}000$ uncorrelated bivariate samples. The desired number of uncorrelated samples is obtained by repeatedly generating bivariate samples $(X_i, Z_i), i = 1, \ldots, 50$ and keeping only those for which the correlation between X and Z is less than 0.001.

Then, the power function of $T_n(h)$ is approximated by

$$P(T_n(h) > \text{cut} - \text{off}) = \frac{\#T_n(h) > \text{cut} - \text{off}}{m}, \tag{2.5}$$

for $m = 100$ replications. Specifically, 40 equidistant correlation levels, ρ, between 0 and 1 are determined. For each ρ, 100 independent (i.e., $\text{corr}(X, Z) < 0.001$) bivariate samples $(X_i, Z_i), i = 1, \ldots, 50$ are drawn. The actual samples used by $T_n(h)$ are $(X_i, Y_i), i = 1, \ldots, 50$ where the Y_i's are obtained by the transformation

$$Y = \rho X + Z\sqrt{1 - \rho^2}.$$

The provision of drawing samples with $\text{corr}(X, Z) < 0.001$ is sought because this ensures that the above transformation will return samples (X_i, Y_i) with the desired level of correlation. Then for each ρ, $T_n(h)$ is calculated 100 times using the (X_i, Y_i) samples and the empirical power is calculated by (2.5).

The tests of [2] (noted as B_n) and [6] (noted as D_n) are used for benchmarking $T_n(h)'s$ behavior. The B_n test is calculated as described in [9, p. 43]. Its power function is approximated by (2.5) with $T_n(h)$ replaced by B_n and with cut-off points found in Table 2 of [9]. The D_n test is implemented by the hoeffd function of R (package Hmisc) which also returns its p-value for the given sample. It's power function is approximated by $\{\#\text{of}D_n\text{p} - \text{values} < a\}/m$.

Now, three examples are presented next to exhibit the test's power behavior and asses, its practical performance. For each distribution utilized in each example, the power functions presented result by an average of 40 power functions calculated as described above. Further, all three tests are always calculated on the same samples.

The first example (Fig. 2.1) utilizes the bivariate distribution with p.d.f.

$$f_1(x, y) = \exp(-(x + y)), \quad x > 0, \; y > 0.$$

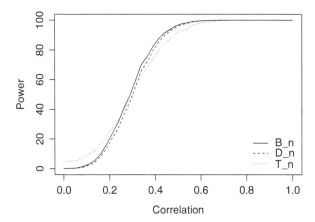

Fig. 2.1 Empirical test powers for T_n (*dotted line*), B_n (*solid line*), and D_n (*dashed line*), $n = 50$ with data from $f_1(x, y)$

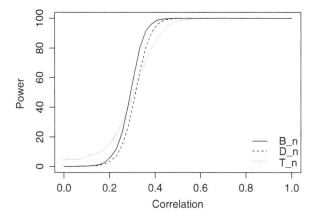

Fig. 2.2 Empirical test powers for T_n (*dotted line*), B_n (*solid line*), and D_n (*dashed line*), $n = 50$ with data from $f_2(x, y)$

In implementing the power function of $T_n(h)$, the bandwidth factor $c = 0.32$ has been found optimal. The significance level is $a = 1\%$.

For the second example (Fig. 2.2) the bivariate distribution with p.d.f.

$$f_2(x, y) = 2, \quad 0 \leq x \leq y \leq 1,$$

is employed, $a = 5\%$ and the bandwidth factor for $T_n(h)$ is found to be $c = 1.34$.

The last example (Fig. 2.3) uses the bivariate normal distribution and $a = 5\%$. $T_n(h)$ is implemented with bandwidth factor $c = 0.7$.

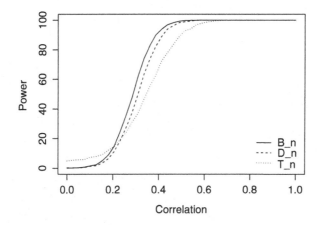

Fig. 2.3 Empirical test powers for T_n (*dotted line*), B_n (*solid line*), and D_n (*dashed line*), $n = 50$ with data from the bivariate normal distribution

References

1. Basset, G., Koenker, R.: An empirical quantile function for linear models with i.i.d. errors. J. Am. Stat. Assoc. **77**, 407–415 (1982)
2. Blum, J.R., Kiefer, J., Rosenblatt, M.: Distribution free tests for independence based on the sample distribution function. Ann. Math. Stat. **32**, 485–498 (1961)
3. Chan, E.: Testing constancy of regression quantiles using generalised sign test. Ph.D. thesis, University of Birmingham (2007)
4. Einmahl, J., McKeauge, I.: Empirical likelihood based hypothesis testing. Bernoulli **9**, 267–290 (2003)
5. Feurverger, A.: A consistent test for bivariate dependence. Int. Stat. Rev. **61**, 419–433 (1993)
6. Hoeffding, W.: A nonparametric test for independence. Ann. Math. Stat. **19**, 546–557 (1948)
7. Hyndman, R., Fan, Y.: Sample quantiles in statistical packages. Am. Statist. **50**, 361–365 (1996)
8. Li, Q., Wang, S.: A simple consistent bootstrap test for parametric regression function. J. Econ. **87**, 145–165 (1998)
9. Mudholkar, G.S., Wilding, G.E.: On the conventional wisdom regarding two consistent tests of bivariate independence. Statistician **52**, 41–57 (2003)
10. Patil, P., Sengupta, D.: On testing constancy of regression quantiles in non parametric regression models via kernel smoothing. Research Report CMA-SRR24-94, Centre for Mathematics and its Applications. The Australian National University (2003)
11. Rosenblatt, M.: A quadratic measure of deviation of two-dimensional density estimates and a test of independence. Ann. Stat. **3**, 1–14 (1975)
12. Rosenblatt, M., Wahlen, B.: A nonparametric measure of independence under a hypothesis of independent components. Stat. Probab. Lett. **15**, 245–252 (1992)
13. Ruppert, D., Sheather, S.J., Wand, M.P.: An effective bandwidth selector for local least squares regression. J. Am. Stat. Assoc. **90**, 1257–1270 (1995)
14. Tjostheim, D.: Measures of dependence and tests of independence. Statistics **28**, 249–284 (1996)

Chapter 3
Empirical φ*-Divergence Minimizers for Hadamard Differentiable Functionals

Patrice Bertail*, Emmanuelle Gautherat, and Hugo Harari-Kermadec

Abstract We study some extensions of the empirical likelihood method, when the Kullback distance is replaced by some general convex divergence or φ-discrepancy. We show that this generalized empirical likelihood method is asymptotically valid for general Hadamard differentiable functionals.

Keywords Generalized empirical likelihood • Empirical process • Hadamard differentiability.

3.1 Introduction

Empirical likelihood [20–22] is now a classical method for testing or constructing confidence regions for the value of some parameters in nonparametric or semi-parametric models. A possible interpretation of empirical likelihood is to see it as the

This research was partially developed within the MME-DII Center of Excellence (ANR-11-LABX-0023-01).

*The first author acknowledges the support of the French Agence Nationale de la Recherche (ANR) under grant ANR-13-BS-01-0005 (project SPADRO).

P. Bertail (✉)
MODAL'X, University Paris-Ouest, Nanterre, France

CREST-LS, Malakoff, France
e-mail: patrice.bertail@gmail.com

E. Gautherat
CREST-LS, Malakoff, France

REGARDS, University of Reims, Reims, France
e-mail: gauthera@ensae.fr

H. Harari-Kermadec
ENS-Cachan, Cachan, France
e-mail: hugo.harari@ens-cachan.fr

© Springer Science+Business Media New York 2014   21
M.G. Akritas et al. (eds.), *Topics in Nonparametric Statistics*, Springer Proceedings in Mathematics & Statistics 74, DOI 10.1007/978-1-4939-0569-0_3

minimization of the Kullback divergence, say K, between the empirical distribution of the data $\mathbb{P}_n$ and a measure (or a probability measure) $\mathbb{Q}_n$ dominated by $\mathbb{P}_n$, under linear or non-linear constraints imposed on $\mathbb{Q}_n$ by the model (see [2,19,26]). Related results may be found in the probabilistic literature about divergence or the method of entropy in mean (see [7,8,10,12,17,18]). Some generalizations of the empirical likelihood method have also been obtained by using Cressie-Read discrepancies [1, 9] and led to some econometric extensions known as "generalized empirical likelihood" [19]. Bertail et al. [4] have shown that Owen's original method in the case of the mean can be extended to any regular convex statistical divergence or φ^*-discrepancy (where φ^* is a regular convex function) under weak assumptions. The purpose of this paper is to show that this general method remains asymptotically valid for a large class of non-linear parameters, mainly Hadamard differentiable parameters in the same spirit as [2].

The layout of this paper is the following. In Part 2, we first recall some basic facts about convex integral functionals and their dual representation. As a consequence, we briefly state the asymptotic validity of the corresponding "empirical φ^*-discrepancy" method in the case of M-estimators. In part 3, we then extend this method to general Hadamard differentiable functionals, $T(\mathbb{P})$ in $\mathbb{R}^q$. An interesting interpretation of this method is that the image by T of the ball centered at $\mathbb{P}_n$ with radius $\chi_q^2(1 - \alpha)/2n$ is, in regular cases, a confidence region asymptotically with coverage probability $1 - \alpha$, for any "regular" φ^*-discrepancies.

3.2   Empirical φ*-Discrepancy Minimizers

3.2.1   φ*-Discrepancy Minimizers and Duality

We consider a measured space $(\mathcal{X}, \mathcal{A}, \mathcal{M})$ where $\mathcal{M}$ is a space of signed measures. Let f be a measurable function defined from $\mathcal{X}$ to $\mathbb{R}^r$, $r \geq 1$. For any signed measure $\mu \in \mathcal{M}$, we write $\mu f = \int f d\mu$ and if μ is a density of probability, $\mu f = \mathbb{E}_\mu(f(X))$. In the following, we consider φ, a convex function whose support $d(\varphi)$, defined as $\{x \in \mathbb{R}, \varphi(x) < \infty\}$, is assumed to be non-void (φ is said to be proper). We denote respectively $\inf d(\varphi)$ and $\sup d(\varphi)$, the extremes of this support. For every convex function φ, its convex dual or Fenchel-Legendre transform is given by

$$\varphi^*(y) = \sup_{x \in \mathbb{R}}\{xy - \varphi(x)\}, \ \forall \ y \in \mathbb{R}.$$

Recall that φ^* is then a semi-continuous inferiorly (s.c.i.) convex function. We define by $\varphi^{(i)}$ the derivative of order i of φ when it exists. From now on, we will assume the following assumptions for the function φ.

H1 φ is strictly convex and $d(\varphi)$ contains a neighborhood of 0;
H2 φ is twice differentiable on a neighborhood of 0;
H3 (renormalization) $\varphi(0) = 0$ and $\varphi^{(1)}(0) = 0$, $\varphi^{(2)}(0) = 1$, which implies that φ has an unique minimum at zero;

H4 φ is differentiable on $d(\varphi)$, that is to say differentiable on int$\{d(\varphi)\}$, with right and left limits on the respective endpoints of the support of $d(\varphi)$, where int$\{.\}$ is the topological interior;

H5 φ is twice differentiable on $d(\varphi)$;

H6 $\varphi^{(1)}$ is itself convex on his domain.

Assumptions **H4–H6** will be useful in studying the generalized empirical likelihood. Notice that **H6** implies that the second order derivative of φ is bounded from below by some constant $m > 0$ on $\mathbb{R}^+ \cap d(\varphi)$, an assumption required in [4]. Let φ satisfies the hypotheses **H1, H2, H3**. Then, the Fenchel dual transform φ^* of φ also satisfies these hypotheses. Under **H1–H6**, φ^* is convex, with a minimum at 0, $\varphi^*(0) = 0$, nonnegative thus invertible on $d(\varphi^*) \cap \mathbb{R}^+$ $\varphi^{*(2)}$ is nonincreasing on $d(\varphi^*) \cap \mathbb{R}^-$. The φ^*-discrepancy I_{φ^*} between $\mathbb{Q}$ and $\mathbb{P}$, where $\mathbb{Q}$ is a signed measure and $\mathbb{P}$ a positive measure, is defined as follows:

$$I_{\varphi^*}(\mathbb{Q}, \mathbb{P}) = \begin{cases} \int_{\mathscr{X}} \varphi^* \left(\frac{d\mathbb{Q}}{d\mathbb{P}} - 1 \right) d\mathbb{P} \text{ if } \mathbb{Q} \ll \mathbb{P} \\ +\infty \hspace{3cm} \text{else.} \end{cases} \tag{3.1}$$

For details on φ^*-discrepancies or divergences Csiszàr [4, 7, 10] and some historical comments, see [16, 18, 27, 28]. For us, the main interest of φ^*-discrepancies lies on the following duality representation, which follows from results of [6] on convex functional integrals (see also [4, 8, 17]).

Theorem 1. *Let $\mathbb{P} \in \mathscr{M}$ be a probability measure with a finite support and f be a measurable function on $(\mathscr{X}, \mathscr{A}, \mathscr{M})$. Let φ be a convex function satisfying assumptions $H1$–$H3$. If the following constraints qualification holds,*

$$\text{Qual}(\mathbb{P}): \exists \mu \in \mathscr{M}, \mu f = \theta_0 \text{ and } \inf d(\varphi^*) < \inf_{\mathscr{X}} \frac{d\mu}{d\mathbb{P}} \leq \sup_{\mathscr{X}} \frac{d\mu}{d\mathbb{P}} < \sup d(\varphi^*), \mathbb{P}-a.s.,$$

then, we have the dual equality:

$$\inf_{\mathbb{Q} \in \mathscr{M}, (\mathbb{Q}-\mathbb{P})f=\theta_0} \{I_{\varphi^*}(\mathbb{Q}, \mathbb{P})\} = \sup_{\lambda \in \mathbb{R}^r} \left\{ \lambda'\theta_0 - \int_{\mathscr{X}} \varphi(\lambda' f) d\mathbb{P} \right\}. \tag{3.2}$$

If φ satisfies $H4$, then the supremum on the right hand side of (3.2) is achieved at a point λ^* and the infimum on the left hand side at $\mathbb{Q}^*$ is given by $\mathbb{Q}^* = (1 + \varphi^{(1)}(\lambda^{*\prime} f))\mathbb{P}$.

3.2.2 Empirical Optimization of φ^*-Discrepancies

Let $X_1, \ldots X_n$ be i.i.d. r.v.'s defined on $\mathscr{X}$ with common probability measure $\mathbb{P}$. Consider the empirical probability measure $\mathbb{P}_n = \frac{1}{n} \sum_{i=1}^n \delta_{X_i}$, where δ_{X_i} is the Dirac measure at X_i. We will first consider that the parameter of interest $\theta_0 \in \mathbb{R}^q$ is the solution of some M-estimation problem $\mathbb{E}_{\mathbb{P}} f(X, \theta_0) = 0$, where f is a regular differentiable function from $\mathscr{X} \times \mathbb{R}^q \to \mathbb{R}^r$. For simplicity, we now assume that f takes its value in $\mathbb{R}^q$, that is $r = q$ and that there is no over-identification problem. The over-identified case can be treated similarly by first reducing the problem to the strictly identified case (see [3, 26]). Denote $\mathscr{M}_n = \{\mathbb{Q}_n \in \mathscr{M}$ with $\mathbb{Q}_n \ll \mathbb{P}_n\} = \{\mathbb{Q}_n = \sum_{i=1}^n q_i \, \delta_{X_i}, \, (q_i)_{1 \leq i \leq n} \in \mathbb{R}^n\}$. Considering this set of measures, instead of a set of probabilities, can be partially explained by Theorem 1, to establish the existence of a solution for the dual problem.

For a given φ, we define, by analogy to [21, 22], the quantity

$$\forall \theta \in \mathbb{R}^q, \quad \beta_n(\theta) = \inf_{\mathbb{Q}_n \in \mathscr{M}_n, \, \mathbb{Q}_n f(.,\theta)=0} I_{\varphi^*}(\mathbb{Q}_n, \mathbb{P}_n)$$

We define the corresponding random confidence region

$$\mathscr{C}_n(r) = \{\theta \in \mathbb{R}^q \,|\, \exists \mathbb{Q}_n \in \mathscr{M}_n \text{ with } \mathbb{Q}_n f(., \theta) = 0 \text{ and } n I_{\varphi^*}(\mathbb{Q}_n, \mathbb{P}_n) \leq r\},$$

where $r = r(\alpha)$ is a quantity such that $P(\theta_0 \in \mathscr{C}_n(r)) = 1 - \alpha + o(1)$.

The underlying idea of empirical likelihood and its extensions is actually a plug-in rule. Consider the functional defined by

$$\forall \theta \in \mathbb{R}^q, \quad \beta(\mathbb{P}, \theta) = \inf_{\mathbb{Q} \in \mathscr{M}, \, \mathbb{Q} \ll \mathbb{P}, \, \mathbb{Q} f(.,\theta)=0} I_{\varphi^*}(\mathbb{Q}, \mathbb{P})$$

that is, the minimization of a contrast under the constraints imposed by the model. This can be seen as a projection of $\mathbb{P}$ on the model of interest for the given pseudo-metric I_{φ^*}. If the model is true at $\mathbb{P}$, that is, if $\mathbb{E}_{\mathbb{P}} f(X, \theta_0) = 0$ at the true underlying probability $\mathbb{P}$, then clearly $\beta(\mathbb{P}, \theta_0) = 0$. A natural estimator of $\beta(\mathbb{P}, \theta)$ for fixed θ is given by the plug-in estimator $\beta(\mathbb{P}_n, \theta)$, which is $\beta_n(\theta)$. This estimator can then be used to test $\beta(\mathbb{P}, \theta) = 0$ or, in a dual approach, to build confidence region for θ_0 by inverting the test.

For $\mathbb{Q}_n$ in $\mathscr{M}_n$, the constraints can be rewritten as $(\mathbb{Q}_n - \mathbb{P}_n) f(., \theta) = -\mathbb{P}_n f(., \theta)$. Using Theorem 1, we get the dual representation

$$\begin{aligned}
\beta_n(\theta) &:= \inf_{\mathbb{Q}_n \in \mathscr{M}_n, \, (\mathbb{Q}_n - \mathbb{P}_n) f(.,\theta)=-\mathbb{P}_n f(.,\theta)} I_{\varphi^*}(\mathbb{Q}_n, \mathbb{P}_n) \\
&= \sup_{\lambda \in \mathbb{R}^q} \mathbb{P}_n \left(-\lambda' f(., \theta) - \varphi(\lambda' f(., \theta)) \right).
\end{aligned} \tag{3.3}$$

Notice that $-x - \varphi(x)$ is a strictly concave function and that the function $\lambda \to \lambda' f$ is also concave. The parameter λ can be simply interpreted as the Kuhn and Tucker coefficient associated to the original optimization problem. From this representation of $\beta_n(\theta)$, we can now derive the usual properties of the empirical likelihood and its generalization. In the following, we will also use the notations

$$\overline{f}_n = \frac{1}{n} \sum_{i=1}^{n} f(X_i, \theta), \quad S_n^2 = \frac{1}{n} \sum_{i=1}^{n} f(X_i, \theta) f(X_i, \theta)' \text{ and} S_n^{-2} = (S_n^2)^{-1}.$$

The following theorem states that generalized empirical likelihood essentially behaves asymptotically like a self-normalized sum. Links to self-normalized sum for finite n have been investigated in [3,5].

Theorem 2. *Let X, $X_1, \ldots,$ X_n be in $\mathbb{R}^p$, i.i.d. with probability $\mathbb{P}$ and $\theta_0 \in \mathbb{R}^q$ such that $\mathbb{E}_{\mathbb{P}} f(X, \theta_0) = 0$. Assume that $S^2 = \mathbb{E}_{\mathbb{P}} f(X, \theta_0) f(X, \theta_0)'$ is of rank q and that φ satisfies the hypotheses **H1–H4**. Assume that the qualification constraints $\text{Qual}(\mathbb{P}_n)$ hold. For any α in $]0, 1[$, set $r = \frac{\chi_q^2(1-\alpha)}{2}$, where $\chi_q^2(.)$ is the χ^2 distribution quantile. Then $\mathscr{C}_n(r)$ is an asymptotic confidence region with*

$$\lim_{n \to \infty} P(\theta_0 \in \mathscr{C}_n(r)) = \lim_{n \to \infty} P(n\beta_n(\theta_0) \leq r)$$

$$= \lim_{n \to \infty} P\left(n\overline{f}_n' S_n^{-2} \overline{f}_n \leq \chi_q^2(1-\alpha)\right)$$

$$= 1 - \alpha.$$

The proof of this theorem starts from the convex dual-representation and follows the main arguments of [4] and [22] for the case of the mean. It is left to the reader.

Remark 1. If φ is finite everywhere then the qualification constraints are not needed (this is for instance the case for the χ^2 divergence). In this case, the empty set problem emphasized by [14] is solved. For empirical-likelihood, the $\text{Qual}(\mathbb{P})$ constraint qualification simply says that there should be at least a solution which belongs to the support of the discrepancy. For the case of the mean, it boils down to assuming that θ belongs to the convex hull of the points.

3.3 Empirical φ^*-Discrepancy Minimizers for Hadamard Differentiable Functionals

We now extend the preceding results to general functional parameter $\theta_0 = T(\mathbb{P})$ defined on the space of signed measures $\mathscr{M}$ taking their value in $\mathbb{R}^q$. The empirical φ^*-discrepancy minimizers evaluated at θ are now defined by

$$\forall \theta \in \mathbb{R}^q, \quad \beta_n(\theta) = \inf_{\mathbb{Q}_n \in \mathscr{M}_n, \, T(\mathbb{Q}_n) = \theta} I_{\varphi^*}(\mathbb{Q}_n, \mathbb{P}_n). \tag{3.4}$$

For any $r > 0$, define the empirical ball center at $\mathbb{P}_n$ with radius r by

$$\mathscr{M}_n(r) = \{\mathbb{Q}_n \in \mathscr{M}_n, \, n I_{\varphi^*}(\mathbb{Q}_n, \mathbb{P}_n) \leq r\}$$

By using the arguments of [2, 21], the confidence region given by

$$\mathcal{C}_n(r) = \{\theta \in \mathbb{R}^q, \exists \mathbb{Q}_n \in \mathcal{M}_n(r), \theta = T(\mathbb{Q}_n)\}, \tag{3.5}$$

with $r = \frac{\chi_q^2(1-\alpha)}{2}$ has asymptotically coverage probability $1 - \alpha$. This was the main motivation of [2] for proving that empirical likelihood of general Hadamard differentiable functionals is still valid asymptotically. It means that the image by T of the ball with respect to I_{φ^*}, centered at $\mathbb{P}_n$ with radius $\frac{\chi_q^2(1-\alpha)}{2n}$ for any pseudo-metric I_{φ^*} is always an asymptotically $1 - \alpha$ confidence region for $T(\mathbb{P})$. If T is a convex functional (in particular if it is linear) then the corresponding region is automatically convex (see also [15]).

3.3.1 Hadamard Differentiability

For this we consider the following abstract empirical process framework (see [30] for details). $\mathcal{F}$ is a subset of functions of $L^2(\mathbb{P}) = \{h, \mathbb{P}h^2 < \infty\}$ endowed with $||f||_{2,\mathbb{P}} = (\mathbb{P}(f^2))^{1/2}$. We assume that $\mathcal{L}_\infty(\mathcal{F})$ is equipped with the uniform norm

$$||\tilde{\mathbb{Q}} - \mathbb{Q}||_{\mathcal{F}} = d_{\mathcal{F}}(\tilde{\mathbb{Q}}, \mathbb{Q}) = \sup_{h \in \mathcal{F}} |(\tilde{\mathbb{Q}} - \mathbb{Q})h()|.$$

We assume that expectations (resp. measures) are outer expectations (resp. outer measures) so that weak convergence is defined as Hoffman-Jörgensen convergence. This avoids measurability problems. For the same reason, we will also assume that $\mathcal{F}$ is image admissible Suslin (see [11, 30]). This ensures that the classes of the square functions and difference of square functions are $\mathbb{P}$-measurable. Assume in addition that

H7 $\mathcal{F}$ is a Suslin–Donsker Class of functions with envelop H (without loss of generality such that $H \geq 1$) such that $0 < \mathbb{P}H^2(.) < \infty$.

Recall that expectation should be understood as outer expectation. Under **H7**, the empirical process $n^{1/2}(\mathbb{P}_n - \mathbb{P})$ indexed by $\mathcal{F}$ converges (as an element of $\mathcal{L}_\infty(\mathcal{F})$) to a limit $G_\mathbb{P}$, which is a tight Borel measurable element of $\mathcal{L}_\infty(\mathcal{F})$ such that the sample paths $f \to G_\mathbb{P}(f)$ are uniformly $|| . ||_{2,\mathbb{P}}$ continuous.

Denote the covering number—the minimal number of ball of radius ε for any seminorm $||.||$ needed to cover $\mathcal{F}$—by $\mathcal{N}(\varepsilon, \mathcal{F}, ||.||)$.

H8 The following usual uniform entropy condition holds

$$\int_0^\infty \sup_{\tilde{\mathbb{P}} \in \mathcal{D}} \sqrt{\log(N(\varepsilon||H||_{2,\tilde{\mathbb{P}}}, \mathcal{F}, ||.||_{2,\tilde{\mathbb{P}}}))} d\varepsilon < \infty,$$

where $\mathcal{D}$ is the set of all discrete finitely probability measures $\tilde{\mathbb{P}}$ with $0 < \tilde{\mathbb{P}}H^2(.) < \infty$.

Define now $B(\mathscr{F}, \mathbb{P})$, the subset of $\mathscr{L}_\infty(\mathscr{F})$ which are $\|\ \|_{2,\mathbb{P}}$-uniformly continuous and bounded. We recall the following definition of Hadamard differentiability tangentially to $B(\mathscr{F})$ adapted from [24]. Notice that differentiation taken tangentially to $B(\mathscr{F}, \mathbb{P})$ weakens the notion of differentiation and makes it easier to check in statistical problems (see examples in [13, 24], Chap. 3.9 of [30] and Chap. 20 of [29]). Our result may be applied for instance the well known functional $\int F dG$, see p. 298 of [29]. The empirical counterpart of this functional yields the Mann–Whitney statistic. It is known that this functional is Hadamard differentiable tangentially to some appropriate sets.

The functional T from $\mathscr{M} \subset \mathscr{L}_\infty(\mathscr{F})$ to $\mathbb{R}^q$ is said to be Hadamard differentiable at $\mathbb{P} \in \mathscr{M}$ tangentially to $B(\mathscr{F}, \mathbb{P})$, say T is $HDT_{\mathscr{F}} - \mathbb{P}$, iff there exists a continuous linear mapping $dT_{\mathbb{P}} : \mathscr{M} \to \mathbb{R}^q$, such that for every sequence $\mu_n \to \mu \in B(\mathscr{F}, \mathbb{P})$, for every sequence $t_n \to 0$, as $n \to \infty$

$$\frac{T(\mathbb{P} + t_n \mu_n) - T(\mathbb{P})}{t_n} - dT_{\mathbb{P}}.\mu \underset{n\to\infty}{\to} 0.$$

For a Hadamard differentiable functional, $T^{(1)}(., \mathbb{P})$ is the canonical gradient or influence function, that is any function from $\mathscr{X}$ to $\mathscr{B}_1$ such that $dT_{\mathbb{P}}\mu = \mu T^{(1)}(., \mathbb{P})$, with the normalization $\mathbb{P}T^{(1)}(., \mathbb{P}) = 0$.

The following theorem establishes the validity of generalized empirical likelihood for Hadamard differentiable functionals.

Theorem 3. *Assume H1 to H8. If T defined on $\mathscr{M}$ is $HDT_{\mathscr{F}} - \mathbb{P}$ with gradient $T^{(1)}(., \mathbb{P})$ and $\mathbb{P}(T^{(1)}(., \mathbb{P}))^2 < \infty$ of rank q, for all $\alpha \in [@; 1]$, for $r = \frac{\chi_q^2(r-\alpha)}{2}$, we have*

$$\lim_{n\to\infty} \mathbb{P}\left(\theta_0 \in \mathscr{C}_n(r)\right) = 1 - \alpha.$$

An interesting example of Hadamard differentiability is given in [24] in the framework of two-dimensional censored survival times, with applications to tests of independence between duration data (see [25]). The idea is to show that the two-dimensional cumulative hazard function is Hadamard differentiable tangentially to a well chosen class of functions (given in [24]). The same kind of results may be obtained directly for real Hadamard functionals of the cumulative hazard function by the chain rule. Recall that Hadamard differentiability is fairly the weakest form of differentiability which ensures the validity of the chain rule (see [30]). Note that it is not needed to construct an empirical likelihood version adapted to the censored data as done for instance in [23] for univariate censored data. The censored structure is directly taken into account into the constraints. Comparisons between the two approaches would be of interest and will be the subject of future applied works. Other examples of interest may be found in [23] and may be treated by using Hadamard differentiability. In this framework, the choice of an adequate divergence is also a crucial issue which requires some extensive works.

3.4   Proofs of the Main Results

The following lemma generalizes a result obtained by Bertail [2] for the Kullback
divergence. It shows that the set $\mathcal{M}_n(r)$ is small and controlled by the number of
weights close to $1/n$. For a signed measure for $\mathbb{Q}_n = \sum_{i=1}^{n} q_i \delta_{X_i}$, we define $I_n =$
$\{i = 1, \ldots, n, q_i > \frac{1}{n}\}$, $I_n^+ = \{i = 1, \ldots, n, 0 < q_i \leq \frac{1}{n}\}$, and $I_n^- = \{i =$
$1, \ldots, n, q_i \leq 0\}$. We denote respectively N, N^+, N^- the cardinals of I_n, I_n^+, I_n^-.
The following lemma is interesting by itself and control the size and number of
weights q_i which can be negative.

Lemma 1. *Let **H1–H6** hold. For any $r > 0$ then, for $\mathbb{Q}_n$ in $\mathcal{M}_n(r)$ we have*

$$\forall i \in I_n, \quad \frac{1}{n} \leq q_i \leq \frac{\varphi^{*-1}(r) + 1}{n}$$

$$\forall i \in I_n^+, \quad \frac{1 - \sqrt{2r}}{n} \mathbb{I}_{r < \frac{1}{2}} \leq q_i \leq \frac{1}{n}$$

$$\text{For } r \geq \frac{1}{2}, \quad \max_{i \in I_n^+} q_i \geq \frac{1}{n} \max\left\{\left(1 - \sqrt{\frac{2r}{N^+}}\right); 0\right\} \text{ and } \frac{(N^+ - 2r)^2}{4N^+ n^2} \leq \sum_{i \in I_n^+} q_i^2$$

$$\forall i \in I_n^-, \quad r \geq \frac{1}{2} \quad \frac{1 - \sqrt{2r}}{n} \leq q_i \leq 0, \quad \text{and finally } N^- \leq 2r \mathbb{I}_{r \geq \frac{1}{2}}.$$

Proof. For $r > 0$ and $\mathbb{Q}_n \in \mathcal{M}_n(r)$, we have $\sum_{i=1}^{n} \varphi^*(nq_i - 1) \leq r$. φ^* is non-
negative, so we have for all $i \in \{1, \ldots, n\}$, $\varphi^*(nq_i - 1) \leq r$. Under **H1–H3**, φ^*
is invertible on $\mathbb{R}^+ \cap d(\varphi^*)$, so that for $r > 0$, for $i \in I_n$, $\frac{1}{n} < q_i \leq \frac{\varphi^{*-1}(r)+1}{n}$.
Now, consider $i \in I_n^+ \cup I_n^-$, $\varphi^{*(2)}$ is strictly decreasing on $\mathbb{R}^-$ so that $\varphi^{*(2)}(x) \geq$
$\varphi^{*(2)}(0) = 1$ for $x \leq 0$. Using Taylor-Lagrange expansion, we have

$$\frac{1}{2} \sum_{i \in I_n^+ \cup I_n^-} \varphi^{*(2)}(n\tilde{q}_i - 1)(nq_i - 1)^2 = \sum_{i \in I_n^+ \cup I_n^-} \varphi^*(nq_i - 1) \leq \sum_{i=1}^{n} \varphi^*(nq_i - 1) \leq r$$

with $n\tilde{q}_i - 1 \in]nq_i - 1; 0[$. Thus, $n\tilde{q}_i - 1 \leq 0$ and we have $\sum_{i \in I_n^+ \cup I_n^-}(nq_i - 1)^2 \leq 2r$
In particular, for any $i \in I_n^+$ and $r < \frac{1}{2}$, $\frac{1 - \sqrt{2r}}{n} \leq q_i \leq \frac{1}{n}$. Moreover, for any $r > 0$,
we get $\frac{1 - \sqrt{\frac{2r}{N^+}}}{n} \leq \max_{i \in I_n^+} q_i \leq \frac{1}{n}$. Now, for any $i \in I_n^-$, $1 - nq_i > 1$,

$$\sqrt{N^-} \leq \sqrt{N^-} \min_{i \in I_n^-}(1 - nq_i) \leq \sqrt{2r},$$

yielding $N^- \le 2r$. And, for any $i \in I_n^-$ and $r \ge \frac{1}{2}$, $\frac{1-\sqrt{2r}}{n} \le q_i \le 0$. Now, we also have $\sum_{i \in I_n^+}(nq_i - 1)^2 \le 2r$ so that $-2n\sum_{i \in I_n^+} q_i + N^+ \le 2r$ yielding

$$\frac{N^+ - 2r}{N+2n} \le \frac{1}{N^+}\sum_{i \in I_n^+} q_i \le \left(\frac{1}{N^+}\sum_{i \in I_n^+} q_i^2\right)^{1/2}.$$

We finally get $\frac{(N^+ - 2r)^2}{4N+n^2} \le \sum_{i \in I_n^+} q_i^2$.

Lemma 2. *Let $r > 0$,*

$$\left(\sum_{i=1}^n q_i^2\right)^{-1/2}(\mathbb{Q}_n - \mathbb{P}) \overset{w}{\underset{n \to \infty}{\to}} G_\mathbb{P} \text{ in}\mathcal{L}_\infty(\mathcal{F}),$$

uniformly over $\mathcal{M}_n(r)$ with $\mathbb{Q}_n = \sum q_i \delta_{X_i} \in \mathcal{M}_n(r)$ and $G_\mathbb{P}$ is a gaussian process in $\mathcal{L}_\infty(\mathcal{F})$.

Proof. We modify the proof of [2] which is itself an adaptation of theorem 2.11.1 and example 2.11.7 of [30]. To obtain the uniform convergence on $\mathcal{M}_n(r)$ of the weighted empirical process, it is sufficient to check that (i) $\lim_{n \to +\infty} \max_{1 \le i \le n} \frac{|q_i|}{\sqrt{\sum_{i=1}^n q_i^2}}$ and (ii) for $\delta > 0$, $\mathcal{F}_{\delta,\mathbb{P}} = \{g - h; g, h \in \mathcal{F}, \|g - h\|_{2,\mathbb{P}} \le \delta\}$ satisfies an uniform equicontinuity condition, with $\|g-h\|_{2,\mathbb{P}}^2 = \mathbb{E}_\mathbb{P}(g(X)-h(X))^2$. The domination condition in example 2.11.8 of [30] is clearly satisfied.

We start by proving? (i). For $0 < r < \frac{1}{2}$, from lemma 1,

$$\max_{i=1,\dots,n} |q_i| \le \frac{\varphi^{*-1}(r) + 1}{n} \quad \text{and} \quad \sum_{i=1}^n q_i^2 \ge \frac{1}{n^2}N + (n - N)\frac{(1 - \sqrt{2r})^2}{n^2}.$$

So for any $c \in [0, 1]$, with $\lim_{n \to \infty} N/n = c$, (i) is true.

Now, for $r > \frac{1}{2}$, we have

$$\max_{i=1,\dots,n} |q_i| \le \max\left(\frac{\varphi^{*-1}(r) + 1}{n}, \frac{1}{n}, \frac{\sqrt{2r} - 1}{n}\right) \quad \text{and} \quad \sum_{i=1}^n q_i^2 \ge \max\left(\frac{1}{n^2}N, \frac{(N^+ - 2r)^2}{4N^+n^2}\right).$$

So, we have

$$\frac{\max_{i=1,\dots,n} |q_i|}{(\sum_{i=1}^n q_i^2)^{\frac{1}{2}}} \le \frac{\max\left(\frac{\varphi^{*-1}(r)+1}{n}; \frac{\sqrt{2r}-1}{n}\right)}{\sqrt{\max\left(\frac{1}{n^2}N, \frac{(N^+-2r)^2}{4N+n^2}\right)}}.$$

Since by lemma N^- is always finite, and for any $c \in [0, 1]$, with $\lim_{n \to \infty} N/n = c$, (i) is follows.

Now, we prove (ii) that is, for all $\tilde{\varepsilon} > 0$

$$\lim_{\delta \to 0} \limsup_{n \to \infty} \sup_{q_n \Subset \mathcal{M}_n(r)} P\left(\sup_{\|f-g\|_{2,\mathbb{P}} < \delta} |G_{n,q_n}(f) - G_{n,q_n}(g)| > \tilde{\varepsilon} \right) = 0 \qquad (3.6)$$

where $G_{n,q_n}(f) = \sum_{i=1}^{n} \frac{q_i}{(\sum_{i=1}^{n} q_i^2)^{\frac{1}{2}}} (\delta_{X_i} - \mathbb{P})(f)$ and $q_n \Subset \mathcal{M}_n(r)$ is used

for $\mathbb{Q}_n = \sum_{i=1}^{n} q_i \delta_{X_i} \in \mathcal{M}_n(r)$. We define $\|f\|_{2,q_n}^2 = \sum_{i=1}^{n} \frac{q_i^2}{\sum_{i=1}^{n} q_i^2} f(X_i)^2$, $\|f\|_{2,\mathbb{P}_n}^2 = \frac{1}{n} \sum_{i=1}^{n} f(X_i)^2$.

We now follow the proof of [2] (except for controlling the maximal inequality with the covering numbers because we do not have the equivalence between the two norms $\|f - g\|_{2,q_n}^2$ and $\|f - g\|_{2,\mathbb{P}_n}^2$). To check (ii), it is sufficient to prove that, for any $\epsilon > 0$ and $\delta_n \to 0$ with n, $\Delta_n = \mathbb{P}(\sup_{\|f-g\|_{2,q_n} < \delta_n} |G_{n,q_n}(f) - G_{n,q_n}(g)| > \tilde{\varepsilon})$ tends to 0 with n. By Markov inequality and symmetrization arguments there exists a constant $C > 0$ such that

$$\Delta_n \leq \frac{2C}{\tilde{\varepsilon}} \mathbb{E}_{\mathbb{P}} \left(\|H\|_{2,q_n} \int_0^{\frac{a_n}{\|H\|_{2,q_n}}} \sqrt{\ln(N(\epsilon \|H\|_{2,q_n}, \mathscr{F}_{\delta,\mathbb{P}}, \|.\|_{2,q_n}^2))} d\epsilon \right),$$

with $a_n = \sup_{f \in \mathscr{F}_{\delta,\mathbb{P}}} \|f(X_i)\|_{2,q_n}$. Using Cauchy–Schwartz inequality, there exists a constant $C_1 > 0$ such that

$$\Delta_n^2 \leq C_1 \mathbb{P} \|H\|_{2,q_n}^2 \mathbb{E}_{\mathbb{P}} \left(\int_0^{\frac{a_n}{\|H\|_{2,q_n}}} \sqrt{\ln(N(\epsilon \|H\|_{2,q_n}, \mathscr{F}_{\delta,\mathbb{P}}, \|.\|_{2,q_n}^2))} d\epsilon \right)^2.$$

Besides we have

- $H > 1$, that is $\|H\|_{2,q_n} \geq 1$, and $\frac{a_n}{\|H\|_{2,q_n}} \leq a_n$.
- $\mathbb{P} \|H\|_{2,q_n}^2 = \mathbb{P}(H^2) < \infty$.
- Moreover, there exists some non-negative constant A such that, for any $q_n \Subset \mathcal{M}_n(r)$, $\|f - g\|_{2,q_n} \leq A \|f - g\|_{2,\mathbb{P}_n}$. So we have $a_n = \sup_{f \in \mathscr{F}_{\delta,\mathbb{P}}} \|f(X_i)\|_{2,q_n} \leq A \sup_{f \in \mathscr{F}_{\delta,\mathbb{P}}} \|f(X_i)\|_{2,\mathbb{P}_n} = \tilde{a}_n$. That is due to Lemma 1, which implies

$$\text{for } r \leq \frac{1}{2}, \forall i, \quad \frac{q_i^2}{\sum_{i=1}^{n} q_i^2} \leq \frac{(\varphi^{*-1}(r) + 1)^2}{(n - N)(1 - \sqrt{2r})^2 + N}$$

$$\text{for } r > \frac{1}{2}, \forall i, \quad \frac{q_i^2}{\sum_{i=1}^{n} q_i^2} \leq \frac{\max((\varphi^{*-1}(r) + 1)^2; 1; (1 - \sqrt{2r})^2}{N + \max\left(1 - \sqrt{\frac{2r}{n-N}}; 0\right)^2 + (1 - \sqrt{2r})^2 N^-}.$$

- Denote $Q_n^+ = \sum_{i=1}^{n} |q_i| \delta_{X_i}$ positive measure. Remark that $\|.\|_{2,q_n} = \|.\|_{2,q_n^+}$ and $Q_n^+ \in \mathscr{D}$ set of finitely discrete probability measure.

It follows by Cauchy–Schwartz inequality, there exists a constant $C_1 > 0$, such that,

$$\Delta_n^2 \leq C_1 \mathbb{P} H^2 \mathbb{E}_{\mathbb{P}} \left(\int_0^{\tilde{a}_n} \sup_{\mu \in \mathbb{D}} \sqrt{\ln(N(\epsilon \|H\|_{2,\mu}, \mathscr{F}_{\delta,\mathbb{P}}, \|.\|_{2,\mu}^2))} d\epsilon \right)^2.$$

We achieve this proof by the same arguments as [2].

Proof of Theorem 3. Let $r > 0$ and define $\mathscr{C}_n(r) = \{\theta \in \mathbb{R}^q \,|\, \exists \mathbb{Q}_n \in \mathcal{M}_n(r)$ with $\mathbb{Q}_n f(., \theta) = 0\}$ with $f(x, \theta) = T(\mathbb{P}) - \theta + T^{(1)}(x, \mathbb{P})$. Then, on the one hand, for $r = \frac{\chi_q^2(1-\alpha)}{2}$ we have by theorem 2, $\lim_{n\to\infty} \mathbb{P}(\mathscr{C}_n(r)) = 1 - \alpha$. On the other hand, we can write

$$\mathscr{C}_n(r) = \left\{ \theta \in \mathbb{R}^q \,|\, \exists \mathbb{Q}_n \in \mathcal{M}_n(r), \theta = T(\mathbb{P}) + \frac{1}{\mathbb{Q}_n(1)} (\mathbb{Q}_n - \mathbb{P})(T^{(1)}(., \mathbb{P})) \right\}.$$

We show that it is the linearized and renormalized part of

$$\tilde{\mathscr{C}}_n(r) = \left\{ \theta \in \mathbb{R}^q \,|\, \exists \mathbb{Q}_n \in \mathcal{M}_n(r), \theta = T(\mathbb{P}) + (\mathbb{Q}_n - \mathbb{P})(T^{(1)}(., \mathbb{P})) + R_n(\mathbb{Q}_n, \mathbb{P}) \right\},$$

where $R_n(\mathbb{Q}_n, \mathbb{P})$ is the remainder in the Hadamard-differentiability of $T(\mathbb{Q}_n)$ around $T(\mathbb{P})$ because T is $HDT_{\mathscr{F}} - \mathbb{P}$ with canonical gradient $T^{(1)}(., \mathbb{P})$. Lemma 2 gives that

$$\left(\sum_{i=1}^n q_i^2 \right)^{-1/2} (\mathbb{Q}_n - \mathbb{P})(T^{(1)}(., \mathbb{P})) \xrightarrow[n\to\infty]{w} G_{\mathbb{P}}(T^{(1)}(., \mathbb{P})).$$

and yields the uniform convergence of $\left(\sum_{i=1}^n q_i^2 \right)^{-1/2} (\mathbb{Q}_n - \mathbb{P})$ on $\mathcal{M}_n(r)$. Thus by Theorem 20.8 of [29] we have the control $R(\mathbb{Q}_n, \mathbb{P}) = o_{\mathbb{P}}((\sum_{i=1}^n q_i^2)^{1/2})$ uniformly over $\mathcal{M}_n(r)$. Moreover $\mathbb{Q}_n(1)$ converges in probability to $\mathbb{P}(1)$ with n. It follows that $\mathscr{C}_n(r)$ and $\tilde{\mathscr{C}}_n(r)$ are asymptotically equivalent.

References

1. Baggerly, K.A.: Empirical likelihood as a goodness-of-fit measure. Biometrika **85**, 535–547 (1998)
2. Bertail, P.: Empirical likelihood in some semi-parametric models. Bernoulli **12**, 299–331 (2006)
3. Bertail, P., Gautherat, E., Harari-Kermadec, H.: Exponential bounds for quasi-empirical likelihood. Working Paper n34, CREST. http://www.crest.fr/images/doctravail//2005-34.pdf (2005). Cited 12 Dec 2012
4. Bertail, P., Harari-Kermadec, H., Ravaille, D.: φ-divergence empirique et vraisemblance empirique généralisée. Annales d'Économie et de Statistique **85**, 131–158 (2007)

5. Bertail, P., Gautherat, E., Harari-Kermadec, H.: Exponential bounds for multivariate self-normalized sums. Electron. Commun. Probab. **13**, 628–640 (2008)
6. Borwein, J.M., Lewis, A.S.: Duality relationships for entropy like minimization problem. SIAM J. Optim. **29**, 325–338 (1991)
7. Broniatowski, M., Keziou, A.: Minimization of φ-divergences on sets of signed measures. Studia Sci. Math. Hungar. **43**(4), 403–442 (2006)
8. Broniatowski, M., Keziou, A.: Parametric estimation and tests through divergences and the duality technique. J. Multivariate Anal. **100**(1), 16–36 (2009)
9. Corcoran, S.: Bartlett adjustment of empirical discrepancy statistics. Biometrika, **85**, 967–972 (1998)
10. Csiszár, I.: On topology properties of f-divergences. Studia Sci. Math. Hungar. **2**, 329–339 (1967)
11. Dudley, R.M.: A Course on Empirical Processes. Ecole d'été de probabilité de Saint Flour. Lecture Notes in Mathematics, vol. 1097, pp. 2–241. Springer, New York (1984)
12. Gamboa, F., Gassiat, E.: Bayesian methods and maximum entropy for ill-posed inverse problems. Ann. Stat. **25**, 328–350 (1996)
13. Gill, R.D.: Non- and semiparametric maximum likelihood estimators and the von Mises method. Scand. J. Stat. **16**, 97–128 (1989)
14. Grendar, M., Judge, G.: Empty set problem of maximum empirical likelihood methods, Electron. J. Stat. **3**, 1542–1555 (2009)
15. Hall, P., La Scala, B.: Methodology and algorithms of empirical likelihood. Int. Stat. Rev. **58**, 109–127 (1990)
16. Léonard, C.: Convex conjugates of integral functionals. Acta Mathematica Hungarica. **93**, 253–280 (2001)
17. Leonard, C.: Minimizers of energy functionals. Acta Math. Hungar. **93**, 281–325 (2001)
18. Liese, F., Vajda, I.: Convex Statistical Distances. Teubner, Leipzig (1987)
19. Newey, W.K., Smith, R.J.: Higher order properties of GMM and generalized empirical likelihood estimators. Econometrica, **72**, 219–255 (2004)
20. Owen, A.B.: Empirical likelihood ratio confidence intervals for a single functional. Biometrika. **75** (2), 237–249 (1988)
21. Owen, A.B.: Empirical likelihood ratio confidence regions. Ann. Stat. **18**, 90–120 (1990)
22. Owen, A.B.: Empirical Likelihood. Chapman and Hall/CRC, Boca Raton (2001)
23. Pan, X.R., Zhou, M.: Empirical likelihood in terms of cumulative hazard function for censored data. J. Multivariate Anal. **80**, 166–188 (2002)
24. Pons, O., Turckheim, E.: Von Mises method, Bootstrap and Hadamard differentiability. Statistics **22**, 205–214 (1991)
25. Pons, O., Turckheim E.: Tests of independence for bivariate censored data based on the empirical joint hazard function. Scand. J. Stat. **18**(1), 21–37 (1991)
26. Qin, J., Lawless, J.: Empirical likelihood and general estimating equations. Ann. Stat. **22**, 300–325 (1994)
27. Rockafeller, R.: Integrals which are convex functionals. Pacific J. Math. **24**, 525–339 (1968)
28. Rockafellar, R.T.: Convex Analysis. Princeton University Press, Princeton (1970)
29. van der Vaart, A.W.: Asymptotic Statistics. Cambridge Series in Statistical and Probabilistic Mathematics. Cambridge University Press, Cambridge (1998)
30. van der Vaart, A.W., Wellner, J.A.: Weak Convergence and Empirical Process: With Applications to Statistics. Springer, New York (1996)

Chapter 4
Advances in Permutation Tests for Covariates in a Mixture Model for Preference Data Analysis

Stefano Bonnini and Francesca Solmi

Abstract The rating problem arises very often in statistical surveys and a new approach for this problem is represented by the combination of uniform and binomial models. In the present work a simulation study is presented to prove the good power behavior of a permutation test on the covariates of a complex model, with more than one covariate. Moreover a discussion on the minimum sample size needed to perform such permutation test is also given.

Keywords Permutation test • Mixture model • Combination based test • Covariate effect

4.1 Introduction

Ordinal data are typical of many application fields like marketing, clinical studies, performance analysis, and many others. In this framework a new approach is represented by the combination of uniform and binomial models (briefly CUB models), introduced by [3,6], and [7] and then generalized by [8] and [5]. It assumes that the judgment process follows a psychological mechanism which takes into account the feeling towards the object under judgment and an uncertainty generated by the presence of multiple alternatives. CUB models are generated by a class of discrete probability distributions obtained as a mixture of a shifted Binomial and a Uniform random variable. One of the most interesting proposals of mixture models for discrete data in biomedical studies is that of [2], related to the distribution of the number of times a given event has been identified by the registration system of a specific disease.

S. Bonnini (✉)
Department of Economics and Management, University of Ferrara, Via Voltapaletto 11,
35020 Ferrara, Italy
e-mail: stefano.bonnini@unife.it

F. Solmi
Department of Statistical Sciences, University of Padova, Padova, Italy
e-mail: francesca.solmi@unipd.it

© Springer Science+Business Media New York 2014
M.G. Akritas et al. (eds.), *Topics in Nonparametric Statistics*, Springer Proceedings
in Mathematics & Statistics 74, DOI 10.1007/978-1-4939-0569-0_4

In the case of CUB models, characteristics of subjects (consumers, patients,...) and objects (products, drugs,...) can be also introduced. Let us assume n people are rating a given item, and hence we observe the sample data $y = (y_1, y_2, \ldots, y_n)$. For example let us consider an international marketing survey performed in Italy, Sweden, and Great Britain where a sample of n people is asked to give a vote (satisfaction level) to a new wine after testing it. Value y_i is the vote (integer number from 1 to $m = 10$) given by the i-th evaluator. Moreover let x_i and w_i ($i = 1, \ldots, n$) be row-vectors of subject's covariates for explaining feeling and uncertainty respectively. In our marketing example let us consider for the feeling the binary covariates *gender* (1: male; 0: female), *italian* (1: Italian; 0: non Italian), and *swedish* (1: Swedish; 0: non Swedish) and for the uncertainty the numerical covariate *age*. As a matter of fact males usually appreciate wine more than females, hence the feeling toward wine of males is supposed to be greater than that of females. Similarly, for enological culture and traditions, Italians are supposed to have a greater feeling than people from Great Britain while Swedes are supposed to have less feeling than British. Moreover younger people are supposed to be less experienced and therefore more uncertain in their evaluations than older people, thus *age* might be considered a covariate negatively related to uncertainty. Therefore for the i-th evaluator the row-vector of size 3 x_i represents gender and nationality (covariates for feeling), and the variable w_i (row-vector of size 1) represents age (covariate for uncertainty).

According to the CUB model theory, a specific score given by an evaluator may be interpreted as consequence of a psychological mechanism such that the evaluation is the result of pairwise comparisons between all the possible scores. For instance, if an evaluator choose the vote 5, this evaluation is better than other 4 evaluations (1, 2, 3, and 4) and worse than other 5 evaluations (6, 7, 8, 9, and 10), hence we can say that we have 4 successes and 5 failures. Therefore if ξ_i is the probability that the chosen score is worse than another one in a pairwise comparison (failure) and $1 - \xi_i$ is the probability that the chosen score is better than another one (success), probability of observing the score y_i might be represented by a shifted binomial distribution with parameter ξ_i. The value $1 - \xi_i$ can be considered a measure of feeling.

The shifted binomial distribution may represent the psychological mechanism of the choices of an evaluator only in the absence of uncertainty. If the evaluator were completely uncertain, the probability of observing the score y_i could be the same for each possible value in the set $1, 2, \ldots, m$, and the uniform distribution would be suitable. In real situations uncertainty is neither null nor maximum. If π_i denotes the probability of non-uncertainty of i-th evaluator, the probability of observing the score y_i might be computed with the mixture model defined by Eq. (4.1). The value $1 - \pi_i$ can be considered a measure of uncertainty.

The general formulation of a CUB(p, q) model (i.e., with p covariates for uncertainty and q covariates for feeling) is expressed by

$$\Pr(Y_i = y | x_i, w_i) = \pi_i \binom{m-1}{y-1} (1 - \xi_i)^{y-1} \xi_i^{m-y} + (1 - \pi_i) \left(\frac{1}{m} \right), \quad (4.1)$$

with $y = 1, 2, \ldots, m$, $\pi_i = [1 + \exp(-x_i\beta)]^{-1}$, and $\xi_i = [1 + \exp(-w_i\gamma)]^{-1}$, where $\psi = (\beta', \gamma')'$ are the coefficients of the relation between the parameters $\pi = (\pi_1, \ldots, \pi_n)$ and $\xi = (\xi_1, \ldots, \xi_n)$ and the respective covariates. In the marketing example $\beta = (\beta_1, \beta_2, \beta_3)'$ represents the vector of coefficients which describe how the covariates *gender*, *italian*, and *swedish* affect the feeling toward wine and γ represents the coefficient which describe how the covariate *age* affects the uncertainty.

Inference on CUB models has been developed both in a parametric framework, via maximum likelihood and asymptotic theory (see [6]), and via nonparametric inference (see [1]). The permutation test on the effect of covariates in a CUB model proposed by [1] works according to the following procedure: (i) calculate the observed value of a suitable test statistic T function of the observed dataset; (ii) permute the rows of only the columns of the dataset related to the tested covariates, keeping fixed the remaining columns of the dataset (perform constrained permutations within blocks, in which the nontested covariates are constant); (iii) calculate the value of the test statistic corresponding to the permuted dataset; (iv) repeat steps (ii) and (iii) B times, obtaining the permutation distribution of the test statistic; (v) calculate the p-value of the test as usual, according to the permutation distribution of T.

Three different test statistics are proposed in [1]: (a) the classical likelihood-ratio statistic (hereafter T_{lrt}); (b) a linear combination of the *Wald* type test statistics for the partial tests on the single coefficients (T_{wald}); (c) a nonparametric combination of p-values of the partial tests on the single coefficients (T_{npc}). The methods are competitive alternatives to the classical parametric test for small sample sizes, since their rejection rates respect α under H_0 and assume greater values under H_1 (higher estimated powers).

In Sect. 4.2 the main results of a Monte Carlo simulation study are shown, to prove the good performances of the permutation solutions in specific situations characterized by more than one covariate, to study how the number of covariates in the model affects the performance of the tests. Section 4.3 is devoted to a discussion about the minimum sample size needed to perform this powerful permutation test. In Sect. 4.4 the test is applied to a real problem related to a customer satisfaction survey of a ski school. The conclusions of the study are reported in Sect. 4.5.

4.2 Simulation Study

Considering dichotomous covariates and assuming nonnegative coefficients, that is $\beta_j \geq 0$ for $j = 1, \ldots, p$ and $\gamma_s \geq 0$ for $s = 1, \ldots, q$, a measure of the "distance" between the model under the null hypothesis of no covariates' effect, namely the CUB$(0, 0)$ model, and the CUB(p, q) model under the alternative hypothesis, in terms of feeling and uncertainty, can be given by the values $\delta_{\pi(x_1, \ldots, x_q)} = \pi_{(1, \ldots, 1)} - \pi_{(0, \ldots, 0)}$ and $\delta_{\xi(w_1, \ldots, w_q)} = \xi_{(1, \ldots, 1)} - \xi_{(0, \ldots, 0)}$, where $\pi_{x_1, \ldots, x_p}$ and $\xi_{w_1, \ldots, w_q}$ are the values of π and ξ respectively for the i-th subject/object, computed as function of the observed covariates. In the present simulation study a fixed marginal effect of the only covariate taken into account for the uncertainty parameter is considered: $\delta_{\pi(x_1)} = \pi_{(1)} - \pi_{(0)} = -0.6$. For the parameter of feeling three alternative values for the fixed marginal effect of each of the three considered covariates are taken into account: $\delta_{\xi(w_1)} = \xi_{(1,0,0)} - \xi_{(0,0,0)} = \delta_{\xi(w_2)} = \xi_{(0,1,0)} - \xi_{(0,0,0)} = \delta_{\xi(w_3)} = \xi_{(0,0,1)} - \xi_{(0,0,0)} = 0.05, 0.10$ or 0.20.

In each setting the marginal effects of the covariates of the feeling parameter ξ (and thus the coefficients $\gamma_1, \ldots, \gamma_q$) are simulated to be equal. Under the null CUB$(0,0)$ model the values $\pi = \pi_{(0,\ldots,0)} = 0.80$ and $\xi = \xi_{(0,\ldots,0)} = 0.10$, very common in real problems, are assumed. Let us consider that the value of the parameter of feeling as function of the covariates is $\xi_i = [1 + \exp(-w_i\gamma)]^{-1} = [1 + \exp(-\gamma_0 - \sum_{s=1}^{q} w_{is}\gamma_s)]^{-1}$, where w_{is} is the value of s-th covariate for i-th evaluator. Thus $\xi_{(0,\ldots,0)} = [1 + \exp(-\gamma_0)]^{-1}$. The overall effects $\delta_{\xi(w_1,\ldots,w_q)}$ are computed according to the following steps:

1. compute the value γ_0 as function of $\xi_{(0,\ldots,0)}$: $\gamma_0 = ln[\xi_{(0,\ldots,0)}/(1 - \xi_{(0,\ldots,0)})]$. If $\xi_{(0,\ldots,0)} = 0.10$ then $\gamma_0 = -2.197$.
2. Compute the coefficient $\gamma_s = \gamma$ as function of the marginal effect $\delta_{\xi(w_s)} = \delta$, because $\delta_{\xi(w_s)} = [1 + \exp(-\gamma_0 - \gamma_s)]^{-1} - \xi_{(0,\ldots,0)} \Rightarrow \gamma_s = \gamma = ln[(\delta + \xi_{(0,\ldots,0)})/(1 - \delta - \xi_{(0,\ldots,0)})] - \gamma_0$. For example if $\delta = 0.05$ then $\gamma = 0.462$.
3. Compute $\delta_{\xi(w_1,\ldots,w_q)} = \xi_{(1,\ldots,1)} - \xi_{(0,\ldots,0)} = [1 + \exp(-\gamma_0 - \sum_{s=1}^{q} \gamma_s)]^{-1} - [1 + \exp(-\gamma_0)]^{-1} = [1 + \exp(-\gamma_0 - q\gamma)]^{-1} - [1 + \exp(-\gamma_0)]^{-1}$. Thus for example if $\gamma = 0.462$ and $q = 2$ then $\delta_{\xi(w_1,\ldots,w_q)} = \delta_{\xi(w_1,w_2)} = 0.12$.

The considered scenarios, that is the models in the alternative hypothesis and the marginal and joint effects of the covariates on π and ξ are reported in Table 4.1. Clearly we wish to study the performance of the testing procedures in the less favorable conditions under H_0 and under H_1. Under the null hypothesis we are interested to study the accuracy of the tests, that is to determine whether the

Table 4.1 Simulation settings: models considered under the alternative hypothesis and marginal and overall effects of the covariates on π and ξ

Setting	Model under H_1	$\delta_{\pi(x_1)}$	$\delta_{\xi(w_s)}$ $s = 1,\ldots,q$	$\delta_{\xi(w_1,\ldots,w_q)}$
1	CUB$(0,2)$	–	0.05	0.12
	CUB$(0,3)$	–	0.05	0.21
	CUB$(1,2)$	–0.6	0.05	0.12
	CUB$(1,3)$	–0.6	0.05	0.21
2	CUB$(0,2)$	–	0.10	0.26
	CUB$(0,3)$	–	0.10	0.46
	CUB$(1,2)$	–0.6	0.10	0.26
	CUB$(1,3)$	–0.6	0.10	0.46
3	CUB$(0,2)$	–	0.20	0.52
	CUB$(0,3)$	–	0.20	0.76
	CUB$(1,2)$	–0.6	0.20	0.52
	CUB$(1,3)$	–0.6	0.20	0.76

Table 4.2 Estimated rejection probabilities for $m = 7$, $n = 50$, and $\alpha = 0.05$

True model	H_0	H_1	T_{lrt}	T_{wald}	T_{npc}	$T_{\text{par-lrt}}$
H_0	CUB(0,0)	CUB(3, 3)	0.059	0.049	0.064	0.119
H_1, Setting 1	CUB(0, 0)	CUB(0, 2)	0.235	0.229	0.220	0.501
	CUB(0, 0)	CUB(0, 3)	0.257	0.275	0.232	0.671
	CUB(0, 0)	CUB(1, 2)	0.507	0.064	0.195	0.919
	CUB(0, 0)	CUB(1, 3)	0.394	0.044	0.111	0.891
H_1, Setting 2	CUB(0, 0)	CUB(0, 2)	0.688	0.686	0.646	0.740
	CUB(0, 0)	CUB(0, 3)	0.806	0.816	0.681	0.849
	CUB(0, 0)	CUB(1, 2)	0.437	0.167	0.188	0.743
	CUB(0, 0)	CUB(1, 3)	0.468	0.213	0.333	0.761
H_1, Setting 3	CUB(0, 0)	CUB(0, 2)	0.985	0.983	0.958	0.992
	CUB(0, 0)	CUB(0, 3)	0.986	0.981	0.965	0.999
	CUB(0, 0)	CUB(1, 2)	1.000	0.999	1.000	1.000
	CUB(0, 0)	CUB(1, 3)	0.999	0.999	1.000	1.000

rejection rates are near α and do not exceed α. The CUB(3, 3) model presents 6 more coefficients than the CUB(0, 0) model, hence some of the considered testing procedures, like T_{wald} and T_{npc} should be disadvantaged because they are based on multiple tests and the power is expected to increase with the number of partial tests (i.e., with the number of tested coefficients). Hence a CUB(3, 3) model in the alternative hypothesis when the null hypothesis is true is less favorable than the other models considered in the alternative hypothesis when the null hypothesis is false and the alternative is true. Conversely when the alternative hypothesis is true and the greater the rejection rates the better the performance of the tests, the less favorable conditions are represented by simpler models with a lower number of tested coefficients.

The permutation solutions T_{lrt}, T_{wald}, and T_{npc} and the parametric likelihood-ratio test, hereafter labeled as $T_{\text{par-lrt}}$ are compared. A number of $B = 1,000$ permutations and $CMC = 1,000$ Monte Carlo replications are considered. Tables 4.2 and 4.3 summarize the results of the simulations, in terms of estimated rejection probabilities of the compared tests, respectively for sample size $n = 50$ and $n = 100$. Notice how the power of all the procedures tends to increase with the sample size n. The only exception to this general rule is represented by setting three in presence of one covariate for π (last two lines of Tables 4.2 and 4.3) in correspondence to T_{lrt}, T_{wald}, and $T_{\text{par-lrt}}$. However the power behavior of the testing methods in these cases is very similar, the rejection rates are very near 1, and the differences between the rejection rates in the case $n = 50$ and in the case $n = 100$ are very small especially when compared to the other settings.

It is evident that the parametric solution does not control the type I error for small sample sizes (specifically $n = 50$). A better performance can be appreciated for $n = 100$. As a matter of fact a curious power behavior can be observed for all the testing

Table 4.3 Estimated rejection probabilities for $m = 7$, $n = 100$, and $\alpha = 0.05$

True model	H_0	H_1	T_{lrt}	T_{wald}	T_{npc}	$T_{par-lrt}$
H_0	CUB(0, 0)	CUB(3, 3)	0.039	0.035	0.037	0.063
H_1, Setting 1	CUB(0, 0)	CUB(0, 2)	0.455	0.466	0.420	0.476
	CUB(0, 0)	CUB(0, 3)	0.626	0.635	0.528	0.785
	CUB(0, 0)	CUB(1, 2)	0.874	0.393	0.589	0.959
	CUB(0, 0)	CUB(1, 3)	0.810	0.294	0.571	0.964
H_1, Setting 2	CUB(0, 0)	CUB(0, 2)	0.972	0.968	0.915	0.965
	CUB(0, 0)	CUB(0, 3)	0.999	0.997	0.965	0.998
	CUB(0, 0)	CUB(1, 2)	0.853	0.623	0.718	0.978
	CUB(0, 0)	CUB(1, 3)	0.866	0.628	0.708	0.991
H_1, Setting 3	CUB(0, 0)	CUB(0, 2)	1.000	1.000	1.000	1.000
	CUB(0, 0)	CUB(0, 3)	1.000	1.000	1.000	1.000
	CUB(0, 0)	CUB(1, 2)	0.946	0.926	1.000	0.987
	CUB(0, 0)	CUB(1, 3)	0.954	0.939	1.000	0.998

procedures: when a significant covariate is introduced for the uncertainty parameter, the estimated rejection probabilities tend to decrease. This behavior seems to be more evident for the permutation solutions based on the nonparametric combination of partial permutation tests (T_{wald} and T_{npc}). This result is due to the anomalous behavior of the partial test on the coefficient of the covariate of the uncertainty parameter π: a possible reason is that when more uncertainty is introduced in a part of the observations (one of the two sub-groups identified by the covariate affecting the uncertainty) it is actually more difficult for the tests to recognize the influence of the other covariates on the feeling parameter. On the other side, if the likelihood-ratio tests (both parametric and permutation versions) are considered, there is a loss of power but usually, when a new covariate is added into the model under H_1 and consequently the "distance" between null and alternative model grows, there is a power improvement of the tests. However it must to be noticed that, for fixed values of $\delta_{\pi(x_1)}$, the power of all the procedures is an increasing function of $\delta_{\xi(w_s)}$ because it grows passing from Setting 1 to Setting 3.

4.3 Minimum Sample Size for the Application of the Permutation Test

The application of a permutation test is not possible when the minimum available p-value of the test is greater than the significance level α, according to the distribution of the test statistic. This problem may occur when the number of possible permutations, and consequently the cardinality of the permutation space, is too low: when all the B possible permutations are considered, the p-value of a permutation test when the null hypothesis is rejected for high values of the test statistic T is given by

$$\lambda = \frac{\sharp(T^* \geq T_{\text{obs}})}{B} = \frac{\sum_{b=1}^{B} I(T_b^* \geq T_{\text{obs}})}{B},$$

where T^* is the random variable which represents the test statistic according to the permutation distribution, T_{obs} is the observed value of T, T_b^* is the value of T^* corresponding to the b-th permutation of the dataset, B is the total number of permutations, and $I(T_b^* \geq T_{\text{obs}})$ represents the indicator function of the event $T_b^* \geq T_{\text{obs}}$, i.e., it takes value 1 when the inequality is true and 0 otherwise. The p-value satisfies the constraints $\lambda_{\min} = 1/B \leq \lambda \leq 1 = \lambda_{\max}$. Hence a necessary condition for the applicability of the method is $\lambda_{\min} = 1/B < \alpha$ or equivalently $B > 1/\alpha$. If $\alpha = 0.05$ then the minimum number of total distinct permutations should be $B = \text{int}(1/\alpha)+1 = 21$. Similarly if $\alpha = 0.10$ then the minimum number of permutations should be $B = 11$; if $\alpha = 0.01$ then the minimum number of permutations should be $B = 101$; etc.

Let us consider the permutation test of the problem under study and let us indicate the number of subjects/evaluators belonging to the j-th group defined by a given combination of values/levels of the nontested covariates with n_j, $j = 1, \ldots, M$, where M indicates the number of groups in the sample. The total number of distinct permutations is given by

$$B = \prod_{j=1}^{M} n_j!,$$

while the *all-cell-permutation* condition, i.e., the condition that at least two data are observed for each combination of values of the nontested covariates (see [1]), necessary in order to apply the permutation solution, implies that the minimum number of permutations is equal to $B = 2^M$. The permutation strategy should take into account that under H_0 exchangeability exists between populations with identical distributions. In [4] it was proved that for tests of hypotheses of no effect, when the two or more compared populations present nonidentical distributions the permutation test can have a greater type I error rate, even under the null hypothesis. Hence no large sample sizes are needed for the application of the test. For example when $\alpha = 0.05$ and $M \geq 5$ the all-cell-permutation condition allows the application of the test, as $B \geq 2^5 = 32 > 21$.

Tables 4.4 and 4.5 report the minimum sample sizes (and the optimal partition of evaluators in the groups) for some values of M and for two alternative values of the significance level $\alpha = 0.05$ and 0.10. Notice that the minimum required sample sizes appear to be quite low. As expected, they increase with M, i.e., with the number of nontested covariates (or the number of possible combinations of levels of nontested covariates). In real case applications a common problem consists in testing the influence of a covariate on the response in presence of two, three, or four nontested dichotomous covariates: in these cases we would have $M = 4, 8$ and 16 respectively.

Table 4.4 Minimum sample sizes (and optimal partition of subjects in the groups) for $\alpha = 0.05$ and some values of M

M	$n_{\min}$	$n_j, j = 1, \ldots, M$
1	$n = 4$	$n_1 = 4 \ (B = 24)$
2	$n = 6$	$n_1 = 2; n_2 = 4 \ (B = 48)$ or $n_1 = n_2 = 3 \ (B = 36)$
3	$n = 7$	$n_1 = n_2 = 2; n_3 = 3 \ (B = 24)$
4	$n = 4$	$n_1 = n_2 = n_3 = 2; n_4 = 3 \ (B = 48)$
$M \geq 5$	$n = 2M$	$n_j = 2; j = 1, \ldots, M \ (B = 2^M)$

Table 4.5 Minimum sample sizes (and optimal partition of subjects in the groups) for $\alpha = 0.10$ and some values of M

M	$n_{\min}$	$n_j, j = 1, \ldots, M$
1	$n = 4$	$n_1 = 4 \ (B = 24)$
2	$n = 5$	$n_1 = 2; n_2 = 3 \ (B = 12)$
3	$n = 7$	$n_1 = n_2 = 2; n_3 = 3 \ (B = 24)$
4	$n = 2M = 8$	$n_1 = n_2 = n_3 = n_4 = 2 \ (B = 16)$
$M \geq 5$	$n = 2M$	$n_j = 2; j = 1, \ldots, M \ (B = 2^M)$

4.4 Real Application: Survey on the Ski School of Sesto

In winter 2010/11 a customer satisfaction survey on the ski courses for young children (up to 13 years old) promoted by the Ski School of Sesto, near Bolzano, in Italy, was performed by the University of Padova. A sample of 135 customers was asked to give a vote (on a 1–10 scale) about some aspects of the service. The respondents were the parents of the kids who took part in the courses. Among the monitored aspects let us consider the "ease of learning" of the children and the "helpfulness of the teachers" and let us apply the proposed permutation test on two $CUB(2, 2)$ models to verify if and how the covariates "nationality" and "first participation" affect uncertainty and feeling of the respondents. The use of the permutation solution based on the NPC combination allows to adjust the p-values of the partial tests on the single coefficients controlling the multiplicity, for attributing the eventual significance of the global tests to the single coefficients. The two covariates are dichotomous variables taking value 1 when the respondent is italian and when the child/children take part to the evaluated course for the first time respectively and value 0 otherwise.

Table 4.6 shows that both the covariates have a significant effect in the model for the "ease of learning" (global p-value $= 0.022$, $B = 1,000$ permutations). Specifically they affect parameter π: "nationality" has a positive effect on the parameter (hence a negative effect on the uncertainty) and "first participation" an opposite effect. In other words italian customers are less uncertain and people who take part to the ski courses for the first time present greater uncertainty when they evaluate the "ease of learning." Similarly the covariates globally have an influence on the evaluations about the "helpfulness of the teacher" (global p-value $= 0.071$,

Table 4.6 Ease of learning: coefficient estimates, permutation tests on covariates at $\alpha = 0.10$ (partial adjusted p-values in brackets), and sign of the relation between covariates and uncertainty and/or feeling in the $CUB(2, 2)$ model

Parameter	Nationality	First participation
π	$\hat{\beta}_1 = 3.776$	$\hat{\beta}_2 = -1.710$
	(0.023)	(0.085)
	–	+
ξ	$\hat{\gamma}_1 = n.s.$	$\hat{\gamma}_2 = n.s.$
	(0.518)	(0.518)
global p-value 0.022		

Table 4.7 Helpfulness of the teacher: coefficient estimates, permutation tests on covariates at $\alpha = 0.10$ (partial adjusted p-values in brackets), and sign of the relation between covariates and uncertainty and/or feeling in the $CUB(2, 2)$ model

Parameter	Nationality	First participation
π	$\hat{\beta}_1 = 2.372$	$\hat{\beta}_2 = n.s.$
	(0.072)	(0.421)
	–	+
ξ	$\hat{\gamma}_1 = n.s.$	$\hat{\gamma}_2 = n.s.$
	(0.955)	(0.312)
global p-value 0.071		

$B = 1,000$ permutations). In this case only "nationality" produces a significant effect on the uncertainty and also in this case the Italians are less uncertain (see Table 4.7). No covariate affects the feeling for any of the considered responses.

4.5 Conclusions

In this paper some new properties of a permutation solution to test for covariates' effect on ordinal responses in a combination of uniform and shifted binomial model (CUB model), recently proposed in [1], are discussed. The simulation study proves that the permutation test controls the type I error in the presence of more than one tested covariate also when the sample size is not large and seems to be powerful under H_1. The test based on the likelihood ratio test statistic seems to be a very well performing alternative to the classical parametric counterpart when the sample size is not large even in presence of more than one covariate both for feeling and uncertainty. The minimum sample sizes and the best partitions of subjects in the M different combinations of the nontested-covariates' levels have been presented.

The CUB model is very useful because it is parsimonious in terms of number of parameters and it can be used for several types of applications when the evaluators are people (performance analysis, customer satisfaction, studies on the effects of drugs or therapies,...). This unique and interesting approach allows to distinguish between the two components which mainly affect the final choice of the evaluators: uncertainty and feeling. An interesting aspect consists in the possibility of testing whether and how some individual characteristics of the evaluators affect uncertainty and feeling. In this framework the wald test and likelihood ratio test present two main limits: they are not suitable solutions for small sample sizes (small number of evaluators) and they cannot be applied for multivariate CUB models. The permutation test proposed by [1] allows to overcome these limits because no assumption about the null distribution of the test statistic is needed, thus the asymptotic distribution under H_0 is not required, and the dependence among component variables in the multivariate case can be implicitly taken into account with a suitable permutation strategy and through the nonparametric combination (NPC) methodology. The present work methodologically (study of power behavior) and from the application point of view (minimum sample size determination, application example) contributes to study utility, performances and applicability, conditions of the cited nonparametric solution, in fact providing a new solution to complex application problems in the CUB framework, where performance and applicability of such methodology had not yet been proved.

Acknowledgements Authors wish to thank the University of Padova (CPDA092350/09) and the Italian Ministry for University and Research MIUR project PRIN2008 -CUP number C91J10000000001 (2008WKHJPK/002) for providing the financial support for this research. They also wish to thank Professor Luigi Salmaso and Doctor Rosa Arboretti Giancristofaro (University of Padova, Italy) and Mr. Herbert Summerer (Director of the Sky School of Sesto) for providing the data used in the real case application.

References

1. Bonnini, S., Piccolo, D., Salmaso, L., Solmi, F.: Permutation inference for a class of mixture models. Commun. Stat. A **41**, 2879–2895 (2012)
2. Bohningen, D., Ekkehart, D., Kuhnert, R., Schon, D.: Mixture models for capture-recapture count data. Stat. Methods Appl. **14**, 29–43 (2005)
3. D'Elia, A., Piccolo, D.: A mixture model for preference data anaysis. Comput. Stat. Data Anal. **49**, 917–934 (2005)
4. Huang, Y., Xu, H., Calian, V., Hsu, J.C.: To permute or not to permute. Bioinformatics **22**(18), 2244–2248 (2006)
5. Iannario, M., Piccolo, D.: CUB models: Statistical methodsand empirical evidence. In: Kennett, R.S., Salini, S. (eds.) Modern Analysis of Customer Surveys with applications using R, pp. 231–258. Wiley, Chichester (2011)
6. Piccolo, D.: On the moments of a mixture of uniform and shifted binomial random variables. Quad. Stat. **5**, 85–104 (2003)
7. Piccolo, D.: Observed information matrix for MUB models. Quad. Stat. **8**, 33–78 (2006)
8. Piccolo, D., D'Elia, A.: A new approach for modeling consumer's preferences. Food Qual. Prefer. **19**, 247–259 (2008)

Chapter 5
Semiparametric Generalized Estimating Equations in Misspecified Models

Francesco Bravo

Abstract This paper proposes a two-step procedure for estimating semiparametric models defined by a set of possibly misspecified overidentified generalized estimating equations. The paper shows that the resulting estimator is asymptotically normal with a covariance matrix that crucially depends on the weight matrix used in the estimation process. The paper also considers efficient semiparametric classical minimum distance estimation. An example provides an illustration of the applicability of the results of this paper.

Keywords Misspecification • Two-step estimation

5.1 Introduction

Generalized estimating equations (GEE henceforth) are extensions of generalized linear models [6, 7] and quasi-likelihood methods (see for example Wedderburn [10] and McCullagh [5]), that are often used in the context of nonnormal correlated longitudinal data as originally suggested by Liang and Zeger [4]. Zieger [11] provides a comprehensive review of GEE. In this paper we consider possibly overidentified GEE in which we allow for infinite dimensional parameters as well as possible global misspecification. These models are theoretically interesting and empirically relevant since there are many situations where misspecification in GEE might arise including responses missing not completely at random and/or covariates measured with errors.

Under global misspecification it is not possible to define the notion of a "true" unknown parameter, yet we can assume the existence of a parameter—often called "pseudo-true" using the terminology originally suggested by Sawa [9]— that uniquely minimizes the limit objective function; this parameter becomes the parameter of interest. Note however that as opposed to the case of correctly specified models the pseudo-true value needs not be the same for different choices of the weight matrix used in the estimation process.

F. Bravo (✉)
Department of Economics, University of York, York YO10 5DD, UK
e-mail: francesco.bravo@york.ac.uk

© Springer Science+Business Media New York 2014
M.G. Akritas et al. (eds.), *Topics in Nonparametric Statistics*, Springer Proceedings in Mathematics & Statistics 74, DOI 10.1007/978-1-4939-0569-0_5

In this paper we assume that the parameter of interest is finite dimensional whereas the nuisance parameter is infinite dimensional. The estimation procedure we suggest consists of two steps: in the first step the infinite dimensional parameter is consistently estimated by a nonparametric estimator. In the second step the finite dimensional parameter is estimated using the profile objective function based on the first step estimator. We make two main contributions: first we show that the proposed two step estimator is asymptotically normal with a covariance matrix that depends in a complicated way on the weight matrix used in the estimation. This result generalizes to the semiparametric case results obtained by Hall and Inoue [3] in the context of generalized method of moment estimators for misspecified models. It also generalizes the quadratic inference function approach originally suggested by Qu et al. [8] in the context of correctly specified GEE models.

Second we propose a novel semiparametric classical (efficient) minimum distance estimator. This result generalizes the classical result of Ferguson [2] and can be used for example to estimate partially linear longitudinal data models with unobserved effects using Chamberlain's [1] linear projection approach..

The rest of the paper is structured as follows: next section introduces the model and the estimators. Section 5.3 contains the main results and an illustrative example. An Appendix contains sketches of the proofs.

The following notation is used throughout the paper: "′" indicates transpose, "⁻" denotes the generalized inverse of a matrix, "$\otimes$" denotes Kronecker product, "*vec*" is the vec operator, and finally for any vector v $v^{\otimes 2} = vv'$.

5.2   The Model and Estimator

Let $\{Z_i\}_{i=1}^n$ denote a random sample from the distribution of $Z \in \mathscr{Z} \subset \mathbb{R}^{d_z}$, $\theta \in \Theta \subset \mathbb{R}^k$ denote the unknown finite dimensional parameters of interest, Θ is a compact set, and $h_0(Z) := h_0 \in \mathscr{H} = \mathscr{H}_1 \times \ldots \times \mathscr{H}_m$ is a vector of unknown infinite dimensional nuisance parameters, and $\mathscr{H}$ is a pseudo-metric space. The statistical model we consider is

$$E\left[g\left(Z, \theta, h_0\right)\right] = \mu\left(\theta, h_0\right) \quad \text{for all } \theta \in \Theta, \tag{5.1}$$

where $g\left(\cdot\right): \mathscr{Z} \times \Theta \times \mathscr{H} \to \mathbb{R}^l$ $(l \geq k)$ is a vector-valued known (smooth) function, and $\inf_{\theta \in \Theta} \|\mu\left(\theta, h_0\right)\| > 0$ is the indicator of misspecification.

Let $g\left(Z_i, \theta, h\right) := g_i\left(\theta, h\right)$, $g\left(Z, \theta, h\right) := g\left(\theta, h\right)$, $\hat{g}\left(\theta, h\right) = \sum_{i=1}^n g_i\left(\theta, h\right)/n$, and let $\hat{h} := \hat{h}\left(Z_i\right)$ denote the first-step nonparametric estimator of h_0. The estimator we consider is defined as

$$\hat{\theta} = \arg\min_{\theta \in \Theta} \hat{Q}_{\hat{W}}\left(\theta, \hat{h}\right), \tag{5.2}$$

where

$$\hat{Q}_{\hat{W}}\left(\theta, h\right) = \hat{g}\left(\theta, h\right)' \hat{W} \hat{g}\left(\theta, h\right)$$

denote the GEE objective function and $\hat{W}$ is a possibly random positive semidefinite weighting matrix, that may also depend on a preliminary estimator, say $\hat{\theta}_p$.

In the case of correctly specified model $\mu(\theta, h_0) = 0$ at a unique θ_0. As an example of this situation we discuss how a semiparametric extension to the classical minimum distance (CMD) estimation can be cast into the framework of this paper. Suppose that θ_0 is related to $\pi_0 := m(\theta_0, h_0)$ where $m(\cdot) : \Theta \times \mathcal{H} \to \mathbb{R}^l$ is a vector-valued known (smooth) function, and that there exists an estimator $\hat{\pi} := \hat{\pi}\left(\hat{h}\right)$ for π_0 based on a first-step estimator $\hat{h}$. Then a semiparametric CMD estimator is defined as

$$\hat{\theta} = \arg\min_{\theta \in \Theta} \left(\hat{\pi} - m\left(\theta, \hat{h}\right)\right)' \hat{W} \left(\hat{\pi} - m\left(\theta, \hat{h}\right)\right), \qquad (5.3)$$

which is as in (5.2) with $\hat{g}\left(\theta, \hat{h}\right) = \hat{\pi} - m\left(\theta, \hat{h}\right)$.

5.3 Main Results

We consider three different estimators that correspond to three different choices of $\hat{W}$: first $\hat{W}$ is a nonstochastic positive definite matrix W; second $\hat{W}$ is stochastic but it does not depend on a preliminary estimator $\hat{\theta}_p$. Finally $\hat{W}$ is allowed to depend on $\hat{\theta}_p$. This distinction is important in the context of misspecified overidentified GEE models, because it implies that the pseudo true value $\theta_* := \theta_*(W)$ minimizing the limit objective function (see assumption GEE1(i) below) might be different for different choices of $\hat{W}$. Note however that for notational simplicity in what follows we shall ignore this important point and use θ_* to denote possibly different pseudo-true parameters.

Assume that

GEE1 (i) there exists a θ_* such that $Q_W(\theta, h) := E[g(\theta, h_0)]' WE[g(\theta, h_0)]$ is minimized for all $\theta \in \Theta \backslash \theta_*$, (ii) $\theta_* \in int(\Theta)$, (iii) $g(\theta, h)$ is twice continuously differentiable in a neighborhood of $\theta_* a.s$,

GEE2 (i) $\sup_{\theta \in \Theta} \left\| \hat{g}\left(\theta, \hat{h}\right) - E[g(\theta, h_0)] \right\| = o_p(1)$, (ii) $\left\| \hat{W} - W \right\| = o_p(1)$,

GEE3 (i) the empirical processes $v_n(h) = \sum_{i=1}^n [g_i(\theta_*, h) - E(g(\theta_*, h))] / n^{1/2}$ and $V_n(h) = \sum_{i=1}^n [G_i(\theta_*, h) - E(G(\theta_*, h))] / n^{1/2}$ are stochastically equicontinuous at h_0, where $G_i(\theta, h) := \partial g_i(\theta, h) / \partial \theta'$, (ii) $E\left[g\left(\theta_*, \hat{h}\right)\right] = o_p\left(n^{-1/2}\right)$, (iii) $n^{1/2}\left(\hat{W} - W\right)\mu_* \xrightarrow{d} N(0, \Sigma_W)$, (iv)

$$\sup_{\theta \in \Theta} \left\| \hat{G}\left(\theta, \hat{h}\right) - E[G(\theta, h_0)] \right\| = o_p(1),$$

$$\sup_{\theta \in \Theta} \left\| \partial vec \left(\hat{G} \left(\theta, \hat{h} \right) \right) / \partial \theta' - E \left[\partial vec \left(G \left(\theta, h_0 \right) \right) / \partial \theta' \right] \right\| = o_p (1),$$

GEE4 (i) the matrix $K (\theta_*, h_0, W)$ is nonsingular, where

$$K (\theta, h, W) = E [G (\theta, h)]' W E [G (\theta, h)]$$
$$+ \mu_*' W \otimes I_k E \left[\partial vec \left(G \left(\theta, h \right) \right) / \partial \theta' \right],$$

(ii) the matrix $\Xi (\theta_*, h_0, W)$ is positive definite, where

$$\Xi (\theta, h, W) = E \left\{ (g (\theta, h) - \mu_*)', \left[(G (\theta, h) - E [G (\theta, h)])' W \mu_* \right]', \right.$$
$$\left. \left[\left(\hat{W} - W \right) \mu_* \right]' \right\}'^{\otimes 2},$$

with $\Xi_{jk} (\theta, h, W)$ $(j, k = 1, 2, 3)$ denoting its block components. Let $\hat{\theta}_W$ and $\hat{\theta}_{\hat{W}}$ denote the GEE estimator based on $\hat{W}$ and W, respectively. Let $\hat{\theta}_{p+1}$ denote the $(p + 1)$th iterated GEE estimator based on a preliminary consistent pth GEE estimator $\hat{\theta}_p$ for $p = 1, 2, \ldots, k$, where $\hat{\theta}_1$ corresponds to either $\hat{\theta}_W$ or $\hat{\theta}_{\hat{W}}$; accordingly let $W (\theta_p)$ denote the probability limit of $\hat{W}$ that depends on $\hat{\theta}_p$.

Theorem 1. *Under GEE1–GEE4*

$$n^{1/2} \left(\hat{\theta}_W - \theta_* \right) \xrightarrow{d} N \left(0, K (\theta_*, h_0, W)^{-1} \Psi_{(0)} (\theta_*, h_0, W) K (\theta_*, h_0, W)^{-1} \right),$$

$$n^{1/2} \left(\hat{\theta}_{\hat{W}} - \theta_* \right) \xrightarrow{d} N \left(0, K (\theta_*, h_0, W)^{-1} \Psi_{(1)} (\theta_*, h_0, W) K (\theta_*, h_0, W)^{-1} \right),$$

$$n^{1/2} \left(\hat{\theta}_{p+1} - \theta_* \right) \xrightarrow{d} N \left(0, K (\theta_*, h_0, W)^{-1} \Psi_{(p+1)} (\theta_*, h_0, W, \theta_p) K (\theta_*, h_0, W)^{-1} \right)$$

where

$$\Psi_{(0)} (\theta_*, h_0, W) = E [G (\theta_*, h_0)]' W \Xi_{11} (\theta_*, h_0, W) W E [G (\theta_*, h_0)]$$
$$+ E [G (\theta_*, h_0)]' W \Xi_{12} (\theta_*, h_0, W) + \Xi_{12} (\theta_*, h_0, W)'$$
$$W E [G (\theta_*, h_0)] + \Xi_{22} (\theta_*, h_0, W),$$

$$\Psi_{(1)} (\theta_*, h_0, W) = \Psi_{(0)} (\theta_*, h_0, W) + E [G (\theta_*, h_0)]' \Xi_{33} (\theta_*, h_0, W) E [G (\theta_*, h_0)]$$
$$+ E [G (\theta_*, h_0)]' W \Xi_{13} (\theta_*, h_0, W) E [G (\theta_*, h_0)]$$
$$+ E [G (\theta_*, h_0)]' \Xi_{13} (\theta_*, h_0, W)' W E [G (\theta_*, h_0)]$$
$$+ \Xi_{23} (\theta_*, h_0, W) E [G (\theta_*, h_0)] + E [G (\theta_*, h_0)]' \Xi_{23} (\theta_*, h_0, W),$$

$$\Psi_{(p+1)}\left(\theta_*, h_0, W, \theta_p\right) = \Psi_{(p)}\left(\theta_*, h_0, W, \Xi_{j3}\left(\theta_p, h_0, W\right)\right) \quad j = 1, 2, 3,$$

and the $\Xi_{j3}\left(\theta_p, h_0, W\right)$ terms are to emphasize the dependency of the distribution of the $(p+1)$th iterated estimator on that of the pth one through the $\Xi_{j3}\left(\theta_p, h_0, W\right)$ matrices.

The expression for $\Psi_{(p+1)}\left(\theta_*, h_0, W, \theta_p\right)$ is in general very complicated; see the Appendix for the case $p = 2$ with $\hat{W} = \left(\sum_{i=1}^n g_i\left(\hat{\theta}_I\right)^{\otimes 2} / n\right)^{-1}$ as weight matrix and $\hat{\theta}_I$ is the first step estimator based on $W = I$.

We now consider correctly specified overidentified GEE; the following theorem shows the asymptotic normality of the efficient CMD (also known as minimum χ^2) estimator, that is obtained by using as weight matrix a consistent estimator of $\Omega\left(\theta_0, h_0\right)^{-1}$, where $\Omega\left(\theta_0, h_0\right) = E\left[g\left(\theta, h\right)^{\otimes 2}\right]$ and $g\left(\theta, h\right) = \pi - m\left(\theta, h\right)$. Assume that

CMD1 (i) there exists a unique θ_0 such that $E\left[g\left(\theta, h\right)\right] = 0$, (ii) $\theta_0 \in int\left(\Theta\right)$, (iii) $m\left(\theta, h\right)$ is continuously differentiable in a neighborhood of θ_0,

CMD2 (i) $\sup_{\theta \in \Theta}\left\|\hat{g}\left(\theta, \hat{h}\right) - E\left[g\left(\theta, h_0\right)\right]\right\| = o_p\left(1\right)$, (ii) $\left\|\hat{\Omega}\left(\hat{\theta}_p, \hat{h}\right)^{-1} - \Omega\left(\theta_0, h_0\right)^{-1}\right\| = o_p\left(1\right)$ for some preliminary consistent $\hat{\theta}_p$

CMD3 (i) the empirical process $v_n\left(h\right) = \sum_{i=1}^n\left[g_i\left(\theta_0, h\right) - E\left(g\left(\theta_0, h\right)\right)\right] / n^{1/2}$ is stochastically equicontinuous at h_0, (ii) $E\left[g\left(\theta_0, \hat{h}\right)\right] = o_p\left(n^{-1/2}\right)$, (iii) $M\left(\theta, h\right) := \partial m\left(\theta, h\right) / \partial\theta'$ is continuous at θ_0, h_0, (iv) $rank\left(M\left(\theta_0, h_0\right)\right) = k$, (v) $\Omega\left(\theta_0, h_0\right)$ is positive definite.

Theorem 2. *Under CMD1–CMD3*

$$n^{1/2}\left(\hat{\theta} - \theta_0\right) \xrightarrow{d} N\left(0, \left[M\left(\theta_0, h_0\right)' \Omega\left(\theta_0, h_0\right)^{-1} M\left(\theta_0, h_0\right)\right]^{-1}\right).$$

5.3.1 An Example

The statistical model we consider is

$$E\left[M\left(Z_2\right)\rho\left(Z, \theta, h_0\right)\right] = \mu_M\left(\theta\right), \tag{5.4}$$

where $Z = \left[Z_1', Z_2'\right]'$, $M\left(\cdot\right) : \mathscr{Z}_2 \to \mathbb{R}^l$ is a vector-valued known function often called in the econometric literature instruments, and $\rho\left(\cdot\right) : \mathscr{Z} \times \Theta \times \mathscr{H} \to \mathbb{R}$ is a "residual" scalar function, such as for example the unobservable error in a regression model. Models such as (5.4) arise often in economics as a result of the conditional moment restriction $E\left[\rho\left(Z, \theta, h_0\right)|Z_2\right] = \mu\left(\theta\right) a.s.$ that might be implied by a misspecified economic model. We assume that:

R1 $(\mu_M(\theta_*) - \mu_M(\theta))' W (\mu_M(\theta_*) - \mu_M(\theta))$ is negative definite for all $\theta \in \Theta \backslash \theta_*$,

R2 $\rho(Z, \theta, h)$ is three times differentiable with respect to θ and h with third derivative Lipshitz continuous with respect to θ and h,

R3 (i) $E \sup_{\theta \in \Theta} \left\| \partial^3 \rho(Z, \theta, h) / \partial h \partial h' \partial h_j \right\|_{\mathscr{H}} < \infty j = 1, \ldots, m$, (ii) $E \sup_{\theta \in \Theta} \| \rho(Z, \theta, h_0) \| < \infty$, $E \sup_{\theta \in \Theta} \left\| \partial^2 \rho(Z, \theta, h_0) / \partial \theta \partial \theta' \right\| < \infty$,

R4 (i) $E [\partial \rho(Z, \theta, h_0) / \partial h | Z_2] = 0$ a.s., (ii) $\left\| \hat{h} - h_0 \right\|_{\mathscr{H}} = o_p \left(n^{-1/4} \right)$,

R5 (i) $rank (E [M(Z_2) \partial \rho(Z, \theta*, h_0) / \partial \theta']) = k$,

 (ii) $rank (E [\partial vec (M(Z_2) \partial \rho(Z, \theta*, h_0) / \partial \theta') / \partial \theta']) = k$.

R1 is an identification condition that implies GEE1(i); R2 suffices for the stochastic equicontinuity assumption GEE3(i), whereas R3 suffices for GEE2(i) and GEE3(iv) by the uniform law of large numbers. R4(i)–(ii) imply GEE3(ii) while R5 implies GEE4(i). Thus assuming further that GEE2(ii), GEE3(iii), and GEE4 hold, the conditions of Theorem 1 are satisfied; thus the distribution of the semiparametric nonlinear instrumental variable estimator

$$\hat{\theta} = \arg\min_{\theta \in \Theta} \left(\sum_{i=1}^{n} \rho \left(Z_i, \theta, \hat{h} \right) M (Z_{2i})' / n \right) \hat{W} \sum_{i=1}^{n} M (Z_{2i}) \rho \left(Z_i, \theta, \hat{h} \right) / n$$

is as that given in Theorem 1.

Appendix

Throughout the Appendix "CMT" and "CLT" denote Continuous Mapping Theorem and Central Limit Theorem, respectively.

Proof (Proof of Theorem 1). The consistency of any of the three estimators follows by the identification condition GEE1(i), and the uniform convergence of $\hat{Q}_{\hat{W}} \left(\theta, \hat{h} \right)$ which follows by GEE2(i)–(ii). The asymptotic normality of $\hat{\theta}_W$ and $\hat{\theta}_{\hat{W}}$ follows by standard mean value expansion of the first order conditions

$$0 = \hat{G} \left(\hat{\theta}, \hat{h} \right)' W \hat{g} \left(\hat{\theta}, \hat{h} \right),$$

$$0 = \hat{G} \left(\hat{\theta}, \hat{h} \right)' \hat{W} \hat{g} \left(\hat{\theta}, \hat{h} \right),$$

which hold with probability approaching to 1 by GEE1(ii). We consider $\hat{\theta}_{\hat{W}}$ and note that

$$0 = \hat{G} \left(\theta_*, \hat{h} \right)' \hat{W} n^{1/2} \hat{g} \left(\theta_*, \hat{h} \right) + \hat{G} \left(\hat{\theta}, \hat{h} \right)' \hat{W} \hat{G} \left(\overline{\theta}, \hat{h} \right) n^{1/2} \left(\hat{\theta} - \theta_* \right),$$

hence by GEE2(i)–(ii), GEE3(iii)–(iv), CMT and some algebra it follows that

$$n^{1/2}\left(\hat{\theta}_{\hat{W}} - \theta_*\right) = -\left(I - P\left(\theta_*, h_0\right)^{-1}\mu'_* W \otimes I_k E\left[\partial vec\left(G\left(\theta_*, h_0\right)\right)/\partial\theta'\right]\right)^{-1}$$

(5.5)

$$\left\{E\left[G\left(\theta_*, h_0\right)'\right]WE\left[G\left(\theta_*, h_0\right)\right]\right\}n^{1/2}\left\{\hat{G}\left(\theta_*, h_0\right)'Wn^{1/2}\left(\hat{g}\left(\theta_*, h_0\right) - \mu_*\right)\right.$$

$$\left. + \overline{G}\left(\theta_*, h_0\right)'W\mu_* + E\left[G\left(\theta_*, h_0\right)\right]'\left(\hat{W} - W\right)\mu_*\right\} + o_p\left(1\right),$$

where $P\left(\theta, h\right) = \left\{E\left[G\left(\theta, h_0\right)'\right]WE\left[G\left(\theta, h\right)\right]\right\}$ and $\overline{G}\left(\theta, h\right) = \hat{G}\left(\theta, h\right)$ $- E\left[G\left(\theta, h\right)\right]$. Note that

$$\left(I - P\left(\theta_*, h_0\right)^{-1}\mu'_* W \otimes I_k E\left[\partial vec\left(G\left(\theta_*, h_0\right)\right)/\partial\theta'\right]\right)^{-1}$$

$$= K\left(\theta_*, h_0, W\right)^{-1}P\left(\theta_*, h_0\right)^{-1},$$

and that by CLT

$$n^{1/2}\begin{bmatrix}\hat{g}\left(\theta_*, h_0\right) - \mu_* \\ \overline{G}\left(\theta_*, h_0\right)'W\mu_* \\ \left(\hat{W} - W\right)\mu_*\end{bmatrix} \xrightarrow{d} N\left(0, \begin{bmatrix}\Xi_{11}\left(\theta_*, h_0, W\right) & \Xi_{12}\left(\theta_*, h_0, W\right) & \Xi_{13}\left(\theta_*, h_0, W\right) \\ \Xi_{12}\left(\theta_*, h_0, W\right)' & \Xi_{22}\left(\theta_*, h_0, W\right) & \Xi_{23}\left(\theta_*, h_0, W\right) \\ \Xi_{13}\left(\theta_*, h_0, W\right)' & \Xi_{23}\left(\theta_*, h_0, W\right)' & \Xi_{33}\left(\theta_*, h_0, W\right)\end{bmatrix}\right),$$

(5.6)

so that the conclusion follows by CMT and some algebra. For $\hat{\theta}_W$ the conclusion follows noting that $\Xi_{j3}\left(\theta_*, h_0, W\right)$ $(j = 1, 2)$ and Σ_W are all 0. Finally we consider the iterated semiparametric GEE estimator $\hat{\theta}_p$. We first consider the second-step estimator $\hat{\theta}_2$ based on either $\hat{\theta}_W$ or $\hat{\theta}_{\hat{W}}$ as preliminary (first step) consistent estimator. Assume that the weight matrix $\hat{W}$ is given by $\hat{\Xi}_{11}\left(\hat{\theta}_W, \hat{h}, W\right)^{-1}$ or $\hat{\Xi}_{11}\left(\hat{\theta}_{\hat{W}}, \hat{h}, \hat{W}\right)^{-1}$; the same argument as that used to obtain (5.5) can be used to show that $n^{1/2}\left(\hat{\theta}_2 - \theta_*\right)$ has the same influence function as that given in (5.5) (with $\Xi_{11}\left(\theta_*, h_0, W\right)^{-1}$ replacing W) except for the last term, which in this case needs a further expansion. First note that

$$n^{1/2}\left(\hat{\Xi}_{11}\left(\hat{\theta}_\times, \hat{h}, W\right)^{-1} - \Xi_{11}\left(\theta_*, h_0, W\right)^{-1}\right)\mu_*$$

$$= n^{1/2}\left(\mu'_* \otimes \hat{\Xi}_{11}\left(\hat{\theta}_\times, \hat{h}, W\right)^{-1}\right)\Xi_{11}\left(\theta_*, h_0, W\right)^{-1}$$

$$\otimes I_l vec\left(\hat{\Xi}_{11}\left(\hat{\theta}_\times, \hat{h}, W\right) - \Xi_{11}\left(\theta_*, h_0, W\right)\right),$$

and by GEE(ii) (with $\hat{W} = \hat{\Xi}_{11}\left(\hat{\theta}_{\times}, \hat{h}, W\right)^{-1}$), the triangle inequality and a mean value expansion we have

$$
\left(\mu_*' \Xi_{11}(\theta_*, h_0, W)^{-1} \otimes \Xi_{11}(\theta_*, h_0, W)^{-1}\right) n^{1/2}\left\{ vec\left(\hat{\Xi}_{11}\left(\theta_*, \hat{h}, W\right) - \Xi_{11}(\theta_*, h_0, W)\right)\right.
$$

$$
\left. + \frac{\partial vec \hat{\Xi}_{11}\left(\overline{\theta}_{\times}, \hat{h}, W\right)}{\partial \theta'}\left(\hat{\theta}_{\times} - \theta_*\right)\right\} + o_p(1), \tag{5.7}
$$

which shows that the asymptotic distribution of $n^{1/2}\left(\hat{\theta}_2 - \theta_*\right)$ crucially depends also on that of $n^{1/2}\left(\hat{\theta}_{\times} - \theta_*\right)$, as we now illustrate for $\hat{\theta}_{\times} = \hat{\theta}_I$. Let

$$
L\left(\theta_*, h_0, \Xi_{11}(\theta_*, h_0, W)^{-1}\right) = \mu_*' \Xi_{11}(\theta_*, h_0, W)^{-1} \otimes \Xi_{11}(\theta_*, h_0, W)^{-1}
$$

$$
\times \left[I, E\left(\frac{\partial vec \Xi_{11}(\theta_*, h_0, W)}{\partial \theta'}\right)\right]
$$

and

$$
\hat{S}\left(\theta_* h_0, \Xi_{11}(\theta_*, h_0, W), \theta_I\right) = \begin{bmatrix} vec\left(\hat{\Xi}_{11}\left(\hat{\theta}_I, \hat{h}, W\right) - \Xi_{11}(\theta_*, h_0, W)\right) \\ K(\theta_*, h_0, I)^{-1}\left\{G(\theta_*, h_0)'\left(\hat{g}(\theta_*, h_0) - \mu_*\right) + \overline{G}(\theta_*, h_0)'\mu_*\right\} \end{bmatrix}
$$

so that

$$
n^{1/2}\left(\hat{\Xi}_{11}\left(\hat{\theta}_I, \hat{h}, W\right)^{-1} - \Xi_{11}(\theta_*, h_0, W)^{-1}\right)\mu_* = L\left(\theta_*, h_0, \Xi_{11}(\theta_*, h_0, W)^{-1}\right)
$$

$$
\times n^{1/2}\hat{S}(\theta_* h_0, \Xi_{11}(\theta_*, h_0, W), \theta_I).
$$

Then

$$
\Xi_{13}(\theta_*, h_0, W(\theta_I)) = nCov\{\hat{g}(\theta_*, h_0) - \mu_*,
$$

$$
\hat{S}(\theta_* h_0, \Xi_{11}(\theta_*, h_0, W), \theta_I)' L\left(\theta_*, h_0, \Xi_{11}(\theta_*, h_0, W)^{-1}\right)'\Big\},
$$

$$
\Xi_{23}(\theta_*, h_0, W(\theta_I)) = nCov\left\{\left(\hat{G}(\theta_*, h_0) - E[G(\theta_*, h_0)]\right)' \Xi_{11}(\theta_*, h_0, W)^{-1}\mu_*,\right.
$$

$$
\hat{S}(\theta_* h_0, \Xi_{11}(\theta_*, h_0, W), \theta_I)' L\left(\theta_*, h_0, \Xi_{11}(\theta_*, h_0, W)^{-1}\right)'\Big\},
$$

$$
\Xi_{33}(\theta_*, h_0, W(\theta_I)) = nVar\left\{L\left(\theta_*, h_0, \Xi_{11}(\theta_*, h_0, W)^{-1}\right)\hat{S}(\theta_* h_0, \Xi_{11}(\theta_*, h_0, W), \theta_I)\right\},
$$

and

$$
n^{1/2}\left(\hat{\theta}_2 - \theta_*\right) \xrightarrow{d} N\left(0, K\left(\theta_*, h_0, \Xi_{11}(\theta_*, h_0, W)^{-1}\right)^{-1}\Psi_{(2)}(\theta_*, h_0, W, \theta_p)\right)
$$

$$K\left(\theta_*, h_0, \Xi_{11}\left(\theta_*, h_0, W\right)^{-1}\right)^{-1}\right),$$

where

$$\Psi_2\left(\theta_*, h_0, \Xi_{11}\left(\theta_*, h_0, W\left(\theta_I\right)\right)^{-1}\right) = \Psi_{(0)}\left(\theta_*, h_0, \Xi_{11}\left(\theta_*, h_0, W\left(\theta_I\right)\right)^{-1}\right)$$
$$+ E\left[G\left(\theta_*, h_0\right)\right]' \Xi_{33}\left(\theta_*, h_0, W\left(\theta_I\right)\right) E\left[G\left(\theta_*, h_0\right)\right]$$
$$+ E\left[G\left(\theta_*, h_0\right)\right]' \Xi_{11}\left(\theta_*, h_0, W\left(\theta_I\right)\right)^{-1} \Xi_{13}\left(\theta_*, h_0, W\left(\theta_I\right)\right) E\left[G\left(\theta_*, h_0\right)\right]$$
$$+ E\left[G\left(\theta_*, h_0\right)\right]' \Xi_{13}\left(\theta_*, h_0, W\left(\theta_I\right)\right)' \Xi_{11}\left(\theta_*, h_0, W\left(\theta_I\right)\right)^{-1} E\left[G\left(\theta_*, h_0\right)\right]$$
$$+ \Xi_{23}\left(\theta_*, h_0, W\left(\theta_I\right)\right) E\left[G\left(\theta_*, h_0\right)\right] + E\left[G\left(\theta_*, h_0\right)\right]' \Xi_{23}\left(\theta_*, h_0, W\left(\theta_I\right)\right).$$

The general distribution of the $(p+1)$th iterated GEE estimator can be computed using a recursive argument.

Proof (Proof of Theorem 2). The consistency of $\hat{\theta}$ follows as in the proof of Theorem 1 by CMD1(i), CMD2(i)–(ii). The first order conditions for $\hat{\theta}$ are

$$M\left(\hat{\theta}, \hat{h}\right)' \hat{\Omega}\left(\hat{\theta}_p, \hat{h}\right)^{-1} \left(\hat{\pi} - m\left(\hat{\theta}, \hat{h}\right)\right) = 0,$$

which hold with probability approaching 1 by CMD1(ii). By a standard mean value expansion

$$n^{1/2}\left(\hat{\pi} - m\left(\theta_0, \hat{h}\right) - M\left(\bar{\theta}, \hat{h}\right)\left(\hat{\theta} - \theta_0\right)\right) + o_p(1) \tag{5.8}$$

and by CMD3(i)–(ii), CMD1 (i), and CMD(iv)

$$n^{1/2}\left(\hat{\pi} - m\left(\theta_0, \hat{h}\right)\right) \xrightarrow{d} N\left(0, \Omega\left(\theta_0, h_0\right)\right) + o_p(1)$$

by CLT. The conclusion follows by CMD3(iii)–(iv), (5.8), and CMT.

References

1. Chamberlain, G.: Multivariate regression models for panel data. J. Econom. **18**, 5–46 (1982)
2. Ferguson, T.: A method for generating best asymptotically normal estimates with application to the estimation of bacterial densities. Ann. Math. Stat. **29**, 1046–1062 (1958)
3. Hall, A., Inoue, A.: The large sample behaviour of the generalized method of moment estimator in misspecified models. J. Econom. **114**, 361–394 (2003)
4. Liang, K., Zeger, S.: Longitudinal data analysis using generalized linear models. Biometrika **73**, 12–22 (1986)
5. Mc Cullagh, P.: Quasi-likelihood functions. Ann. Stat. **11**, 59–67 (1983)

6. Mc Cullagh, P., Nelder, J.: Generalized Linear Models, 2nd edn. Chapman and Hall, New York (1989)
7. Nelder, J., Wedderburn, R.: Generalized linear models. J. R. Stat. Soc A **135**, 370–384 (1972)
8. Qu, A., Lindsay, B., Li, B.: Improving generalised estimating equations using quadratic inference functions. Biometrika **87**, 823–836 (2000)
9. Sawa, T.: Information criteria for discriminating among alternative regression models. Econometrica **46**, 1273–1291 (1978)
10. Wedderburn, R: Quasi-likelihood functions, generalized linear models, and the Gauss—Newton method. Biometrika **61**, 439–447 (1974)
11. Ziegler A.: Generalized estimating equations. Lecture Notes in Statistics, vol. 211. Springer, New York (2011)

Chapter 6
Local Covariance Estimation Using Costationarity

Alessandro Cardinali

Abstract In this paper we propose a novel estimator for the time-varying covariance of locally stationary time series. This new approach is based on costationary combinations, that is, time-varying deterministic combinations of locally stationary time series that are second-order stationary. We show with a simulation example that the new estimator has smaller variance than other approaches exclusively based on the evolutionary cross-periodogram, and can therefore be appealing in a large number of applications.

Keywords Local stationarity • Costationarity • Wavelets • Time-varying covariances

6.1 Introduction

Loosely speaking, a stationary time series is one whose statistical properties remain constant over time. A locally stationary (LS) time series is one whose statistical properties can change slowly over time. As a consequence, such a series can appear stationary when examined close up, but appear non-stationary when examined on a larger scale. Priestley [9, 10] provide a comprehensive review of locally stationary processes and their history, and Nason and von Sachs [5] provide a more recent review. The methods described in this article can be applied to locally stationary time series, being triangular stochastic arrays defined in the rescaled time t/T, where T represents the sample size.

Based on this setup, Dahlhaus [3] proposed *locally stationary Fourier* (LSF) processes whose underlying pseudo-spectral structure is defined in terms of Fourier basis. The *locally stationary wavelet* (LSW) model due to Nason et al. [6], instead, decomposes the local structure of the process among different scales through a set of non-decimated wavelets used as basis functions. We refer the interested reader to [7] for computational details pertaining LSW processes. In the following, we will consider the latter family of processes defined as

A. Cardinali (✉)
University of Plymouth, Plymouth, UK
e-mail: alessandro.cardinali@plymouth.ac.uk

© Springer Science+Business Media New York 2014
M.G. Akritas et al. (eds.), *Topics in Nonparametric Statistics*, Springer Proceedings in Mathematics & Statistics 74, DOI 10.1007/978-1-4939-0569-0_6

$$X_{t;T} = \sum_{j=1}^{\infty} \sum_{k=-\infty}^{\infty} W_j\left(\frac{k}{T}\right) \psi_j(t-k)\epsilon_{j,k}, \tag{6.1}$$

where $\{\psi_j(t-k)\}_j$ is a family of discrete non-decimated wavelet filters, with local support spanned by the index k and including the neighborhood of t. The parameter j is integer valued and represents the scale of the corresponding wavelet. The function $W_j(k/T)$ is a time-localized amplitude of bounded variation, referring to dyadic scales indexed by $j = 1, 2, .., J_T$, with $J_T = [\log_2 T]$. Finally $\epsilon_{j,k}$ is a sequence of doubly indexed i.i.d. standardized random variables. This setup allows the definition of a time-varying generalization of the classical spectra, having a well defined limit in the rescaled time $z \in (0, 1)$, defined as

$$S_j(z) = \lim_{T\to\infty} \left| W_j\left(\frac{[zT]}{T}\right) \right|^2, \tag{6.2}$$

where we have set $k = [zT]$, and $[x]$ is the integer part of x. This multiscale LS framework has proven to be useful in order to estimate the time-varying association between non-stationary time series. The estimation of time-localized covariances is relevant in a wide range of disciplines such as climatology, neuroscience, and economics, where the underlying phenomena are inherently characterized by regime changes that cannot be appropriately taken into account by classical stationary models.

Ombao and Van Bellegem [8] and Sanderson et al. [11] propose methodologies based, respectively, on the cross-spectra of LSF and LSW models. In this paper we propose an alternative methodology to estimate time-localized covariances, which is based on the existence of time-varying linear combinations of LS processes which are (co)stationary. More details concerning costationarity can be found in Cardinali and Nason [1], whereas [2] illustrate the related computational aspects. The contribution of this paper is therefore twofold. We first propose our new estimation methodology, and then show this to be statistically efficient in comparison to the method proposed in Sanderson et al. [11]. We illustrate a theoretical example and then validate the comparison by means of simulations.

The article is structured as follows. Section 6.2 reviews the second order properties of LS time series models and briefly describes the concept of costationarity. Section 6.3 introduces our new covariance-based estimator, and illustrates its relative efficiency over alternative estimators using a theoretical and simulation example.

6.2   Local Covariance and Costationarity

When processes are not stationary in the wide sense, covariance operators may have complicated time-varying properties. With non-stationarity their estimation is typically difficult, since a canonical spectral structure does not exist. Meyer

[4] showed that although we cannot find, in general, bases which diagonalize complicated integral operators, it is nevertheless possible to find well structured bases which compress them. This means that time-varying covariance operators can be well represented by sparse matrices with respect to such bases. For $n = 1, 2, \ldots, N$, the locally stationary behavior of LS processes $X_{t;T}^{(n)}$ is characterized by a local auto-covariance function

$$\gamma_T(t, \tau) = \gamma_{n,n;T}(t, \tau) = \mathrm{Cov}\left(X_{t;T}^{(n)}, X_{t+\tau;T}^{(n)}\right).$$

Using the approximation derived in Sanderson et al. [11] this is representable as

$$\gamma_T(t, \tau) = \sum_{j=1}^{[\log_2 T]} S_j\left(\frac{t}{T}\right) \Psi_j(\tau) + \mathcal{O}(T^{-1}) \tag{6.3}$$

where $S_j(t/T) = S_j(t/T)^{(n)}$ is the local spectra for $X_{t;T}^{(n)}$, and $\Psi(\tau) = \sum_t \psi_j(t)\psi_j(t + \tau)$ is the *autocorrelation wavelet*, see Nason et al. [6] for details. Similarly, using another approximation derived in Sanderson et al. [11], for $n, m = 1, 2, \ldots, N$, the local cross-covariance between two locally stationary processes $X_{t;T}^{(n)}$ and $X_{t;T}^{(m)}$ can be defined as

$$\gamma_{n,m;T}(t, \tau) = \sum_{j=1}^{[\log_2 T]} S_j^{(n,m)}\left(\frac{t}{T}\right) \Psi_j(\tau) + \mathcal{O}(T^{-1}), \tag{6.4}$$

where $S_j^{(n,m)}(t/T)$ is the local cross-spectra defined as

$$S_j^{(n,m)}(t/T) = W_j^{(n)}\left(\frac{t}{T}\right) W_j^{(m)}\left(\frac{t}{T}\right), \tag{6.5}$$

where $W_j^{(n)}(t/T)$ and $W_j^{(m)}(t/T)$ are the local amplitude functions for the processes $X_{t;T}^{(n)}$ and $X_{t;T}^{(m)}$ respectively. Note that these functions, along with the local spectra and cross-spectra, are defined in the rescaled time, and their limits (for $T \to \infty$) are well defined as $\gamma(z, \tau)$, $\gamma_{n,m}(z, \tau)$, $S_j(z)$ and $S_j^{(n,m)}(z)$ respectively, for $z \in (0, 1)$. Moreover, note also that for univariate (globally) stationary time series the spectra $S_j(t/T)$ is time invariant, i.e., $S_j(t/T) = S_j \forall j$, which also implies a Toeplitz covariance operator $\gamma(t, \tau) = \gamma(\tau)$. Finally, the time-localized covariance between two LS processes can be obtained as a particular case of Eq. (6.4) by setting $\tau = 0$, and is therefore defined as

$$\gamma_{n,m;T}(t, 0) = \sum_{j=1}^{[\log_2 T]} S_j^{(n,m)}\left(\frac{t}{T}\right) + \mathcal{O}(T^{-1}), \tag{6.6}$$

since the properties of wavelet filters imply $\Psi_j(0) = \sum_t \psi_j^2(t) = 1$, for all j. The covariances (6.6) can be estimated by

$$\hat{\gamma}_{n,m;T}(t,0) = \sum_{j=1}^{[\log_2 T]} \hat{S}_{j,t}^{(n,m)}, \tag{6.7}$$

where $\hat{S}_{j,t}^{(n,m)}$ is an asymptotically unbiased smoothed estimator of the time-varying cross spectra, as the one proposed in Sanderson et al. [11].

6.2.1 Costationarity

We give a multivariate extension for the definition of costationary processes originally proposed in Cardinali and Nason [1]. However, we now concentrate on constant piecewise solution vectors.

Definition 1. Let $\mathbf{X}_{t;T} = \left(X_{t;T}^{(1)}, \ldots, X_{t;T}^{(N)} \right)'$ be a vector time series with local auto-covariances and cross-covariances satisfying Eqs. (6.3) and (6.4). Moreover assume

$$\sup_t \left\{ \frac{\left| \mathrm{Cov}\left(X_{t;T}^{(n)}, X_{t;T}^{(m)} \right) \right|}{\mathrm{Var}\left(X_{t;T}^{(n)} \right)^{1/2} \mathrm{Var}\left(X_{t;T}^{(m)} \right)^{1/2}} \right\} < 1,$$

and for $n, m = 1, 2, \ldots, N$. We call $Z_t^{(i)}$ costationary process if there exists a set of bounded piecewise constant functions $\alpha_t^{(i,n)}$ for $t = 1, \ldots, T$, $n = 1, \ldots, N$, and $i = 1, 2, \ldots, I$ such that

$$Z_t^{(i)} = \sum_{n=1}^{N} \alpha_t^{(i,n)} X_{t;T}^{(n)}$$

is a covariance stationary process.

Costationary solutions can be, in general, multiple. This multiplicity is represented here through the index i. Using a vector notation we can also represent the set of costationary solutions as a time-varying linear system

Definition 2. Let $\mathbf{Z}_t = \left(Z_t^{(1)}, \ldots, Z_t^{(I)} \right)'$ and $\mathbf{X}_{t;T}$ as in Definition 1. We define the costationary system as

$$\mathbf{Z}_t = \mathbf{A}_t \, \mathbf{X}_{t;T},$$

where, for each time point t, $\mathbf{A}_t = \left[\alpha_t^{(i,n)} \right]_{i,n}$ is a $(I \times N)$ dimensional matrix of costationary vector entries for time t.

The piecewise constant functions $\alpha_t^{(i,n)}$ are supposed to be measurable on a disjoint sequence of half-opened dyadic intervals. In this paper intervals of dyadic length have been considered for computational convenience, however, in principle the theory we present will apply to intervals of arbitrary *minimal* length T^*, provided that for $T \to \infty$ we have $T^*/T \to \eta$, where $\eta > 0$. For a discussion on segmentation issues and regularity conditions concerning costationary solutions $\alpha_t^{(i,n)}$ we refer again the interested reader to Cardinali and Nason [1]. For an arbitrary time-varying LS combination, the local variance can be represented as

$$\sigma_{Z_i}^2(t) = \sum_{n,m} \alpha_t^{(i,n)} \alpha_t^{(i,m)} \, \gamma_{n,m;T}(t,0), \tag{6.8}$$

for $n, m = 1, \ldots, N$, and it is typically a time-varying quantity. However, when considering costationary combinations $Z^{(i)}(t)$, the quantity defined in Eq. (6.8) becomes time-invariant, so we can simply refer to it as $\sigma_{Z_i}^2$.

6.3 A Costationary Estimator for Local Covariances

A direct estimator for the time-varying covariance of locally stationary processes can be obtained by (6.7). The aim of this section is to introduce an alternative estimator for local covariances which makes use of costationary solutions. We will use integers $n, m, l, h = 1, 2, \ldots, N$ to identify (pairs of) LS processes. By imposing costationarity, the quantity defined in Eq. (6.8) can be estimated by

$$\hat{\sigma}_{Z_i}^2 = \sum_{n,m} \hat{\alpha}_t^{(i,n)} \hat{\alpha}_t^{(i,m)} \, \hat{\gamma}_{n,m;T}(t,0), \tag{6.9}$$

where $\hat{\gamma}_{n,m;T}(t,0)$ can be any asymptotically unbiased and consistent estimator for $\gamma_{n,m;T}(t,0)$, and $\hat{\alpha}_t^{(i,n)}$ can be determined using the **costat** algorithm described in Cardinali and Nason [1]. In practice we can obtain $\hat{\sigma}_{Z_i}^2$ using the sample variance estimator of costationary processes $Z_t^{(i)}$. However, from Eq. (6.9), we can derive an alternative local covariance estimator as

$$\hat{\gamma}_{l,h;T}^{(i)}(t,0) = \frac{\hat{\sigma}_{Z_i}^2 - \sum_{n,m}^{(n,m) \neq (l,h)} \hat{\alpha}_t^{(i,n)} \hat{\alpha}_t^{(i,m)} \, \hat{\gamma}_{n,m;T}(t,0)}{\hat{\alpha}_t^{(i,l)} \hat{\alpha}_t^{(i,h)}}. \tag{6.10}$$

We are particularly interested in assessing the properties of this estimator, in particular the gain in efficiency possibly due to using costationary combinations. This is because estimators based upon such combinations also exploit information about the global relation of the original LS time series (through the costationary solution vectors), and such information is not taken into account by considering covariance estimators exclusively based upon the time-localized cross-periodogram.

Moreover, because (co)stationarity of single solutions $Z_t^{(i)}$, there always exist a projection diagonalizing their covariance matrix. We therefore expect to achieve a greater degree of efficiency by averaging estimators based upon single costationary series, provided that the original LS series are not perfectly correlated. We will therefore consider the estimator

$$\hat{\gamma}_{l,h;T}^*(t,0) = \frac{1}{I} \sum_{i=1}^{I} \hat{\gamma}_{l,h;T}^{(i)}(t,0). \tag{6.11}$$

The following section illustrates the efficiency of this estimator with a theoretical and simulation example.

6.3.1 Theoretical Example and Simulations

For the illustrative purposes of this example we will only consider time-invariant costationary systems, i.e., costationary systems with time-invariant costationary vectors. These systems can be defined, as a particular case of the general form given in Definition 2, as

$$\mathbf{Z}_t = \mathbf{A} \, \mathbf{X}_{t;T}. \tag{6.12}$$

We consider two uncorrelated LS Gaussian processes $X_{t;T}^{(1)}$ and $X_{t;T}^{(2)}$, respectively having local spectra

$$S_j^{(1)}(t/T) = \begin{cases} 2/3, & \text{if } j = j^* \text{ and } t \leq T/2; \\ 1/3, & \text{if } j = j^* \text{ and } t > T/2; \\ 0, & \text{otherwise,} \end{cases} \tag{6.13}$$

and

$$S_j^{(2)}(t/T) = \begin{cases} 1/3, & \text{if } j = j^* \text{ and } t \leq T/2; \\ 2/3, & \text{if } j = j^* \text{ and } t > T/2; \\ 0, & \text{otherwise.} \end{cases} \tag{6.14}$$

Then, for $i = 1, 2, 3, 4$, and time-invariant costationary vectors

$$\boldsymbol{\alpha}^{(1,\cdot)} = (1, 1),$$
$$\boldsymbol{\alpha}^{(2,\cdot)} = (1, -1),$$
$$\boldsymbol{\alpha}^{(3,\cdot)} = (-1, 1),$$
$$\boldsymbol{\alpha}^{(4,\cdot)} = (-1, -1),$$

the linear combinations $Z_t^{(i)}$ are White-Noise processes with variance $\sigma_{Z_i}^2 = 1$. This is an example of multiple costationary solution vectors which are also time-invariant. Other costationary vectors can be found by multiplying $\alpha^{(i,\cdot)}$ by some real scalar. By applying the singular value decomposition $\mathbf{A} = \mathbf{U}\,\mathbf{D}\,\mathbf{V}$ we obtain

$$\mathbf{A} = \begin{pmatrix} 1 & 1 \\ 1 & -1 \\ -1 & 1 \\ -1 & -1 \end{pmatrix} = \begin{pmatrix} -1/2 & 1/2 \\ -1/2 & -1/2 \\ 1/2 & 1/2 \\ 1/2 & -1/2 \end{pmatrix} \begin{pmatrix} 2 & 0 \\ 0 & 2 \end{pmatrix} \begin{pmatrix} -1 & 0 \\ 0 & 1 \end{pmatrix}$$

In this case we therefore have $\text{rank}(\mathbf{A}) = 2$ and we expect the estimator $\hat{\gamma}_{l,h;T}^*(t,0)$ to be more efficient than the direct estimator proposed in Sanderson et al. [11]. This is because the degree of efficiency of $\hat{\gamma}_{l,h;T}^*(t,0)$ is obtained by exploiting the number of non-perfectly correlated LS time series $X_{t;T}^{(n)}$, which are used to produce *synthetic* linearly independent stationary time series.

Remark 1. Note that, even in the general time-varying case of Definition 1 (i.e., when $\mathbf{A}_t$ depends upon time), conditions therein are sufficient to ensure that $\text{rank}(\mathbf{A}_t) = N$, for all $t = 1, \ldots, T$, provided that $I \geq N$. The greater is the number of non-perfectly correlated time series returning costationary solutions, the greater is the expected degree of efficiency of the new estimator.

We use this simple theoretical example to conduct a simulation experiment investigating the relative efficiency of our costationary estimator for the local covariance function when compared with the classical estimator exclusively based on the local cross-periodogram and defined in Eq. (6.7). We simulate pairs of uncorrelated multiscale LS Gaussian processes having local spectra as defined in Eqs. (6.13) and (6.14).

In this experiment we use a (real-valued) wavelet model representation. We choose $j^* = 4$ and simulate $30 \times 29/2 = 435$ pairs of uncorrelated processes of increasing length. For each pair we first estimate time-localized covariances by using the *classical* estimator defined in Eq. (6.7). We then repeat the estimation by using the statistics defined in Eq. (6.11). We use simulations to compute, at each time point t, the Montecarlo variance (variance of simulated samples) of each estimator as well as the ratio of the two variances, respectively denoted as $\text{Var}[\hat{\gamma}(t,0)]$, $\text{Var}[\hat{\gamma}^*(t,0)]$, and $\text{Eff}[\hat{\gamma}(t,0)/\hat{\gamma}^*(t,0)] = \text{Var}[\hat{\gamma}(t,0)]/\text{Var}[\hat{\gamma}^*(t,0)]$. Table 6.1 shows the time averages (over t) of these respective quantities. The analysis is repeated for simulated time series of increasing size $T = 128, 256, 512, 1{,}024, 2{,}048$. The results of our simulations show that the new estimator is substantially more efficient than the estimator exclusively based upon the cross-periodogram. Interestingly, the gain in efficiency is substantial even for very moderate sample sizes. These results suggest that increasing efficiency can be achieved by considering a larger number of time series in costationary combinations. Future work will consider a full theoretical investigation of this approach as well as some applications to economics and financial data.

Table 6.1 Time-averaged Montecarlo variances and relative efficiency ratios for classical and costationary local covariance estimators

Estimates	T = 128	T = 256	T = 512	T = 1,024	T = 2,048
$\text{Var}_T[\hat{\gamma}(\cdot, 0)]$	0.0092	0.0101	0.0103	0.0112	0.0134
$\text{Var}_T[\hat{\gamma}^*(\cdot, 0)]$	0.0029	0.0028	0.0029	0.0030	0.0032
$\text{Eff}_T[\hat{\gamma}(\cdot, 0)/\hat{\gamma}^*(\cdot, 0)]$	3.1710	3.5970	3.5579	3.7321	4.1934

Acknowledgements I am grateful to Guy Nason for reading a preliminary version of the manuscript and for providing useful comments. All errors are my own responsibility.

References

1. Cardinali, A., Nason, G.P.: Costationarity of locally stationary time series. J. Time Ser. Econom. **2**(2), 1–19 (2010)
2. Cardinali, A., Nason, G.P.: Costationarity of locally stationary time series using costat. J. Stat. Softw. **55**(1):1–22 (2013)
3. Dahlhaus, R.: Fitting time series models to nonstationary processes. Ann. Stat. **25**, 1–37 (1997)
4. Meyer, Y.: Wavelets and operators. Proc. Symp. Appl. Math. **47**, 35–57 (1993)
5. Nason, G., von Sachs, R.: Wavelets in time series analysis. Phil. Trans. R. Soc. Lond. A **357**, 2511–2526 (1999)
6. Nason, G., von Sachs, R., Kroisandt, G.: Wavelet processes and adaptive estimation of the evolutionary wavelet spectrum. J. R. Stat. Soc. B **62**, 271–292 (2000)
7. Nason, G.P.: Wavelet Methods in Statistics with R. Springer, New York (2008)
8. Ombao, H., Van Bellegem, S.: Coherence analysis of nonstationary time series: a linear filtering point of view. IEEE Trans. Signal Proc. **56**(6):2259–2266 (2008)
9. Priestley, M.: Spectral Analysis and Time Series. Academic, New York (1983)
10. Priestley, M.: Non-Linear and Non-Stationary Time Series Analysis. Academic, San Diego (1988)
11. Sanderson, J., Fryzlewicz, P., Jones, M.: Estimating linear dependence between nonstationary time series using the locally stationary wavelet model. Biometrika **97**, 435–446 (2010)

Chapter 7
Stable Nonparametric Signal Filtration in Nonlinear Models

Alexander V. Dobrovidov

Abstract A stationary two-component hidden Markov process $(X_n, S_n)_{n \geq 1}$ is considered where the first component is observable and the second one is non-observable. The problem of filtering a random signal $(S_n)_{n \geq 1}$ from the mixture with a noise by observations $X_1^n = X_1, \cdots, X_n$ is solved under a nonparametric uncertainty regarding the distribution of the desired signal. This means that a probabilistic parametric model of the useful signal (S_n) is assumed to be completely unknown. In these assumptions, it is impossible generally to build an optimal Bayesian estimator of S_n explicitly. However, for a more restricted class of static observation models, in which the conditional density $f(x_n | s_n)$ belongs to the exponential family, the Bayesian estimator satisfies the optimal filtering equation which depends on probabilistic characteristics of the observable process (X_n) only. These unknown characteristics can be restored from the observations X_1^n by using stable nonparametric estimation procedures adapted to dependent data. Up to this work, the author investigated the case where the domain of the desirable signal S_n is the whole real axis. To solve the problem in this case the approach of nonparametric kernel estimation with symmetric kernel functions has been used. In this paper we consider the case where $S_n \in 0, \infty)$. This assumption leads to nonlinear models of observation and non-Gaussian noise which in turn requires more complex mathematical constructions including non-symmetric kernel functions. The nonlinear multiplicative observation model with non-Gaussian noise is considered in detail, and the nonparametric estimator of an unknown gain coefficient is constructed. The choice of smoothing and regularization parameters plays the crucial role to build stable nonparametric procedures. The optimal choice of these parameters leading to an automatic algorithm of nonparametric filtering is proposed.

Keywords Nonlinear models • Positive signals • Nonparametric filtration

A.V. Dobrovidov (✉)
Institute of Control Sciences of Russian Academy of Sciences, Moscow, 117997 Russia
e-mail: dobrovid@ipu.ru

© Springer Science+Business Media New York 2014   61
M.G. Akritas et al. (eds.), *Topics in Nonparametric Statistics*, Springer Proceedings
in Mathematics & Statistics 74, DOI 10.1007/978-1-4939-0569-0_7

7.1 Introduction

This paper considers the problem of extracting the useful random signal $S_n \in \mathcal{S} \subseteq \mathbb{R}$ or a known one-to-one function $\vartheta_n = Q(S_n)$ from the mixture $X_n = \varphi(S_n, \eta_n)$ with the noise η_n, where $\varphi(\cdot)$ is a given function and $\mathcal{S}$ is an admissible set. Here, it is assumed that the state model and the distribution of S_n are completely unknown. Such problems often arise in the processing of information in radio and sonar, as well as getting the true information in some problems of financial mathematics (e.g., the information about realized volatility or exchange rates). The problem is studied in discrete time. The goal is to construct an estimator $\hat{\vartheta}_n$ of the process $\vartheta_n = Q(S_n)$ at the moment n by sample $X_1^n = (X_1, \ldots, X_n)$ of the mixture (X_n). If the performance index is the mean squared deviation $\mathsf{E}(Q(S_n) - \hat{\vartheta}_n)^2$, then it is well known that the optimal estimator equals to the posterior mean $\hat{\vartheta}_n^{opt} = \mathsf{E}(Q(S_n)|X_1^n)$. There are several approaches to calculate $\hat{\vartheta}_n^{opt}$ based on different estimation theories. For $Q(S_n) = S_n$, linear observation model $X_n = S_n + \eta_n$ and Gaussian joint distribution of (X_n, S_n), in accordance with the theorem of a normal correlation [15], the posterior mean

$$\hat{S}_n = \mathsf{E}S_n + \mathsf{cov}(S_n, X_1^n)\mathsf{cov}^{-1}(X_1^n, X_1^n)(X_1^n - \mathsf{E}X_1^n) \tag{7.1}$$

can be expressed in terms of the covariances of observations $\mathsf{cov}(X_1^n, X_1^n)$ and mutual covariances $\mathsf{cov}(S_n, X_i), i = 1, \ldots, n$ of the signal and observations. The formula (7.1) admits a recursive representation called the Kalman filter. Emphasize that such filter can only be built when the Gaussian joint distribution of the signal and observation are completely known. In the more general case when the observation model is nonlinear and the joint distribution of (X_n, S_n) is non-Gaussian, the optimal Bayesian estimator can be obtained by applying the theory of conditional Markov processes of Stratonovich [17]. For this the compound process (X_n, S_n) should be a Markov process. The optimal estimator is represented as

$$\hat{\theta}_n^{opt} = \mathsf{E}(Q(S_n)|X_1^n = x_1^n) = \int_{\mathcal{S}} Q(s_n)w_n(s_n|x_1^n)ds_n, \tag{7.2}$$

where $w_n(s_n|x_1^n)$ is a posterior density of the signal S_n under fixed observations $x_1^n = (x_1, \ldots, x_n)$. In accordance with the theory of conditional Markov processes this posterior density can be written in a recursive form which allows one to trace a density evolution from w_{n-1} to w_n. The density w_n is substituted in (7.2) and, after integration, one receives the estimator desired.

There are some other ways to determine the optimal estimator (7.2). However, most of them require precise assignment of the joint distribution of the process (X_n, S_n). What should we do if the distribution of the useful signal (S_n) is unknown, as in the applications mentioned above? In this case it is impossible to observe the "pure" useful signal (S_n) and to collect any statistics about it. The signal can be observed only with the noise, thus, in general, the posterior estimator (7.2) cannot be built directly. Nevertheless, for a more restricted class of static observation models described by the conditional density of observations $f(x_n|s_n)$, belonging to the exponential family of density distributions (one-dimensional case) [9, 11]

$$f(x_n|s_n) = \tilde{C}(s_n)h(x_n)\exp\{T(x_n)Q(s_n)\}, \quad x_n \in \mathbb{R}, s_n \in \mathbb{R}, \quad (7.3)$$

where $h(x_n)$, $Q(s_n)$, and $T(x_n)$ are given Borel functions, one can construct an exact equation [9] for the optimal estimator (7.2)

$$T'(x_n)\hat{\theta}_n^{opt} = \frac{\partial}{\partial x_n}\left(\ln\frac{f(x_n|x_1^{n-1})}{h(x_n)}\right), \quad (7.4)$$

where $T' = dT/dx_n$ and $\tilde{C}(s_n)$ is a normalizing factor.

The main feature of the Eq. (7.4) lies in its explicit independence from the unknown distribution of (S_n). To our best knowledge, this equation was firstly obtained in [4]. In this equation, the optimal estimator $\hat{\theta}_n^{opt}$ is expressed in terms of probability characteristics of observable random variables X_1^n only. Such property of the estimators is the implementation of the well known empirical Bayesian approach of G. Robbins to the problems of signal processing. Empirical Bayesian approach was used in some papers [14, 16] to find a Bayesian estimators of the unknown constant parameters of probability densities from the exponential family in the case of i.i.d. observations. For the time being, the author has not found any papers concerning the application of empirical Bayesian approach to signal processing in which observations have been statistically dependent.

It should be noted that it is linear entering of the parameter $\theta_n = Q(s_n)$ under the exponent of (7.3) (the so called canonical parametrization) allows us to obtain the Eq. (7.4). Therefore we can construct an estimator of S_n as $\hat{s}_n = Q^{-1}(\hat{\theta}_n)$. This will be shown below in the example of the multiplicative model.

Probability characteristics $\frac{\partial}{\partial x_n}\ln(f(x_n|x_1^{n-1})$ entering into the Eq. (7.4) are unknown but they can be restored using nonparametric kernel procedures of estimation by dependent observations. This approach constitutes the content of the nonparametric signal estimation theory [6, 9] where nonparametric methods of filtering, interpolation, and forecasting of signals with unknown distributions were developed.

Despite the nonparametric kernel techniques are designed to restore functional dependencies with a priori unknown form, they contain some unknown parameters that have to be estimated by observed sample. These parameters are smoothing parameters (bandwidths), which differ one from another while estimating densities and their derivatives. A nonparametric estimator of the logarithmic derivative from the Eq. (7.4) belongs to a class of estimators with peculiarities, because they can take extremely geat values which in practice leads to high spikes. One of the methods to remove these spikes is the piece-wise smooth approximation [9, 12, 13], which includes an unknown regularization parameter δ. This parameter is also to be estimated by observed sample. Having the estimators of these two parameters, algorithms of signal processing become automatic in the sense that no further information apart from the sample of observations x_1^n and the conditional density $f(x_n|s_n)$ is required. This approach corresponds to the notion of unsupervised systems, introduced in the 1960s of last century.

Such automatic algorithm of stochastic signal filtration was built in [7, 8] for the linear observation model with a Gaussian noise, where $S_n, X_n \in \mathbb{R}$. In this case it is natural to compare the nonparametric filtering quality with the quality of Kalman filter. In this paper, an automatic nonparametric filtering algorithm for a multiplicative model $X_n = S_n \eta_n$ with positive variables $S_n, X_n \in [0, \infty]$ is constructed. Here the distribution density of noise $p(\eta_n)$ is generated by normalized χ^2-distribution. Note that these linear and nonlinear models are described by different conditional densities $f(x_n|s_n)$ belonging to the exponential family (7.3). The latter allows us to use the same general equation (7.4) to find the optimal estimator (7.2) and its nonparametric counterpart.

In addition to the optimal nonlinear and nonparametric estimators it is of interest to build for a nonlinear model the optimal linear estimate in the case of complete statistical information. This is done in order to understand what is better: the optimality of linear estimator or the possibility of sub-optimal nonparametric estimator to adapt to unknown functional dependence.

In the case of nonlinear models a lot of additional problems arises. First, how to evaluate the accuracy of nonparametric estimators. Indeed, in the nonlinear case "Kalman type" algorithm for the optimal estimation does not exist. Second, to compare estimators numerically one has to generate dependent rvs simulating the desired signal. This problem arises because any multivariate distributions except Gaussian have no simple analytical form or unknown at all. Third, to construct the linear optimal estimators (with complete information) in the nonlinear models it is necessary to calculate some moments of non-Gaussian rv. Fourth, to estimate characteristics on the positive semi-axis, using nonparametric techniques, generally accepted Gaussian kernels are not suitable. Hence, one has to apply, for example, gamma-kernels [3], which are non-symmetric and take only positive values.

The paper is structured as follows. In Sect. 7.2, we consider Stratonovich equation for transformation of posterior probability densities and construct a numerical algorithm of calculating the optimal estimator. Section 7.3 is devoted to the generation of dependent random variables corresponding to the multiplicative observation model. In Sect. 7.4, the optimal linear estimator of the useful signal (S_n) based on its moment characteristics is derived. The nonparametric estimator of the useful signal with unknown distribution in the multiplicative model is presented in Sect. 7.5. Consistent estimators of optimal smoothing and regularization parameters are presented here also. The comparison of quality of estimators by simulated data is given in Sect. 7.6.

7.2   Model Description

As an example of a nonlinear system, we consider the multiplicative model

$$X_n = S_n \eta_n, \quad X_n \geqslant 0, \ S_n \in \mathcal{S} = \mathbb{R}^+, \ n \in \mathbb{N}, \tag{7.5}$$

where $(S_n)_{n \geqslant 1}$ is a stationary Markov random process defined by Rayleigh transition probability density with distribution parameter σ_s and covariance power moment $\mathsf{E}\left[S_i^2 S_{i+1}^2\right] = \rho, \ i \in \mathbb{N}$. The sequence $(\eta_n)_{n \geqslant 1}$ consists of iid rvs of the form $\eta_n = \frac{1}{k}\sum_{i=1}^{k}\xi_{ni}^2$, $k \in \mathbb{N}$, where ξ_{ni} are independent Gaussian rvs from i-th repeated experiment, $\xi_{ni} \sim \mathcal{N}(0, \sigma^2), 1 \leqslant i \leqslant k$. The probability density function (pdf) of the variable η_n is described by the following expression

$$p_{\eta_n} = C(k, \sigma) \cdot \eta_n^{k/2-1} \exp\left(-\frac{k\eta_n}{2\sigma^2}\right), \ C(k, \sigma) = \left(\frac{k}{2\sigma^2}\right)^{k/2} \Gamma^{-1}(k/2), \ \eta_n > 0. \tag{7.6}$$

Here we assume that random processes (S_n) and (η_n) are mutually independent. Markov property of the signal S_n allows us to specify two-dimensional distribution only for its whole description. Our next goal is to construct the optimal nonlinear estimator of the signal S_n using full statistical information and observations x_1^n.

7.3   Optimal Mean-Square Signal Estimation

As a performance index the mean-squared risk

$$J(\hat{S}_n) = \mathsf{E}\left(S_n - \hat{S}_n(X_1^n)\right)^2 \tag{7.7}$$

is considered. Under fixed observations $X_1^n = x_1^n$, the optimal mean-square estimator $\hat{S}_n^{opt} = \mathsf{E}(S_n|X_1^n)$, minimizing (7.7), turns into its realization $\hat{s}_n^{opt}$, described by (7.2), where $Q(S_n) = S_n$. Thus the solution of the estimation problem is reduced to finding the posterior density $w_n(s_n|x_1^n)$. From Stratonovich theory of conditionally Markovian processes [17], it is known the recursive formula for transformation of the posterior density

$$w_n\left(s_n \mid x_1^n\right) = \frac{f\left(x_n \mid s_n\right)}{f\left(x_n \mid x_1^{n-1}\right)} \int_{\mathbb{R}^+} \tilde{p}\left(s_n \mid s_{n-1}\right) w_{n-1}\left(s_{n-1} \mid x_1^{n-1}\right) ds_{n-1}, \ n \geqslant 2. \tag{7.8}$$

This equation links the posterior density w_{n-1} on the previous step and the density w_n on the present step using the transition density $\tilde{p}(s_n|s_{n-1})$ of the signal S_n and the conditional density $f(x_n|s_n)$, generated by the model (7.5). Our aim now is to calculate these two densities.

To get the formula for $\tilde{p}(s_n \mid s_{n-1})$, the bivariate Rayleigh pdf [18]

$$p(x, y) = \frac{xy}{\sigma_s^4(1 - \rho)} \exp\left\{-\frac{1}{1-\rho}\left(\frac{x^2}{2\sigma_s^2} + \frac{y^2}{2\sigma_s^2}\right)\right\} I_0\left(\frac{\sqrt{\rho}xy}{(1-\rho)\sigma_s^2}\right) \quad (7.9)$$

is used, where $I_0(x)$ is a modified Bessel function and $\rho = \dfrac{\mathrm{cov}(X^2, Y^2)}{\sqrt{\mathrm{var}(X^2)\mathrm{var}(Y^2)}}$

is a power correlation. Since $\tilde{p}(s_n \mid s_{n-1}) = \dfrac{p(s_n, s_{n-1})}{p_{S_{n-1}}(s_{n-1})}$, where $p_{S_{n-1}}(x) = \dfrac{x}{\sigma_s^2}e^{-\frac{x^2}{2\sigma_s^2}}$ is a known univariate Rayleigh density, we get the following expression for transitional density

$$\tilde{p}(y|x) = \frac{y}{\sigma_s^2(1-\rho)} \exp\left\{-\frac{1}{1-\rho}\left(\frac{\rho x^2}{2\sigma_s^2} + \frac{y^2}{2\sigma_s^2}\right)\right\} I_0\left(\frac{\sqrt{\rho}xy}{\sigma_s^2(1-\rho)}\right). \quad (7.10)$$

It is used not only in Stratonovich's equation (7.8), but also for generating dependent Rayleigh rvs in modeling experiment. To build the rvs generator we need a transitional distribution function [18]

$$F(y|x) = \int_0^y \tilde{p}(y|x)dy = 1 - Q_1\left(x\sqrt{\frac{\rho}{(1-\rho)\sigma_s^2}}, y\sqrt{\frac{\rho}{(1-\rho)\sigma_s^2}}\right), \quad (7.11)$$

which is expressed in terms of the Marcum Q-function [18]

$$Q_M(a, b) = \int_b^\infty x\left(\frac{x}{a}\right)^{M-1} \exp\left(-\frac{x^2 + a^2}{2}\right) I_{M-1}(ax)dx, \quad M \geqslant 1.$$

The expression for $f(x_n|s_n)$, obtained from (7.5) and (7.6), has a form

$$f(x_n|s_n) = C(k, \sigma) \cdot s_n^{-k/2} x_n^{k/2-1} \exp\left(-\frac{kx_n}{2\sigma^2 s_n}\right), \quad x_n > 0. \quad (7.12)$$

Now everything is ready to solve the Eq. (7.8). As an initial condition we have the posterior density $w_1(s_1|x_1)$, which is calculated from Bayes formula

$$w_1(s_1|x_1) = \frac{p_{S_1}(s_1)f(x_1|s_1)}{\int p_{S_1}(s_1)f(x_1|s_1)ds_1}.$$

It substitutes on the right-hand side of (7.8). Then some grid depending on required accuracy on the semi-axis is selected and the computational transformation process is start up. At each step n the posterior density w_n is calculated and substituted in (7.2), giving the required optimal estimator.

7.4 Linear Estimation of Signal with Full Statistical Information

Besides the optimal Bayesian estimator, it is of interest to build a somewhat simpler optimal linear estimator

$$\hat{S}_n^{lin} = a_0 + a_1 X_{n-1} + a_2 X_n \tag{7.13}$$

for the signal S_n in the multiplicative model (7.5), depending on two latest observations. A linear estimator, as usual, is expressed through the mathematical expectation and covariances of the processes (S_n) and (η_n). In the non-Gaussian case the moments are calculated, using the known multivariate distribution of the process (S_n), which is Rayleigh one in our case. Because of its complexity we could calculate the covariance of two time-adjacent values of the process (S_n) only. This reason explains the choice of dependence in (7.13) from two variables of the observed process (X_n).

Minimizing the criterion $E\left(S_n - \hat{S}_n^{lin}\right)^2$ in a_0, a_1, a_2, one can obtain the following system of linear equations for the optimal values $\tilde{a}_0$, $\tilde{a}_1$, $\tilde{a}_2$ of a_0, a_1, a_2:

$$\begin{cases} \tilde{a}_0 + \tilde{a}_1 ES_{n-1}E\eta_{n-1} + \tilde{a}_2 ES_n E\eta_n - ES_n = 0, \\ \tilde{a}_0 ES_{n-1}E\eta_{n-1} + \tilde{a}_1 ES_{n-1}^2 E\eta_{n-1}^2 + ES_n S_{n-1}E\eta_{n-1}\left(\tilde{a}_2 E\eta_{n-1}\right) = 0, \\ \tilde{a}_0 ES_n E\eta_n + \tilde{a}_1 ES_n S_{n-1}E\eta_n E\eta_{n-1} + \tilde{a}_2 ES_n^2 E\eta_n^2 - ES_n^2 E\eta_n = 0. \end{cases} \tag{7.14}$$

Coefficients of the linear system depend on first and second moments of random variables $S_n, S_{n-1}, \eta_n, \eta_{n-1}$. All of them are easy to calculate using known univariate Rayleigh and χ^2-distribution except one covariance $ES_n S_{n-1}$, where the formula (7.9) of bivariate Rayleigh density is applied. Substituting the modified Bessel function in the form of series $I_0(x) = \sum_{m=0}^{\infty} \frac{1}{(m!)^2}\left(\frac{x}{2}\right)^{2m}$ into (7.9), one get (after some tedious calculations) the following expression for the correlation moment

$$ES_n S_{n-1} = \sigma_s(1-\rho)^2 \sum_{m=0}^{\infty} \rho^m \left(\frac{\Gamma\left(m+\frac{3}{2}\right)}{\Gamma(m+1)}\right)^2 = \frac{1}{2}\sigma_s\sqrt{(1-\rho)}\sqrt{\pi}.$$

Solution to the linear equation (7.14) is found by standard methods.

7.5 Nonparametric Estimators in Multiplicative Observation Model

7.5.1 Equation of Optimal Filtering

It should be noted that the problem of nonparametric estimation of useful signal S_n with unknown distribution for the linear observation model $X_n = S_n + \eta_n$ with Gaussian noise η_n was solved in [6, 9] and published in [5, 7]. In order to evaluate the degree of complexity in the transition to non-linear models and non-Gaussian signals, the multiplicative model (7.5) is considered where the noise η_n is described by the pdf (7.6). This density and the model (7.5) generate the conditional density of observations (7.12). Comparing (7.12) and (7.3), one can see that conditional pdf (7.12) belongs to the exponential family with $h(x_n) = x_n^{k/2-1}$, $T(x_n) = kx_n/2\sigma^2$, and $Q(s_n) = 1/s_n$. Consequently, the canonical parameter $\theta_n = 1/s_n$. In this case the optimal equation (7.4) is solved with regard to θ_n.

To demonstrate the method proposed let us derive the optimal equation for the model (7.5). At the beginning write the conditional density (7.12) in the canonical exponential representation [11]

$$f(x_n|\theta_n) = C(k, \theta_n) \cdot \theta_n^{k/2} x_n^{k/2-1} \exp\left(-\frac{kx_n\theta_n}{2\sigma^2}\right), \quad x_n > 0. \quad (7.15)$$

To estimate the canonical parameter, the standard mean-squared conditional risk $J_x(\hat{\theta}_n) = \mathsf{E}_x(\vartheta_n - \hat{\theta}_n)^2 = \mathsf{E}[(\vartheta_n - \hat{\vartheta}_n)^2|x_1^n]$ is used, which after minimization in $\hat{\theta}_n$ leads to the optimal estimator $\hat{\theta}_n^{opt} = \mathsf{E}_x(\vartheta_n) = \mathsf{E}(\vartheta_n|x_1^n)$. Since $\vartheta_n = Q(S_n) = 1/S_n$, we put $\hat{\theta}_n = Q(\hat{s}_n) = 1/\hat{s}_n$. Then the conditional risk $J_x(\hat{\theta}_n) = \mathsf{E}_x(1/S_n - 1/\hat{s}_n)^2 \doteq G_x(\hat{s}_n)$. Minimization $G_x(\hat{s}_n)$ by $\hat{s}_n$ yields an expression of one estimator through another:

$$\hat{s}_n = \frac{1}{\mathsf{E}_x(1/S_n)} = \frac{1}{\mathsf{E}_x(\vartheta_n)} = \frac{1}{\hat{\theta}_n}. \quad (7.16)$$

It is necessary to note that the performance index for estimating S_n becomes nonstandard. A loss function $L(s_n, \hat{s}_n) = (1/s_n - 1/\hat{s}_n)^2$ corresponding to this criterion becomes very sensitive (grows quickly) when the signal s_n or its estimator $\hat{s}_n$ tends to zero bound of the definition domain. Nevertheless, when it is acceptable, the estimate (7.16) is optimal by the criterion $G_x(\hat{s}_n)$. Thereby, the problem of estimation of useful signal S_n is reduced to the problem of estimation of the canonical random parameter ϑ_n. Since, by assumption, the random process

(S_n) is Markovian, and the process (ϑ_n) is also Markovian. Thus the recursive Stratonovich's equation (7.8) can be applied to it, but in relation to posterior densities of the canonical process (ϑ_n):

$$w_n(\theta_n|x_1^n) = \frac{f(x_n|\theta_n)}{f(x_n|x_1^{n-1})} \int\limits_{\Theta_{n-1}} \tilde{p}(\theta_n|\theta_{n-1})w_{n-1}(\theta_{n-1}|x_1^{n-1})d\theta_{n-1}, \ n \geq 2. \quad (7.17)$$

To obtain the equation in $\hat{\theta}_n$, the conditional density $f(x_n|\theta_n)$ must be substituted from (7.15) into (7.17). Then it is necessary to integrate the Eq. (7.17) by θ_n, to transfer the normalizing constant $f(x_n|x_1^{n-1})$ from right to left and to differentiate the resulting equation by x_n (such approach in general case of the conditionally exponential family of densities is given in [4,6,9]). At the end we obtain the required equation

$$\hat{\theta}_n = \frac{\sigma^2(k-2)}{kx_n} - \frac{2\sigma^2}{k}\frac{\partial}{\partial x_n}\ln f(x_n|x_1^{n-1}). \quad (7.18)$$

Now substituting (7.18) into (7.16) we get the exact equations for the optimal estimator $\hat{s}_n$ which depends on characteristics of the observed random variables only. This estimator is nonrecursive as well as the estimator (7.1), but in contrast to (7.1) the estimator (7.18) does not depend on characteristics of unobserved signal S_n. When the distribution of S_n is unknown, the distribution of the mixture X_n is also unknown, but it can be restored by observations x_1^n using nonparametric kernel methods. To solve the Eq. (7.18) we have to restore the unknown logarithmic density derivative $\partial/\partial x_n \ln f(x_n|x_1^{n-1})$ by dependent observations x_1^n. The expression for the conditional density $f(x_n \mid x_1^{n-1})$ formally depends on all of the observed sample x_1^n. But there is no need to estimate the density of such a great dimensionality. Taking into account the weak dependence in the process (X_n), we conclude that only some latest observations $X_{n-1}, \ldots, X_{n-\tau}$ in condition can influence on X_n, where τ is called the *length of dependence zone*. More often in applications τ is assumed to be equal to 1, 2, 3. The meaning of the parameter τ is equivalent to the notion of connectivity of Markov process that approximates the non-Markov observation process (X_n). One way of selection τ is considered in [9]. This remark allows to substitute the conditional density $f(x_n \mid x_1^{n-1})$ by the "truncated" conditional density $\bar{f}(x_n \mid x_{n-\tau}^{n-1})$ with predefined accuracy. Thus, the logarithmic derivative of the "truncated" conditional density has the form

$$\psi(x_{n-\tau}^n) = \frac{\partial}{\partial x_n}\ln \bar{f}(x_n \mid x_{n-\tau}^{n-1}), \quad (7.19)$$

where the argument of $\psi(\cdot)$ contains $\tau + 1$ variables.

7.5.2 Gamma Kernel Nonparametric Estimator of Logarithmic Density Derivative

The main difficulty in constructing nonparametric estimators in the multiplicative models consists in the condition $X_n \geqslant 0$. The principal tools for constructing estimators are non-symmetric kernels, defined on the positive semi-axis. Examples of such kernels are gamma kernels

$$K_{\rho_b(x),b}(t) = \frac{t^{\rho_b(x)-1}\exp(-t/b)}{b^{\rho_b(x)}\Gamma(\rho_b(x))} , \quad \rho_b(x) = \begin{cases} x/b, & \text{if } x > 2b \\ \frac{1}{4}(x/b)^2 + 1, & \text{if } x \in [0, 2b), \end{cases}$$
(7.20)

recently proposed by Chen [3]. Here b is the smoothing parameter. These kernels consist of two curves smoothly joined at a point $x = 2b$. Nonparametric estimators, built using these kernels, have been generalized in [2] to the case of multidimensional bounded observations and in [1] to the case of weakly dependent observations. Multiplicative gamma kernel version of the multidimensional non-parametric estimator of a density has the form

$$\hat{f}(x_1, \ldots, x_d) = \frac{1}{n} \sum_{i=1}^{n} \prod_{s=1}^{d} K^s_{\rho_{b_s}(x_s),b_s}(X_{is}),$$
(7.21)

where $b_1, \cdots, b_d$ are smoothing parameters, K^s is a gamma kernel for a variable x_s, and X_{is} is the s-th element of the sample at the moment i.

Since the logarithmic derivative (7.19) is $\psi(x_{n-\tau}^n) = \dfrac{f'_{x_n}(x_{n-\tau}^n)}{f(x_{n-\tau}^n)}$, where $f'_{x_n}(x_{n-\tau}^n) = \partial f(x_{n-\tau}^n)/\partial x_n$ is the partial derivative at the point x_n, than it is necessary to construct nonparametric estimator of the density partial derivative on the positive semi-axis. By differentiating (7.21) in x_d, we have

$$\hat{f}'_{x_d}(x_1^d) = n^{-1} \sum_{i=1}^{n} (\ln(X_{id})) - \ln b - \Psi(\rho(x_d)) \prod_{s=1}^{d} K^s_{\rho_{b_s}(x_s),b_s}(X_{is}),$$
(7.22)

where $\rho_{b_s(x_s)}$ and $K^s_{\rho_{b_s}(x_s),b_s}$ are defined in (7.20) and $\Psi(\cdot)$ is a Digamma function. Conditions of convergence in mean-square of density derivative (7.22) one can find in [10]. Nonparametric estimates of true Rayleigh density and its derivative are represented in Fig. 7.1. Now we can write the nonparametric counterpart of the optimal equation (7.18) as

$$\tau \hat{\theta}_n = \frac{\sigma^2(k-2)}{kx_n} - \frac{2\sigma^2}{k} \tau \hat{\psi}(x_{n-\tau}^n),$$
(7.23)

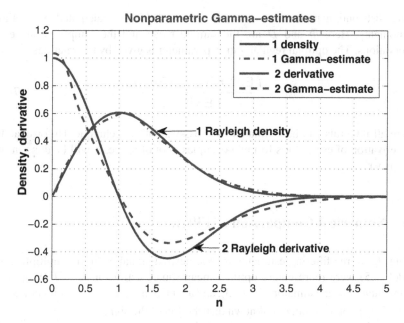

Fig. 7.1 Density, derivative and its Gamma-estimators

where $^{\tau}\hat{\psi}(x^n_{n-\tau}) = \hat{f}'_{x_n}(x^n_{n-\tau})/\hat{f}(x^n_{n-\tau})$ is nonparametric estimate. This statistics is unstable when denominator is near zero. Therefore, we introduce a regularization procedure for obtaining the stable estimator of the form

$$\tilde{\psi}(x^n_{n-\tau}) \doteq \tilde{\psi}_n(x^n_{n-\tau}; \delta_n) = \frac{^{\tau}\hat{\psi}_n(x^n_{n-\tau})}{1 + \delta_n^{\tau}\hat{\psi}_n(x^n_{n-\tau})^4}, \tag{7.24}$$

where the regularization parameter $\delta_n \to 0$ has to be evaluated for a finite n.

7.5.3 Choice of the Smoothing and Regularization Parameters

To solve the Eq. (7.23) the smoothing parameter b and the regularization parameter δ_n have to be chosen, whose optimal values depend on the unknown density $f(x^n_{n-\tau})$, as usual, in kernel estimation procedures. Here the optimality means the minimization of the integrated criterion $\int \mathsf{E}(f - \hat{f})^2$ over b and the integrated criterion $\int \mathsf{E}(\psi - \tilde{\psi})^2$ over δ_n. Since there is no enough space to derive the expressions, we give only the brief results. In univariate case for smoothing parameter, a rough *rule of gamma* with the gamma reference distribution yields an expression

$$b_{rg} = u \left(2\frac{2u - 3v}{3u - 4v} \right)^{2/5} n^{-2/5},$$

where distribution parameters $u = \bar{m}$, $v = \bar{D}/\bar{m}$ are calculated by method of moments. Here $\bar{m}$ and $\bar{D}$ are the sample mean and the sample variance of observations. The optimal regularization parameter is given by the relationship

$$\delta_{opt} = \frac{\int E\hat{\psi}^6 \, f dx - \int \psi E\hat{\psi}^5 \, f dx}{\int E\hat{\psi}^{10} \, f dx}, \qquad (7.25)$$

where all integrals can be estimated by cross-validation technique. Unfortunately, the derivation of estimator's formula occupies a lot of space. It will be done in the next paper.

7.6 Numerical Comparison of Filters

In Fig. 7.2, three filtering estimators of the desired signal S_n in the multiplicative model (7.5) were represented: optimal mean-square estimator $\hat{s}_n^{opt}$ (7.2), optimal mean-square linear estimator $\hat{s}_n^{lin}$ (7.13), and adaptive nonparametric estimator $^{\tau}\hat{s}_n = 1/^{\tau}\hat{\theta}_n$ (7.16) obtained under unknown distribution of the signal S_n.

From these results it can be concluded that the quality of the nonparametric filter is only slightly inferior to the quality of the optimal filter, but it is better

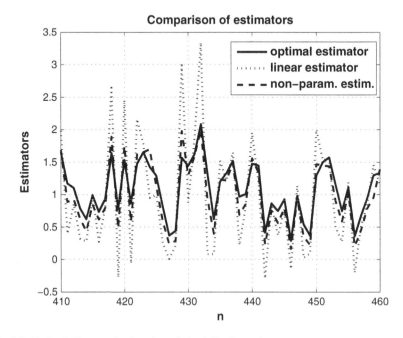

Fig. 7.2 Optimal, linear optimal, and regularized filtering estimators

than the optimal linear filter constructed from the complete statistical information. This means that the nonparametric filter is able to adapt to the unknown structure of functionals of the distributions (i.e., to be adaptive or unsupervised), while the linear filter works within the scope of only two moments and cannot trace the functional form of distributions.

7.7 Conclusion

This paper presents the nonparametric adaptive filter to extract the useful non-negative signal with the unknown distribution. It is shown that the quality of the filter is close to the quality of the Stratonovich nonlinear optimal filter, built using the complete statistical information.

References

1. Bouezmarni, T., Rambouts, J.: Nonparametric density estimation for multivariate bounded data. Core discussion paper, pp 1–31 (2007)
2. Bouezmarni, T., Rambouts, J.V.K.: Nonparametric density estimation for positive time series. Comput. Stat. Data Anal. **54**(2):245–261 (2010)
3. Chen Song Xi.: Probability density function estimation using gamma kernels. Ann. Inst. Statist. Math. **52**, 3, 471–480 (2000)
4. Dobrovidov, A.V.: Nonparametric methods of nonlinear filtering of stationary random sequences. Automat. Remote Control **44**(6):757–768 (1983)
5. Dobrovidov A.V.: Automatic methods of useful signal extraction from noise background under conditions of nonparametric uncertainty. Automat. Remote Control **72**(2):269–282 (2011)
6. Dobrovidov, A.V., Koshkin, G.M.: Nonparametric signal estimation. Moscow: Nauka, Fismatlit, (in Russian) (1997)
7. Dobrovidov, A.V., Koshkin, G.M.: Data-based non-parametric signal filtration. Austrian J. Stat. **1**:15–24 (2011)
8. Dobrovidov, A., Koshkin, G.: Regularized data-based non-parametric filtration of stochastic signals, pp 300–304. World Congress on Engineering 2011, London, UK, 6–8 July 2011
9. Dobrovidov, A.V., Koshkin, G.M., Vasiliev, V.A.: Non-Parametric State Space Models. Kendrick Press, Heber City (2012)
10. Dobrovidov, A.V., Markovich, L.A.: Nonparametric gamma kernel estimators of density derivatives on positive semi-axis. Proceedings of IFAC MIM 2013, 1–6, Petersburg, Russia, 19–21 June 2013
11. Lehmann, E.L.: Testing Statistical Hypotheses. Wiley: New York (1959)
12. Pensky, M.: Empirical Bayes estimation of a scale parameter. Math Methods Stat **5**:316–331 (1996)
13. Pensky, M.: A general approach to nonparametric empirical Bayes estimation. Statistics **29**:61–80 (1997)
14. Pensky, M., Singh, R.S.: Empirical Bayes estimation of reliability characteristics for an exponential family. Can. J. Stat. **27**:127–136 (1999)
15. Shiryaev, A.N.: Probability. Springer, Berlin/Heidelberg/New York (1990)

16. Singh, R.S.: Empirical Bayes estimation in Lebesgue-exponential families with rates near the best possible rate. Ann. Stat. **7**:890–902 (1979)
17. Stratonovich, R.L.: Conditional Markovian Processes and their Application to the Optimal Control Theory. Moscow University Press, Moscow (in Russian) (1966)
18. Tellambura, C., Jayalath, A.D.S.: Generation of Bivariate Rayleigh and Nakagami-m Fadding Envelops. IEEE Commun. Lett. **5**:170–172 (2000)

Chapter 8
Block Bootstrap for the Autocovariance Coefficients of Periodically Correlated Time Series

Anna E. Dudek, Jacek Leśkow, and Sofiane Maiz

Abstract In this paper we propose a new technique of significant frequencies detection for periodically correlated time series. New method is based on bootstrap technique called Generalized Seasonal Block Bootstrap. Bootstrap procedure is applied in the time domain and then Fourier representation of autocovariance function for bootstrap samples is used. Finally, the simultaneous confidence intervals for the absolute values of the Fourier coefficients are calculated. The results are compared in the small simulation study with similar tools based on subsampling methodology and moving block bootstrap for almost periodic processes.

Keywords Autocovariance function • Block bootstrap • Fourier coefficients Periodically correlated time series • Significant frequencies • Simultaneous confidence intervals

8.1 Introduction

In recent years, there is an increased research effort in analyzing nonstationary time series, especially those with cyclic first and second order characteristic. Such time series are quite widely used in signal analysis : a recent paper of Antoni [1] provides an interesting insight into application of such time series into machine diagnostics, while the survey paper of Serpedin et al. [11] gives more than 1,500 different applications of cyclostationary time series and signals in various areas of science and engineering.

A.E. Dudek (✉)
AGH University of Science and Technology, Al. Adama Mickiewicza 30, 30-059 Krakow, Poland
e-mail: aedudek@agh.edu.pl

J. Leśkow
Institute of Mathematics, Cracow University of Technology, Krakow, Poland

S. Maiz
Universite de Saint Etienne, Jean Monnet, F-42000, Saint Etienne, France
LASPI, F-43334, IUT de Roanne, France

© Springer Science+Business Media New York 2014
M.G. Akritas et al. (eds.), *Topics in Nonparametric Statistics*, Springer Proceedings in Mathematics & Statistics 74, DOI 10.1007/978-1-4939-0569-0_8

The most natural way of analyzing first and second order of such nonstationary time series and signals is to estimate and test various components of mean and covariance both in time and in frequency domain. The focus of our research is to provide feasible statistical procedures for estimating Fourier coefficients of covariance function for periodically correlated time series. Before bootstrap and subsampling methods were introduced into the area of nonstationary time series and signals, the statistical inference was based either on direct assumption of gaussianity of the underlying signal or on approximate gaussianity of the estimating procedures. However, the latter approach did not provide practical way of constructing confidence intervals and tests due to a very complicated structure of asymptotic variance-covariance matrix.

In our paper, we focus on providing valid statistical methods of frequency determination for periodically correlated time series and signals. This problem is of critical importance in mechanical signal processing—see, e.g., [1]. We will study a new bootstrap method called Generalized Seasonal Block Bootstrap (GSBB) . This method has been proposed and proven to be consistent for a specific nonstationary time series model in [3]. We will show how GSBB can be applied to the problem of frequency determination.

Section 8.2 of our paper introduces necessary language and notation to be used to study time series and signals with cyclic first and second order characteristics. Moreover, in this section we provide a description of GSBB method. The third section contains a simulation study proving efficiency of GSBB in detecting frequencies for the cyclic autocovariance structure. In Sect. 8.4 we provide an application of our work.

8.2   Problem Formulation

Let $\{X(t), t \in \mathscr{Z}\}$ be a periodically correlated (PC) time series with the known period of the length d, i.e., the considered process has periodic mean and covariance functions

$$E\left(X(t+d)\right) = E\left(X(t)\right) \quad \text{and} \quad \mathrm{Cov}\left(\mathrm{X}(\mathrm{t}+\mathrm{d}), \mathrm{X}(\mathrm{s}+\mathrm{d})\right) = \mathrm{Cov}\left(\mathrm{X}(\mathrm{t}), \mathrm{X}(\mathrm{s})\right).$$

Examples of PC times series can be found, e.g., in [6].

We focus on the Fourier representation of the autocovariance function (see [4] and [5])

$$B_X\left(t, \tau\right) = \sum_{\lambda \in \Lambda_d} a\left(\lambda, \tau\right) \exp\left(i\lambda t\right), \tag{8.1}$$

where

$$a\left(\lambda, \tau\right) = \frac{1}{d} \sum_{t=1}^{d} B_X\left(t, \tau\right) \exp\left(-i\lambda t\right) \tag{8.2}$$

and

$$\Lambda_d = \{\lambda : a(\lambda, \tau) \neq 0\} \subset \{2k\pi/d : k = 0, \ldots, d - 1\}$$

is finite.

Moreover, for all $\lambda \in [0, 2\pi) \setminus \Lambda_d$ coefficients $a(\lambda, \tau)$ are equal to 0.

Assume that we observe the sample $X(1), \ldots, X(n)$ form the considered time series $X(t)$. The estimator of $a(\lambda, \tau)$ is of the form

$$\hat{a}_n(\lambda, \tau) = \frac{1}{n} \sum_{t=1-\min\{\tau,0\}}^{t=n-\max\{\tau,0\}} X(t + \tau) X(t) \exp(-i\lambda t),$$

for more details see [7] and [8].

Very important application of just presented Fourier transformation for the autocovariance function is the significant frequency determination. To construct the test statistic the asymptotic distribution needs to be determined. Unfortunately, the asymptotic variance is of very complicated form, which forces usage of other methods to construct valid confidence intervals. For PC time series the subsampling procedure was used so far (see for example [2] and [9]).

In this paper we propose to use a bootstrap method introduced in [3] called Generalized Seasonal Block Bootstrap (GSBB). The main feature of this method is that it preserves the periodic structure of the time series. Moreover, in contrary to other known block bootstrap methods the block length choice is independent of the period length. Dudek et al. [3] proved the consistency of this method under some moment and α—mixing conditions for the overall and seasonal means of the series. Additionally, the consistent bootstrap simultaneous confidence intervals for the seasonal means were constructed. The performed preliminary simulation study is very encouraging. Actual coverage probabilities are close to nominal ones for the wide spectrum of block length choices. It seems that GSBB is not very sensitive for the block length choice in the simultaneous confidence bands case. Unfortunately, so far there are no consistency results for the second order statistics. However, since subsampling method is computationally very time consuming, no subsampling simultaneous confidence bands for $a(\lambda, \tau)$ were obtained and the subsample length choice has a big impact on the results, and we decided to use GSBB to construct bootstrap versions of $a(\lambda, \tau)$ and simultaneous bands. Our preliminary results should be treated as the beginning of much deeper study that needs to be done.

Below we present the GSBB algorithm proposed in [3]. For the sake of simplicity we focus on its circular version and we assume that the sample size is an integer multiple of the block length b $(n = lb)$ and is also an integer multiple of the period length d $(n = dw)$.

Step 1: Choose a (positive) integer block size $b(< n)$.
Step 2: For $t = 1, b + 1, 2b + 1, \ldots, (l - 1)b + 1$, let

$$(X_t^*, X_{t+1}^*, \ldots, X_{t+b-1}^*) = (X_{k_t}, X_{k_t+1}, \ldots, X_{k_t+b-1})$$

where k_t is iid from a discrete uniform distribution

$$P\left(k_t = t + vd\right) = \frac{1}{w} \quad \text{for} \quad v = 0, 1, \ldots, w - 1.$$

Since we consider the circular version of GSBB, when $t + vd > n$ we take the shifted observations $t + vd - n$.

Step 3: Join the l blocks $(X_{k_t}, X_{k_t+1}, \ldots, X_{k_t+b-1})$ thus obtained together to form a new series of bootstrap pseudo-observations.

We define the bootstrap version of $\hat{a}_n\left(\lambda, \tau\right)$ in the natural way as

$$\hat{a}_n^*\left(\lambda, \tau\right) = \frac{1}{n} \sum_{t=1-\min\{\tau,0\}}^{t=n-\max\{\tau,0\}} X^*\left(t + \tau\right) X^*\left(t\right) \exp\left(-i\lambda t\right),$$

where $X^*(t)$ is a bootstrap version of $X(t)$ obtained by using GSBB.

In the next section we present some simulation study results in which we construct the bootstrap simultaneous equal-tailed confidence interval (same as proposed in [3]) for $|a\left(\lambda, \tau\right)|^2$. We use these results to find the significant frequencies testing if the coefficients are equal to zero.

8.3   Simulation Study

In our study we considered the following model

$$X(t) = a(t) \cos(2\pi f_0 t) + b(t) \cos(2\pi f_1 t),$$

where $a(t)$ and $b(t)$ are two independent stationary random gaussian processes (with zero mean and variance equal to 1). The length of the time series n is equal to 8,192, sine waves frequencies f_0 and f_1 are equal to 0.1 and 0.11 Hz, respectively. In Fig. 8.1 we present surface of estimated absolute values of the autocovariance function. Note that the set of the cyclic significant frequencies is of the form $\{-2f_1, -2f_0, 0, 2f_0, 2f_1\}$. Since the autocorrelation function is symmetric we focused only on the positive part of the real line.

We decided to construct bootstrap simultaneous confidence intervals for $|a\left(\lambda, 0\right)|^2$ using GSBB method, subsampling pointwise confidence intervals using the method proposed in [9] and finally, introduced by Synowiecki in [12] moving block bootstrap (MBB) pointwise confidence intervals for almost periodically correlated (APC) series of the form $X\left(t + \tau\right) X\left(t\right) \exp\left(-i\lambda t\right)$. To be more specific, Lenart et al. in [9] showed the consistency of the standard subsampling technique (see, e.g., [10]) for the estimator of $|a(\lambda, \tau)|$ of a PC time series. On the other hand, Synowiecki in [12] obtained the consistency of MBB for the overall mean of an APC time series and as a possible application pointed out the Fourier coefficients of the autocovariance function.

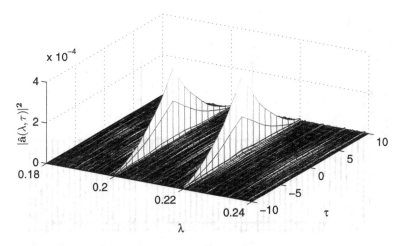

Fig. 8.1 Estimated values of $|a\,(\lambda,\tau)|^2$ for [0.18 Hz, 0.24 Hz] and $\tau \in [-10, 10]$

In our study using GSBB we constructed 95 % equal-tailed bootstrap simultaneous confidence interval. The number of bootstrap samples generated using GSBB is $B = 500$, and the block length b is equal to $\lfloor\sqrt{n}\rfloor$ and $\lfloor\sqrt[3]{n}\rfloor$. In Fig. 8.2 we present the estimated values of $|a\,(\lambda,0)|^2$ together with 95 % simultaneous confidence interval. The only frequencies for which the confidence interval does not include zero value are the true frequencies (for both choices of b).

In Fig. 8.3 the subsampling pointwise confidence intervals are presented. The subsample sizes were taken as previously as $\lfloor\sqrt{n}\rfloor$ and $\lfloor\sqrt[3]{n}\rfloor$. Although it would be enough to cut the lower confidence interval values at zero, we prefer to keep the original values in figures as in the opposite case the subsampling intervals would be very narrow and figures would be very hard to read. In contrary to GSBB, subsampling performance seems to be very dependent on the block length. For b of order $\sqrt{n}$ the significant frequencies were detected properly. But for the shorter block choice the confidence intervals are wider and all contain zero value, which means that no significant frequency was detected. Moreover, note that the most of the subsampling confidence intervals contain strongly negative values. It seems that the bootstrap distribution is very skewed and converges very slowly to the normal distribution. As a result the upper quantile is very high and the lower bound of confidence interval is negative.

Finally, we repeated the simulation study using MBB. Please note that this method is extremely time consuming as for each τ and λ the series $X\,(t+\tau)\,X\,(t)\exp\,(-i\lambda t)$ needs to be constructed. We consider only $\tau = 0$, but wide range of frequencies in the interval [0.18 Hz, 0.24 Hz]. Moreover, there is no method to construct the simultaneous confidence intervals in this case. For MBB we kept all the same parameters as for GSBB. We detected the true significant frequencies for both values of b (see Fig. 8.4). Comparing with other methods the confidence intervals are very narrow when the true frequency is not significant.

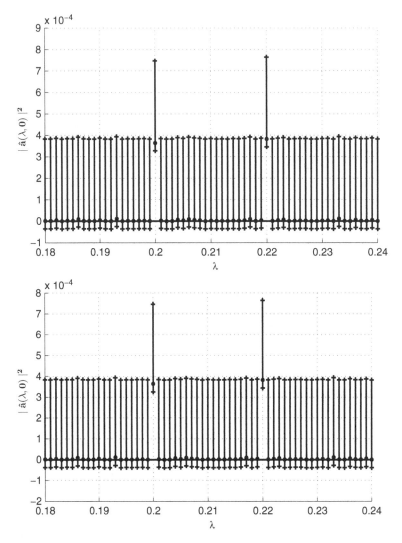

Fig. 8.2 Estimated values of $|a(\lambda, 0)|^2$ (*black dots*) together with 95 % GSBB equal-tailed simultaneous confidence interval (*vertical lines*). Block length b was chosen as $b = \lfloor \sqrt{n} \rfloor$ (*top figure*) and $b = \lfloor \sqrt[3]{n} \rfloor$ (*bottom figure*)

On the other hand those confidence intervals for subsampling method were wide but contained negative values, so finally after removing negative parts of intervals both methods will provide comparable results in this case.

It seems that GSBB is much more powerful than subsampling and MBB in the considered very simple example. Additionally, it is much less computationally time consuming than MBB method and potentially provides simultaneous confidence intervals, which cannot be achieved in MBB case. Although, we are aware that

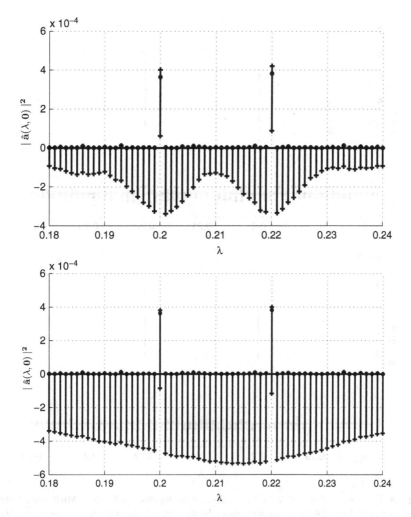

Fig. 8.3 Estimated values of $|a(\lambda, 0)|^2$ (*black dots*) together with 95 % subsampling equal-tailed pointwise confidence intervals (*vertical lines*). Subsample size b was chosen as $b = \lfloor \sqrt{n} \rfloor$ (*top figure*) and $b = \lfloor \sqrt[3]{n} \rfloor$ (*bottom figure*)

much deeper simulation and research study is needed to make any final statement. Also the problem of the optimal choice of b is open for all methods. We decided to use same b values for bootstrap and subsampling only to show that obtained results are not accidental and values are not taken to favor any of the techniques. Performance of GSBB is very promising and research on its consistency for different applications should be continued.

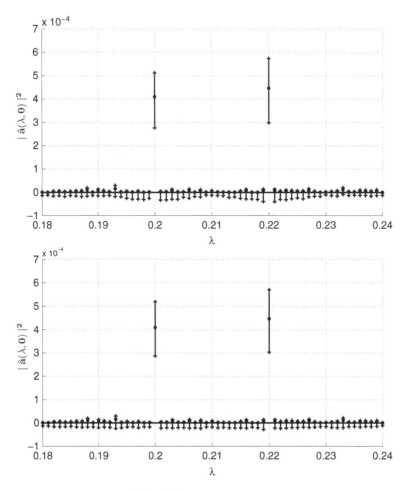

Fig. 8.4 Estimated values of $|a(\lambda, 0)|^2$ (*black dots*) together with 95 % MBB equal-tailed pointwise confidence intervals (*vertical lines*). Block length b was chosen as $b = \lfloor \sqrt{n} \rfloor$ (*top figure*) and $b = \lfloor \sqrt[3]{n} \rfloor$ (*bottom figure*). To indicate the additional significant frequency the x-axis range was reduced in comparison to other cases

8.4 Possible Applications

Just presented preliminary results can be applied in the future in many different settings. Of course, first the consistency theorems are essential. The second order frequency analysis is often necessary in condition monitoring and mechanical systems, rotating machinery, telecommunications, biomechanics signals. Many kind of faults appear and manifest themselves by the presence of second order cyclic frequencies. These frequencies can be the result of a mixture of various random phenomena with some periodic structures related to the studied system. GSBB could be a very useful tool for identifying these frequencies, testing their significance and then making decision about the state of the considered system. The wide range of motivating signal examples can be found in [1].

Acknowledgements Research of Anna Dudek was partially supported by the Polish Ministry of Science and Higher Education and AGH local grant. Research work of Sofiane Maiz was supported by the REGION RHONE ALPES grant.

References

1. Antoni, J.: Cyclostationarity by examples. Mech. Syst. Signal Process. **23**(4), 987–1036 (2009)
2. Dehay, D., Dudek, A., Leśkow, J.: Subsampling for continuous-time nonstationary stochastic processes. J. Stat. Plan. Inf. **150**, 142–158 (2014)
3. Dudek, A., Leśkow, J., Politis, D., Paparoditis, E.: A generalized block bootstrap for seasonal time series. J. Time Ser. Anal. **35**, 89–114 (2014)
4. Hurd, H.: Nonparametric time series analysis for periodically correlated processes. IEEE Trans. Inf. Theory **35**, 350–359 (1989)
5. Hurd, H.: Correlation theory of almost periodically correlated processes. J. Multivar. Anal. **30**, 24–45 (1991)
6. Hurd, H.L., Miamee, A.G.: Periodically Correlated Random Sequences: Spectral Theory and Practice. Wiley, Hoboken (2007)
7. Hurd, H., Leśkow, J.: Estimation of the Fourier coefficient functions and their spectral densities for ϕ-mixing almost periodically correlated processes. Stat. Probab. Lett. **14**, 299–306 (1992a)
8. Hurd, H., Leśkow, J.: Strongly consistent and asymptotically normal estimation of the covariance for almost periodically correlated processes. Stat. Decis. **10**, 201–225 (1992b)
9. Lenart, Ł., Leśkow, J., Synowiecki, R.: Subsampling in testing autocovariance for periodically correlated time series. J. Time Ser. Anal. **29**(6), 995–1018 (2008)
10. Politis, D., Romano, J., Wolf, M.: Subsampling. Springer, New York (1999)
11. Serpedin, E., Panduru, F., Sari, Il, Giannakis, G.: Bibliography on cyclostationarity. Signal Process. **85**, 2233–2300 (2005)
12. Synowiecki, R.: Consistency and application of moving block bootstrap for non-stationary time series with periodic and almost periodic structure. Bernoulli **13**(4), 1151–1178 (2007)

Chapter 9
Bootstrapping Realized Bipower Variation

Gang Feng and Jens-Peter Kreiss

Abstract Realized bipower variation is often used to measure volatility in financial markets with high frequency intraday data. Considering a nonparametric volatility model in discrete time, we propose a nonparametric i.i.d. bootstrap procedure by resampling the noise innovations based on discrete time returns, and a nonparametric wild bootstrap procedure by generating pseudo-noise that imitates correctly the first and second order properties of the underlying noise, in order to approximate the distribution of realized bipower variation. Asymptotic validity of the proposed procedures is proved. Furthermore, the finite sample properties of the proposals are investigated in a simulation study and are also compared with the standard normal approximation.

Keywords Realized bipower variation • Nonparametric volatility estimation • Bootstrap

9.1 Introduction

We begin with a standard continuous-time model for the log-price process (P_t) of a financial asset

$$d \log P_t = \mu_t dt + \sigma_t dW_t,$$

where μ_t denotes the drift, σ_t is a volatility term, and W_t is a standard Brownian motion. Assume that equidistant intraday data with lag $1/n, n \in N$ is observable.

$$X_{i,n} := \log P_{\frac{i}{n}} - \log P_{\frac{i-1}{n}}$$

denotes the intraday log-return over the time interval $[\frac{i-1}{n}, \frac{i}{n}]$.

The integrated volatility (IV) over a day, an important value to quantify the variation of the price process, is defined as

G. Feng (✉) • J.-P. Kreiss
Technische Universität Braunschweig, Institut für Mathematische Stochastik, Pockelsstrasse 14, 38106 Braunschweig, Germany
e-mail: g.feng@tu-bs.de; j.kreiss@tu-bs.de

© Springer Science+Business Media New York 2014   85
M.G. Akritas et al. (eds.), *Topics in Nonparametric Statistics*, Springer Proceedings
in Mathematics & Statistics 74, DOI 10.1007/978-1-4939-0569-0_9

$$\text{IV} := \int_0^1 \sigma_t^2 dt.$$

Encouraged by the increased availability of high frequency data, there exists quite a number of publications that deal with the estimation of integrated volatility in the last few years (see, e.g., Andersen and Bollerslev [1, 3]). The microstructure noise effect of high frequency data on the properties of estimators of integrated volatility is observed, but will not be considered in this paper.

A simple estimator of IV, known as realized volatility (RV) [2], is defined for the model above as

$$\text{RV} := \sum_{i=1}^n X_{i,n}^2.$$

A Central Limit Theorem for $\sqrt{n}(\text{RV} - \text{IV})$ is already given (see, e.g., Barndorff-Nielsen and Shephard [3]).

The more general realized bipower variation (RBV) estimator (see, e.g., Barndorff-Nielsen and Shephard [4]) is defined as

$$\text{RBV}(r, s) := n^{\frac{r+s}{2}-1} \sum_{i=2}^n |X_{i,n}|^r |X_{i-1,n}|^s, \quad r, s \geq 0.$$

It is shown that RBV is robust to finite activity jumps if $r, s < 2$. We concentrate here on RBV(1, 1).

Barndoff-Nielsen et al. [5] showed the following convergence in probability

$$\text{RBV}(r, s) \xrightarrow{p} \mu_r \mu_s \int_0^1 |\sigma_u|^{r+s} du,$$

and under certain assumptions on the stochastic volatility process (σ_t), as $n \to \infty$,

$$T_n^{PZ} := \frac{\sqrt{n} \left(\text{RBV}(r, s) - \mu_r \mu_s \int_0^1 |\sigma_t|^{r+s} dt \right)}{\rho(r, s)} \xrightarrow{d} N(0, 1), \qquad (9.1)$$

where $\mu_r = E(|u|^r)$, $u \sim N(0, 1)$ and

$$\rho^2(r, s) = \left(\mu_{2r} \mu_{2s} + 2\mu_r \mu_s \mu_{r+s} - 3\mu_r^2 \mu_s^2 \right) \int_0^1 |\sigma_t|^{2(r+s)} dt.$$

As an alternative tool to the first-order asymptotic theory, Goncalves and Meddahi [6] primarily introduced two bootstrap methods in the context of realized volatility. Podolskij and Ziggel [8] extended it to realized bipower variation. Podolskij and Ziggel [8] proved first-order asymptotic validity and used Edgeworth expansions and Monte Carlo simulations to compare the accuracy of the bootstrap with existing approaches. It is worth mentioning that Podolskij and Ziggel as well as

Goncalves and Meddahi focus on standardized quantities like (9.1) for their boostrap procedures. We propose in this paper two further bootstrap methods in the context of a nonparametric model and we do not restrict to standardized quantities.

In the following, we consider a discrete-time model for the intraday log-return process $(X_{t,n})$:

$$X_{t,n} := \frac{1}{\sqrt{n}} \sigma \left(\frac{t-1}{n} \right) \varepsilon_t, \quad t = 1, \ldots, n, \tag{9.2}$$

where $n \in N$ is the number of intraday observations.

Assumption A.

- (ε_t) are i.i.d. but not necessarily normally distributed random variables with $E\varepsilon_t = 0$, $E\varepsilon_t^2 = 1$, and $E\varepsilon_t^4 < \infty$.
- σ denotes a spot volatility term. We assume it can be described with a non-stochastic continuous differentiable function $\sigma : [0, 1] \to (0, \infty)$.

The following theorem can be shown.

Theorem 1. *For the discrete-time model (9.2), it holds under assumption A, as $n \to \infty$, that*

$$T_n := \sqrt{n} \left(\mathrm{RBV}(1,1) - (E|\varepsilon_1|)^2 \mathrm{IV} \right) \xrightarrow{d} N \left(0, \tilde{\rho}^2(1,1) \right), \tag{9.3}$$

where

$$\tilde{\rho}^2(1,1) = \left(\left(E|\varepsilon_1|^2 \right)^2 + 2 \left(E|\varepsilon_1| \right)^2 E|\varepsilon_1|^2 - 3 \left(E|\varepsilon_1| \right)^4 \right) \int_0^1 |\sigma_u|^4 du.$$

Proof. We define

$$U_{i,n} := \sigma \left(\frac{i-1}{n} \right) \sigma \left(\frac{i-2}{n} \right) \left(|\varepsilon_i \varepsilon_{i-1}| - (E|\varepsilon_1|)^2 \right).$$

For each $n \in N$, $\{U_{i,n} : i = 1, \cdots, n\}$ are centered and 1-dependent random variables. T_n can be written as follows: $T_n = A_n + B_n$, where

$$A_n := \frac{1}{\sqrt{n}} \sum_{i=2}^n U_{i,n}$$

and

$$B_n := \left[\frac{1}{\sqrt{n}} \sum_{i=2}^n \sigma \left(\frac{i-1}{n} \right) \sigma \left(\frac{i-2}{n} \right) (E|\varepsilon_1|)^2 \right] - \sqrt{n} \, (E|\varepsilon_1|)^2 \mathrm{IV}.$$

To handle A_n we make use of a central limit theorem (CLT) for m-dependent triangular arrays (cf. [7]). It can easily be shown, as $n \to \infty$, that

$$\frac{1}{n} \sum_{i=2}^{n} E\left(U_{i,n}^2\right) \longrightarrow \left(\left(E|\varepsilon_1|^2\right)^2 - \left(E|\varepsilon_1|\right)^4\right) \int_0^1 |\sigma_u|^4 du := c(0),$$

and

$$\frac{1}{n} \sum_{i=2}^{n-1} E\left(U_{i,n} U_{i+1,n}\right) \longrightarrow \left(\left(E|\varepsilon_1|\right)^2 E|\varepsilon_1|^2 - \left(E|\varepsilon_1|\right)^4\right) \int_0^1 |\sigma_u|^4 du := c(1).$$

The function $c(\cdot)$ fullfills

$$c(0) + 2c(1) = \tilde{\rho}^2(1,1).$$

A direct computation furthermore leads to

$$\frac{1}{n^2} \sum_{i=1}^{n} E|U_{i,n}|^4 = o(1).$$

Thus a Ljapunov-condition is also fulfilled and Lemma 8.4, [7], gives

$$A_n \xrightarrow{d} N\left(0, \tilde{\rho}^2(1,1)\right).$$

Additionally it is easy to show that, as $n \to \infty$, $B_n \to 0$, which concludes proof of Theorem 1.   □

Remark 1. Podolskij and Ziggel [8] approximate the finite sample distribution of T_n^{PZ}, which is a standardized statistic. We want to approximate the finite sample distribution of T_n including its (asymptotic) variance. Thus, if we want to construct a confidence interval of RBV, our results will directly lead to confidence intervals without further estimation of a standard deviation ρ (as which has to be done in Podolskij and Ziggel [8]).

9.2 Nonparametric I.I.D. Bootstrap

The i.i.d. bootstrap for realized volatility introduced by Goncalves and Meddahi [6] is motivated from constant volatility, i.e., they used a standard resampling scheme from observed log-returns and showed the asymptotic validity under certain assumptions. Podolskij and Ziggel [8] introduced a bootstrap method with the similar idea in context of bipower variation. In contrast, we propose a nonparametric bootstrap procedure by resampling estimated noise innovations based on discrete time returns, which closely mimics the varying volatility structure of observed log-returns.

9.2.1 Bootstrap Procedure

Let realizations $X_{1,n}, \cdots, X_{n,n}$ be given.

- **Step 1:** Compute $\hat{\sigma}$ via a kernel estimator, e.g.,

$$\hat{\sigma}(u)^2 = \frac{\sum_{t=1}^{n} X_{t,n}^2 K\left(\frac{\frac{t-1}{n}-u}{h}\right)}{\frac{1}{n}\sum_{t=1}^{n} K\left(\frac{\frac{t-1}{n}-u}{h}\right)}, \tag{9.4}$$

  where $h > 0$ denotes the bandwidth and $K(u)$, a probability density (typically with bounded support) called kernel function.

- **Step 2:** Let $\hat{\varepsilon}_t = \dfrac{\sqrt{n}X_{t,n}}{\hat{\sigma}\left(\frac{t-1}{n}\right)}$, $t = 1, \cdots, n$. Standardizing $\{\hat{\varepsilon}_1, \cdots, \hat{\varepsilon}_n\}$ gives $\{\bar{\varepsilon}_1, \cdots, \bar{\varepsilon}_n\}$.

- **Step 3:** Generate the bootstrap intraday returns via

$$X_{t,n}^* = \frac{1}{\sqrt{n}}\hat{\sigma}\left(\frac{t-1}{n}\right)\varepsilon_t^*,$$

  in which $\varepsilon_t^* = \bar{\varepsilon}_{I_t}$, $I_t \sim$ Laplace on $\{1, \cdots, n\}$, i.e., ε_t^* are drawn with replacement from the set $\{\bar{\varepsilon}_1, \cdots, \bar{\varepsilon}_n\}$.

The bootstrap realized bipower variation is defined as:

$$\text{RBV}^*(r, s) := n^{\frac{r+s}{2}-1} \sum_{i=2}^{n} |X_{i,n}^*|^r |X_{i-1,n}^*|^s, \quad r, s \geq 0.$$

9.2.2 Validity of the Bootstrap

Theorem 2. *It holds for a continuous differentiable function* $\sigma : [0, 1] \rightarrow (0, \infty)$ *in model (9.2) and a kernel estimator* $\hat{\sigma}$, *for which* $\sup_{i \in \{0,1,\cdots,n\}} |\hat{\sigma}^2(i/n) - \sigma^2(i/n)| = o_P(1)$ *and* $\sup_{u \in [0,1]} |\hat{\sigma}'(u)| = o_P(\sqrt{n})$, *that as* $n \rightarrow \infty$,

$$T_n^* := \sqrt{n}\left(\text{RBV}^*(1, 1) - \hat{\mu}_1^2 \frac{1}{n}\sum_{t=1}^{n}\hat{\sigma}^2\left(\frac{t-1}{n}\right)\right) \xrightarrow{d} N(0, \tilde{\rho}^2(1, 1)) \tag{9.5}$$

in probability. Here $\hat{\mu}_r = E^*|\varepsilon_1^*|^r = \dfrac{1}{n}\sum_{i=1}^{n}|\bar{\varepsilon}_i|^r$. $\tilde{\rho}^2(1, 1)$ *is defined in Theorem 1. The result implies as* $n \rightarrow \infty$:

$$\sup_{x \in R} |P(T_n^* \leq x) - P(T_n \leq x)| \xrightarrow{P} 0.$$

Proof. Similarly to the proof of Theorem 1, we define for each $n \in N$ centered and 1-dependent random variables $\{U_{i,n}^* : i = 1, \cdots, n\}$ as

$$U_{i,n}^* := \hat{\sigma}\left(\frac{i-1}{n}\right)\hat{\sigma}\left(\frac{i-2}{n}\right)\left(|\varepsilon_i^* \varepsilon_{i-1}^*| - \hat{\mu}_1^2\right),$$

and rewrite T_n^* as: $T_n^* = A_n^* + \tilde{B}_n$, where

$$A_n^* := \frac{1}{\sqrt{n}}\sum_{i=2}^{n} U_{i,n}^*,$$

and

$$\tilde{B}_n := \frac{1}{\sqrt{n}}\left(\sum_{i=2}^{n}\hat{\sigma}\left(\frac{i-1}{n}\right)\hat{\sigma}\left(\frac{i-2}{n}\right)\hat{\mu}_1^2 - \sum_{i=1}^{n}\hat{\sigma}^2\left(\frac{i-1}{n}\right)\hat{\mu}_1^2\right).$$

The CLT for m-dependent triangular arrays (Lemma 8.4, [7]) yields the desired result.   □

9.3 Nonparametric Wild Bootstrap

Based on the wild Bootstrap for realized volatility introduced by Goncalves and Meddahi [6], a wild bootstrap method in context of bipower variation was defined by Podolskij and Ziggel [8]. It uses the same summands as the original realized bipower variation, but the returns are all multiplied by an external random variable. We propose a nonparametric wild bootstrap procedure by generating pseudo-noise that imitates correctly the first and second order properties of the underlying noise.

9.3.1 Bootstrap Procedure

Let realizations $X_{1,n}, \cdots, X_{n,n}$ be given.

- **Step 1:** Compute $\hat{\sigma}$ with an kernel estimator, e.g. (9.4).
- **Step 2:** Generate pseudo-noise $\varepsilon_1^*, \cdots, \varepsilon_n^*$ with ε_t^* i.i.d. such that $E^*\varepsilon_t^* = 0$,

$$E^*|\varepsilon_t^*| = \sqrt{\frac{RBV(1,1)}{\frac{1}{n}\sum_{i=1}^{n}\hat{\sigma}^2\left(\frac{i-1}{n}\right)}} \quad \text{and} \quad E^*|\varepsilon_t^*|^2 = \sqrt{\frac{RBV(2,2)}{\frac{1}{n}\sum_{i=1}^{n}\hat{\sigma}^4\left(\frac{i-1}{n}\right)}}.$$

For example, one can easily define even a two point distribution that matches all three moment conditions.

- **Step 3:** Generate the wild bootstrap intraday returns via

$$X_{t,n}^{\text{WB}} = \frac{1}{\sqrt{n}} \hat{\sigma}\left(\frac{t-1}{n}\right) \varepsilon_t^*.$$

The bootstrap realized bipower variation is defined as:

$$\text{RBV}^{\text{WB}}(r,s) := n^{\frac{r+s}{2}-1} \sum_{i=2}^{n} |X_{i,n}^{\text{WB}}|^r |X_{i-1,n}^{\text{WB}}|^s, \quad r,s \geq 0.$$

9.3.2 Validity of the Bootstrap

Based on the result $\text{RBV}(r,r) \xrightarrow{p} (E|\varepsilon|^r)^2 \int_0^1 \sigma^{2r}(u)du$ for our model (compare [4] for stochastic volatility), we have that

$$E^*|\varepsilon_t^*|^r \xrightarrow{p} E|\varepsilon_t|^r, \quad r = 1,2,$$

so that the first and second order properties of the underlying noise are correctly mimicked.

Theorem 3. *It holds for a continuous differentiable function* $\sigma : [0,1] \to (0,\infty)$ *in model (9.2) and a kernel estimator* $\hat{\sigma}$, *for which* $\sup_{i \in \{0,1,\cdots,n\}} |\hat{\sigma}^2(i/n) - \sigma^2(i/n)| = o_P(1)$ *and* $\sup_{u \in [0,1]} |\hat{\sigma}'(u)| = o_P(\sqrt{n})$, *that as* $n \to \infty$,

$$T_n^{\text{WB}} := \sqrt{n}\left(\text{RBV}^{\text{WB}}(1,1) - (E^*|\varepsilon_1^*|)^2 \frac{1}{n} \sum_{t=1}^{n} \hat{\sigma}^2\left(\frac{t-1}{n}\right)\right) \xrightarrow{d} N(0, \tilde{\rho}^2(1,1))$$

$$(9.6)$$

in probability. $\tilde{\rho}^2(1,1)$ *is defined in Theorem 1. This implies as* $n \to \infty$:

$$\sup_{x \in R} |P(T_n^{\text{WB}} \leq x) - P(T_n \leq x)| \xrightarrow{p} 0.$$

Proof. Similar to the proof of Theorem 2 and therefore omitted. □

9.4 A Simulation Study

Using Monte Carlo simulations, we compare the accuracy of the proposed bootstrap methods with the normal approximation by considering 2.5% and 97.5% quantiles of the finite sample statistic T_n.

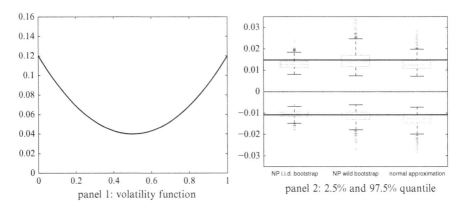

panel 1: volatility function

panel 2: 2.5% and 97.5% quantile

Fig. 9.1 Volatility function and quantiles of nonparametric bootstrap and normal approximation

We choose here a volatility function $\sigma(u) = 0.32(u - 0.5)^2 + 0.04$ (cf. panel 1, Fig. 9.1), a noise innovation of $\varepsilon \sim N(0, 1)$, and a sample size of $n = 200$. The observations are simulated according to model (9.2) and T_n is computed. On the one hand, we compute $\hat{\sigma}$ with the kernel estimator (9.4), generate the bootstrap data, and compute T_n^* and T_n^{WB} [cf. (9.5) and (9.6)]. With 1,000 repetitions of the bootstrap procedures, we get empirical quantiles of the distribution of T_n^* and T_n^{WB}. On the other hand, we computed the desired quantiles via normal approximation with estimated standard deviation $\tilde{\rho}(1, 1)$.

The whole simulation again is repeated 1,000 times to obtain boxplots of sample quantiles of interest. The boxplots on the left side of panel 2 in Fig. 9.1 give the approximations via nonparametric i.i.d. bootstrap, while the ones in the middle give the approximations via nonparametric wild bootstrap. The boxplots on the right side present the results obtained from normal approximation. The true quantiles, indicated as lines in panel 2 of Fig. 9.1, of the finite sample distribution of T_n are also obtained by simulation (100,000 repetitions).

One can see that the median of both bootstrap boxplots nearly hits the true 2.5% quantile. For the 97.5% quantile, both bootstraps perform not as well as for the 2.5% quantile, but at least slightly better than the normal approximation. One reason could be the fact that the kernel volatility estimator (9.4) underestimates the true high volatility at the boundary of the interval. Therefore, the 97.5% quantile, which is strongly related to high volatility, is not so well approximated. The normal approximation of course cannot mimic the skewness of the finite sample distribution of T_n.

With the same setup of n and ε, but another volatility function, namely $\sigma(u) = 0.08 + 0.04 \sin(2\pi u)$ (cf. panel 1, Fig. 9.2), we do the same simulation study as before. The results are displayed in panel 2 of Fig. 9.2.

The situation is quite similar. Bootstrap medians are closer to the true values than the medians of the normal approximation, indicating that bootstrap might be better able to mimic to a certain extent the skewness of the distribution of T_n.

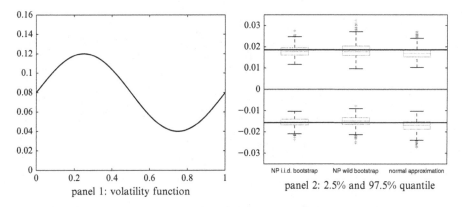

panel 1: volatility function

panel 2: 2.5% and 97.5% quantile

Fig. 9.2 Volatility function and quantiles of nonparametric bootstrap and normal approximation

An underestimation of the 97.5 % quantile does not appear in this case, in which the high volatility is located in a non-border area of the interval. Both bootstraps therefore perform better for the 97.5 % quantile in contrast with the simulation before.

Acknowledgements Both authors gratefully acknowledge the report of an anonymous referee and also the remarks of one of the editors. Their comments lead to a considerable improvement of the manuscript.

References

1. Andersen, T.G., Bollerslev, T.: Answering the skeptics: yes, standard volatility models do provid accurate forecasts. Int. Econ. Rev. **39**, 885–905 (1998)
2. Andersen, T.G., Bollerslev, T.: Parametric and nomparametric measurements of volatility. In: Aït-Sahalia, Y., Hansen, L.P. (eds.) Handbook of Financial Ecnometrics. Elsevier, North-Holland (2002)
3. Barndorff-Nielsen, O.E., Shephard, N.: Econometric analysis of realised volatility and its use in estimating stochastic volatitily models. J. R. Stat. Soc. Ser. B **64**, 253–280 (2002)
4. Barndorff-Nielsen, O.E., Shephard, N.: Power and bipower variation with stochastic volatility and jumps. J. Financ. Econom. **2**(1), 1–37 (2004)
5. Barndorff-Nielsen, O.E., Graversen, S.E., Jacod, J., Podolskij, M., Shephard, N.: A central limit theorem for realized power and bipower variation of continuous semimartingales. In: From Stochastic Analysis to Mathematical Finance, Festschrift for Albert Shiryaev, pp. 33–68. Springer, Berlin (2006)
6. Gonçalves, S., Meddahi, N.: Bootstrapping realized volatility. Econometrica **77**(1), 283–306 (2009)
7. Kreiss, J.-P.: Asymptotical properties of residual bootstrap for autoregression. Technical Report, TU Braunschweig. (1997). Available via: https://www.tu-braunschweig.de/Medien-DB/ stochastik/kreiss-1997.pdf
8. Podolskij, M., Ziggel, D.: Bootstrapping bipower variation. Technical report, Ruhr-University Bochum (2007). Available via: http://www.podolskij.uni-hd.de/files/PZ2007.pdf

Chapter 10
Multiple Testing Approaches for Removing Background Noise from Images

John Thomas White and Subhashis Ghosal*

Abstract Images arising from low-intensity settings such as in X-ray astronomy and computed tomography scan often show a relatively weak but constant background noise across the frame. The background noise can result from various uncontrollable sources. In such a situation, it has been observed that the performance of a denoising algorithm can be improved considerably if an additional thresholding procedure is performed on the processed image to set low intensity values to zero. The threshold is typically chosen by an ad-hoc method, such as 5% of the maximum intensity. In this article, we formalize the choice of thresholding through a multiple testing approach. At each pixel, the null hypothesis that the underlying intensity parameter equals the intensity of the background noise is tested, with due consideration of the multiplicity factor. Pixels where the null hypothesis is not rejected, the estimated intensity will be set to zero, thus creating a sharper contrast with the foreground. The main difference of the present context with the usual multiple testing applications is that in our setup, the null value in the hypotheses is not known, and must be estimated from the data itself. We employ a Gaussian mixture to estimate the unknown common null value of the background intensity level. We discuss three approaches to solve the problem and compare them through simulation studies. The methods are applied on noisy X-ray images of a supernova remnant.

Keywords Image denoising • Background noise • Multiple testing • Multiscale representation • Bayesian approach • Gaussian mixture model

*Research is partially supported by NSF grant number DMS-1106570.

J.T. White
SAS Institute Inc., Cary, NC, USA
e-mail: John.Thomas.White@gmail.com

S. Ghosal (✉)
North Carolina State University, Raleigh, NC, USA
e-mail: sghosal@stat.ncsu.edu

© Springer Science+Business Media New York 2014   95
M.G. Akritas et al. (eds.), *Topics in Nonparametric Statistics*, Springer Proceedings
in Mathematics & Statistics 74, DOI 10.1007/978-1-4939-0569-0_10

10.1   Introduction

In astronomy, distant starts and supernovas are often studied by images taken by X-ray telescopes such as Chandra X-ray observatory, where the data consist of pixel by pixel photon counts (Starck and Murtagh [10]). Typically, these images have very low intensity, where Poisson distribution appropriately model the pixel level photon counts. It is very convenient to analyze such data using a multi-scale framework, where the likelihood function factorizes according to different levels of smoothness (Kolaczyk [6], Kolaczyk and Nowak [7]). Several methods have been developed for denoising such low-intensity images; see [3, 5, 7, 8, 12, 14]. The last reference used a Bayesian approach with a prior that induces ties using a Chinese restaurant process (Pitman [9]) in the values of relative intensities of neighboring pixels, thus allowing structure formation, yet preserving a conditionally conjugate structure for fast computation. At the same time, the Bayesian method gives estimates of the variability in the estimated intensities associated with each pixel.

A common feature of X-ray images is that the area on the screen other than the part occupied by the object of interest shows presence of a faint, roughly constant, background noise, which ideally would have been completely dark (Starck and Murtagh [10]). This noise may come from various other celestial sources and cosmic radiations. Typically, the impact of background noise relative to the radiation from the object of interest is negligibly small for images taken by telescopes reading visible light such as the Hubble telescope. However, X-ray images are typically taken for distant supernova remnants which appear to be extremely faint from the solar system, and so the effect of the background noise becomes much more prominent.

The background noise can be incorporated in the Poisson model for pixel-level photon counts by adding an extra unknown constant term η in the intensity values $\lambda_{(i,j)}$ of each pixel (i, j), making a nugget-like effect as in some Gaussian regression models with measurement errors. However, in the corresponding multiscale framework, the parameter η remains present in all levels thus making it hard to use benefits of the multiscale modeling.

To overcome the difficulty, we take a post-processing approach to remove the effect of the background noise. While the approach may be combined with almost any denoising method for photon limited images, we specifically apply the background noise removal technique on images denoised by the Bayesian method proposed by White and Ghosal [12]. It was shown in [12] that their method outperforms other methods when both accuracy and computational complexity are considered. Moreover their method allows an estimate of variability that will turn out to be useful in our proposed background noise removal approach.

A naive but effective method of background noise removal is a simple thresholding—any estimated intensity below a certain level is set to zero. This introduces more contrast between the intensities of the foreground and background, thus creating sharper boundaries. It has been empirically observed by simulation that in a typical photon limited image, a thresholding below 5 % level of the maximum estimated intensity can often improve distances between processed and true images by 10–20% (White [11]). However, the 5% level of thresholding is chosen in an ad-hoc manner. The processed image can be sensitive to the choice of threshold, and important features may be lost if a too strong threshold is used.

The problem of determining the appropriate threshold can be formulated as a multiple hypotheses testing problem. At a given pixel, the true intensity can be the null value equal to the background intensity or an alternative value which is considerably higher than the null value. We can then test these pair of hypotheses at each pixel using the estimated intensity parameter and the estimated intensity is set to zero only if the null hypothesis is not rejected at that pixel. However, due to a large number of pixels in a frame, which can be between several thousands to a million, a considerable number of false rejections will show up just by the randomness in the estimated intensity values. In order to address the issue, we need to correct for multiplicity. In the next section, we describe three approaches to multiplicity correction. In Sect. 10.3, we present a simulation study to check the effectiveness of the proposed multiple testing method. Some real astronomical images are presented in Sect. 10.4. We then conclude the paper with a discussion of remaining issues and future developments.

10.2   Background Noise Detection by Multiple Testing

Consider an image represented by an array of $N \times N$ pixel-specific intensity values $\lambda_{(i,j)}$, $i, j = 1, \ldots, N$, where N is typically a power of 2, $N = 2^L$. We observe a noisy version of the image as an array $X_{(i,j)}$, $i, j = 1, \ldots, N$. In the low intensity images in X-ray astronomy we are interested in, it is reasonable to model $X_{(i,j)}$ as independent Poisson variables with parameters $\lambda_{(i,j)}$, $i, j = 1, \ldots, N$. Typically, in a considerably large portion of the image, the source of the emission is a background noise with low constant intensity η, that is, $\lambda_{(i,j)} = \eta$ for these pixels, although it is unknown which pixels correspond to the intensity value η. Moreover, the value of η is also unknown. The only information we have, from the intuitive understanding of X-ray astronomical images, is that η is small compared to $\max\{\lambda_{(i,j)} : i, j = 1, \ldots, N\}$ and that for many pixels (i, j), $\lambda_{(i,j)} = \eta$. This is an interesting sparsity structure that will allow improved processing of images. When viewed on a frame, features in the image are much more clearly visible if the intensities at pixels receiving only background noise are set to zero.

To distinguish between signal and background noise, at each pixel (i, j), we can consider a pair of statistical hypotheses, the null $H_{(i,j)} : \lambda_{(i,j)} = \eta$ against an alternative $H'_{(i,j)} : \lambda_{(i,j)} > \eta$. Based on a noisy image data, we have estimates of intensity values $\hat{\lambda}_{(i,j)}$, $(i, j) = 1, \ldots, N$. Below we propose a method of removing the background noise from the processed image using a multiple testing method. We shall illustrate the idea behind the multiple testing technique using images processed by a multi-scale Bayesian method based on a Chinese restaurant process prior recently introduced by White and Ghosal [12]. A brief description of this method is given below. We note that the multiple testing method can be applied on any good Bayesian or non-Bayesian image processing method. For instance, the basic denoising method on which multiple testing is performed can be other commonly used methods such as the translation invariant Haar wavelet method (Kolaczyk [5]), median filtering method, method of Markov random field prior, two-dimensional multi-scale maximum a posteriori method (Nowak and Kolaczyk [8]), multi-scale complexity regularization method (Kolaczyk and Nowak [7]), or the platelet method (Willett and Nowak [14]). The multiple testing method itself may be non-Bayesian or Bayesian. We shall present two variations of the multiple testing method, one of which is based on a Bayesian approach and the other is non-Bayesian.

The multi-scale Bayesian method of [12] is based on a factorization of the Poisson likelihood in different scales and independent assignment of prior distributions on the parameters appearing in each factor such that neighboring values of the parameters are tied with certain probabilities. At intermediate scales $l = 1, \ldots, L - 1$, there are 4^l block-pixels, whose photon counts $X_{l,(i,j)}$ are Poisson distributed with parameter $\lambda_{l,(i,j)}$, say. Let $X^*_{l+1,(i,j)} = (X_{l+1,(i',j')} : i' = 2i - 1, 2i, \, j' = 2j - 1, 2j)$. Then the likelihood function can be factorized as

$$\mathscr{P}(X_{0,(1,1)}|\lambda_{0,(1,1)}) \times \prod_{l=0}^{L-1} \prod_{i=1}^{2^l} \prod_{j=1}^{2^l} \mathscr{M}(X^*_{l+1,(i,j)}|X_{l,(i,j)}, \boldsymbol{\rho}^*_{l+1,(i,j)}), \qquad (10.1)$$

where $\mathscr{P}$ and $\mathscr{M}$ respectively stand for Poisson and multinomial distributions and $\boldsymbol{\rho}^*_{l+1,(i,j)}$ stands for the vector of relative intensities. Independent priors on parameters at different levels will lead to independent posterior distributions and allow easy computation of the posterior means of the original λ-parameters.

To encourage structures formation, White and Ghosal [12] constructed a prior on the relative intensity parameters within each-parent-child group by first conditioning on a randomly formed "configuration" of tied relative intensity parameters, and independently, across different parent-child groups of all possible levels of the multi-scale representation, using Dirichlet distributions of the appropriate dimensions, depending on the number of free parameters. Configurations are formed within each-parent-child group by independent Chinese restaurant processes (Pitman [9]). The Chinese restaurant process creates random partitions dictating which values are to be tied with what. The conjugacy of the Dirichlet distribution for the multinomial likelihood, conditional on the configurations, help represent the posterior distribution analytically. Posterior probabilities of each configuration can be computed using explicit expressions of the integrated likelihood, and hence allow fast analytic computation of posterior mean and variances of intensity parameters.

A test for $H_{(i,j)}$ against $H'_{(i,j)}$ can be based on the estimated intensity $\hat{\lambda}_{(i,j)}$. Ideally, a natural decision is to reject the null hypothesis (i.e., to declare that the source of emission at the location is not just the background image, but is some genuine object of interest) for large values of $(\hat{\lambda}_{(i,j)} - \eta)/\sigma_{(i,j)}$, where $\sigma_{(i,j)}$ stands for the standard deviation of $\hat{\lambda}_{(i,j)}$.

There are several difficulties associated with the testing problem. Firstly, the test statistics $(\hat{\lambda}_{(i,j)} - \eta)/\sigma_{(i,j)}$, $i, j = 1, \ldots, N$, are not observable since the value of the background noise intensity η is unknown. Moreover, the standard deviation $\sigma_{(i,j)}$ of $\hat{\lambda}_{(i,j)}$ is difficult to obtain, and will involve unknown values of true intensities in a very complicated way, so even a plug-in estimator of $\sigma_{(i,j)}$ is not readily available. The distribution of the test statistic is not known too, although a normal approximation may be reasonable. Finally, as there are many pairs of hypotheses (one for each pixel), the multiplicity must be taken into account to provide appropriate controls on testing errors.

Multiple testing issues commonly appear in genomics and other modern statistical applications. Suppose that there are n null hypotheses $H_{10}, \ldots, H_{n0}$, which are to be tested, out of which n_0 are true. Suppose that R of n null hypotheses are rejected if tested individually using a common critical level α. Let V be the (unobserved) number of true null hypotheses that are rejected. Ideally, one would like to control the family-wise error rate (FWER) given by $P(V > 0)$, but in most applications, that turns out to be too conservative. Benjamini and Hochberg [1] considered a different measure, called the false discovery rate (FDR) given by $E(V/ \max(R, 1))$ and devised a procedure to control the FDR. Let $p_{(1)}, \ldots, p_{(n)}$ denote the ordered nominal p-values and $\hat{i} = \max\{i : p_{(i)} < i\alpha/n\}$. Then rejecting all null hypotheses H_{0j} corresponding to p-values $p_j \leq p_{(\hat{i})}$ will control the FDR at level α (Benjamini and Hochberg [1]). If the proportion of true hypothesis n_0/n, is known to be, or can be estimated as w, then an improved procedure is obtained by replacing $i\alpha/n$ by $i\alpha/(nw)$. Often, the Benjamini–Hochberg procedure remains valid even when the test statistics (equivalently, p-values) in different locations are not independent, but are positively related, although it can become substantially conservative. Ghosal and Roy [4] found that modeling p-values or test statistics by nonparametric mixture models can be beneficial, especially if the distributions of the test statistics are hard to obtain and dependence is present across test statistics corresponding to different hypotheses.

In the present context, we are interested in testing the $n = N^2$ hypotheses $\lambda_{(i,j)} = \eta$, $i, j = 1, \ldots, N$, simultaneously. We model the estimated intensity values nonparametrically as a Gaussian mixture model (GMM). The idea is of course motivated from the observation that each $\hat{\lambda}_{(i,j)}$ is approximately normally distributed with mean $\lambda_{(i,j)}$—a consequence of a Bernstein–von Mises theorem that may be established since the Poisson distribution forms a regular finite dimensional family. As finite Gaussian mixtures are quite accurate in approximating a more general Gaussian mixture, it will suffice to consider a relatively few number of components, say, three or five leading to the following mixture model for the marginal density of $\hat{\lambda}_{(i,j)}$:

$$f(\lambda) = \sum_{k=1}^{r} w_k \, \phi(\lambda; \mu_k, \sigma_k), \tag{10.2}$$

where w_k is the weight of the component with mean μ_k and standard deviation σ_k, r is the number of mixture components, and ϕ stands for the normal density. It may be noted that $\hat{\lambda}_{(i,j)}$ will not be independently distributed across pixels as $\hat{\lambda}_{(i,j)}$ involves all $X_{(i',j')}$. However, in image processing applications, the dependence becomes extremely weak for pixels that are far apart, leading to a mixing-type dependence. This will allow estimation of parameters in (10.2). The mixture component with the smallest center (and usually with the highest proportion also) is of interest for our purpose. By studying the histogram of the estimated intensity parameters corresponding to the pixels, we find that three Gaussian components are needed to adequately represent the empirical histogram so that the null component is clearly visible, and hence to allow accurate estimation of this component. If the full density needs to be estimated as in the Bayesian multiple testing method, it appears that five mixture components are needed. The resulting methods appear to be fairly insensitive to the choice of the number of components r, as long as $r \geq 3$ and $r \geq 5$ for the respective problems. Hence we shall use $r = 3$ or 5 depending on the multiple testing methods in the numerical illustrations below. However, a more formal approach for the choice of r may be based on model selection criteria such as AIC or BIC. If desired, completely nonparametric Gaussian mixture models can be used too at the expense of additional computational complexity, but it does not appear to improve accuracy of the multiple testing method. Note that the multiplicity of the testing problem has a benefit, namely, the estimation of the marginal density $f(\lambda)$ is possible only due to the multiplicity, which works as repeated sampling. This together with the assumed sparsity structure allows us to detect the intensity level η of the background noise.

To estimate the parameters in (10.2), we employ an EM-algorithm starting with an initial value given by a k-means algorithm as described by Calinon [2]. Once the parameters are estimated and component with the lowest value of μ_k is identified as the estimated value of η, we can consider three different methods of multiple testing. The first approach employs the classical multiplicity correction based on the Benjamini–Hochberg procedure, where test statistics are normalized by the estimated standard deviation corresponding the null component in the Gaussian mixture model for the post-processing intensity values. The second approach is a minor variation of the first, where the only difference is in the estimate of the standard deviation, namely, a pixel-specific estimated standard deviation is obtained from the posterior distribution, and is applicable only for a Bayesian method such as that of [12]. The third approach uses a Bayesian multiple testing method. Below, we describe these methods in more details:

- Gaussian Mixture Model (GMM) Method: We use the test statistic $T_{(i,j)} = (\hat{\lambda}_{(i,j)} - \hat{\eta})/\hat{\sigma}$, where $\hat{\sigma}$ is the standard deviation of the Gaussian component associated with center $\hat{\eta}$, and compare its value with the one-sided standard normal critical value. Then the Benjamini–Hochberg procedure is applied on $T_{(i,j)}$ to locate pixels corresponding to background intensity η.

- Posterior Variance Method: We use the same idea except that we use a pixel specific estimate of the standard deviation $\hat{\sigma}_{(i,j)}$, obtained from the posterior variance of $\lambda_{(i,j)}$, i.e., $T_{(i,j)} = (\hat{\lambda}_{(i,j)} - \hat{\eta})/\hat{\sigma}_{(i,j)}$.
- Bayes Classification Method: The intensity value for each pixel (i, j) is classified as the null or the alternative component using a standard Bayes classifier, treating the estimated weights as prior probabilities of each component and the component Gaussian distribution as the likelihood. Thus if the "posterior probability" $\hat{w}\phi_{\hat{\mu},\hat{\sigma}}(\hat{\lambda}_{(i,j)})/\hat{f}(\hat{\lambda}_{(i,j)})$, where $\hat{w}$ is the weight of the null Gaussian mixture component and $(\hat{\mu}, \hat{\sigma})$ its parameters estimated using the GMM method, and $\hat{f}$ represents the full density estimate using the GMM method, is less than $1/2$, then we keep the pixel intensity estimate and do not threshold it to 0. As the Bayes classifier is based on posterior probability rather than a p-value, it does not need multiplicity correction.

A fully Bayesian variation of the Bayes classification method can be implemented by using a Gibbs sampling algorithm on the Gaussian mixture model to compute the posterior probability of each observation coming from the null component. This method appears to be computationally more expensive, but it will be interesting to study its performance.

10.3   Empirical Study

In this section, we use a well-known image of Saturn and simulate photon counts by following the Poisson model and adding some background noise to study the relative performance of the proposed multiple testing methods for background noise removal. We consider two situations—light background noise (maximum intensity 10 and background intensity 0.3) and heavy background noise (maximum intensity 5 and background intensity 1). The simulation results are shown in Table 10.1. In the light background noise scenario, the error, computed by the mean absolute deviation (MAD), was reduced by 64% after the usual denoising stage without the background noise correction. The error was further reduced by over 50% using the posterior variance method of removing background. The other two methods, GMM and Bayes, reduce error by about 45% in the second stage. Second stage reduction at the level 20–25% is obtained in terms of the root-mean-squared error (RMSE). In the heavy background noise scenario, the MAD was reduced respectively by 57%, 70%, and 58% in the second stage by the three background noise removal methods respectively, while these figures are respectively 40%, 51%, and 41% in terms of RMSE. Figure 10.1 shows how the image looks like before and after background noise removal in the heavy background noise situation.

Table 10.1 Saturn Image denoising under two different level background images—light and heavy. Figures are given in the scale 10^{-6}. The numbers in the parentheses denote the standard errors of the estimates. Results are based on ten replications

Method	Light background noise		Heavy background noise	
	MAD	RMSE	MAD	RMSE
Observed	5.42 (0.008)	8.55 (0.013)	8.78 (0.009)	11.75 (0.011)
Smoothed	1.95 (0.010)	2.38 (0.014)	6.53 (0.009)	7.20 (0.010)
GMM	1.11 (0.009)	1.89 (0.015)	2.81 (0.024)	4.33 (0.025)
Posterior variance	0.96 (0.008)	1.77 (0.015)	2.02 (0.019)	3.51 (0.026)
Bayes	1.08 (0.009)	1.86 (0.015)	2.73 (0.081)	4.25 (0.074)

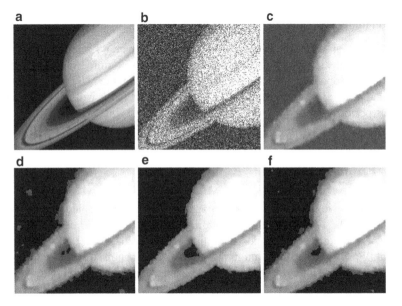

Fig. 10.1 Saturn image with maximum intensity 5 and background intensity 1. (**a**) True. (**b**) Observed. (**c**) Smooth. (**d**) GMM. (**e**) Post. (**f**) Bayes

10.4   Applications

We apply the proposed background removal methods on an X-ray image of the supernova G1.9+0.3 obtained by the Chandra observatory that is publicly available from NASA website. Figure 10.2 shows the smoothed image in the top left. The images after background removal are shown for each algorithm. It appears that the Posterior Variance background noise removal method appears to remove the background noise somewhat more aggressively than the other two methods.

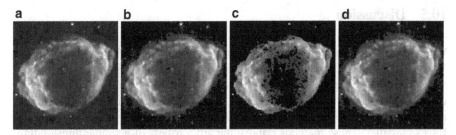

Fig. 10.2 X-ray image of G1.9+0.3 from the Chandra X-ray observatory with denoising and background noise correction. (**a**) Smoothed. (**b**) GMM. (**c**) Post. (**d**) Bayes

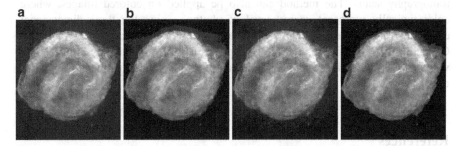

Fig. 10.3 X-ray image of Kepler's supernova remnant from the Chandra X-ray observatory with denoising and background noise correction. (**a**) Smoothed. (**b**) GMM. (**c**) Post. (**d**) Bayes

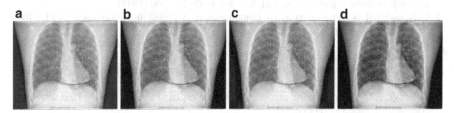

Fig. 10.4 Chest X-ray image with denoising and background noise correction. (**a**) Smoothed. (**b**) GMM. (**c**) Post. (**d**) Bayes

Another X-ray image is shown in Fig. 10.3. This is an image of Kepler's supernova remnant, which has a much larger photon flux due to longer exposure given to it because of its importance. Visually it appears that the Bayes background noise removal algorithm works the best. This may be due to the higher exposure level compared with usual X-ray images.

Finally, we show a medical application of the proposed background noise removal method using a chest X-ray photograph in Fig. 10.4. It appears that the clarity of the method improves when background noise removal technique is applied, especially the Bayes classifier method.

10.5   Discussion

We have proposed a multiple testing method for removing background noise from X-ray images that are given by low photon counts. The method is based on modeling of estimated pixel intensities using a Gaussian mixture model. We considered three approaches to multiple testing, two of which are based on FDR controlling mechanism using two different test statistics and the third one is based on a Bayes classifier. These methods remove the arbitrariness of a simple thresholding procedure and is shown to be effective in further reducing noise from processed images. The method is also applicable to some medical images such as computed tomography scans. The method can also be applied on colored images, where color typically represent energy level of photons, using the three dimensional extension of the method of White and Ghosal [12] proposed recently by White and Ghosal [13]. Although the multiple testing idea is well motivated by intuition and works well in numerical experiments, much of its theoretical justification remains to be established.

References

1. Benjamini, Y., Hochberg, Y.: Controlling the false discovery rate: A practical and powerful approach to multiple testing. J. R. Stat. Soc. Ser. B **57**, 289–300 (1995)
2. Calinon, S.: Robot Programming by Demonstration: A Probabilistic Approach. EPFL/CRC Press, Boca Raton, Springer, Berlin-Heidelberg (2009)
3. Esch, D.N., Connors, A., Karovska, M., van Dyk, D.A.: An image restoration technique with error estimates. Astrophys. J. **610**, 1213–1227 (2004)
4. Ghosal, S., Roy, A.: Predicting false disovery proportion under dependence. J. Am. Stat. Assoc. **106**, 1208–1218 (2011)
5. Kolaczyk, E.D.: Nonparametric estmation of gamma-ray burst intensities using Haar wavelets. Astrophys. J. **483**, 340–349 (1997)
6. Kolaczyk, E.D.: Bayesian multiscale models for Poisson processes. J. Am. Stat. Assoc. **94**, 920–933 (1999)
7. Kolaczyk, E.D., Nowak, R.D.: Multiscale likelihood analysis and complexity penalized estimation. Ann. Stat. **32**, 500–527 (2004)
8. Nowak, R.D., Kolaczyk, E.D.: A statistical multiscale framwork for Poisson inverse problems. IEEE Trans. Inf. Theory **46**, 1811–1825 (2000)
9. Pitman, J.: Exchangeable and partially exchangeable random partitions. Probab. Theory Relat. Fields **102**, 145–158 (1995)
10. Starck, J., Murtagh, F.: Astronomical Image and Data Analysis. Springer, Berlin-Heidelberg (2006)
11. White, J.T.: Bayesian multi-scale smoothing of photon limited images with applications in astronomy and medicine. Ph. D. Thesis, North Carolina State Univeristy (2009)
12. White, J.T., Ghosal, S.: Bayesian smoothing of photon limited images with applications in astronomy. J. R. Stat. Soc. Ser. B **73**, 579–599 (2011)
13. White, J.T., Ghosal, S.: Denoising three-dimensional and colored images using a Bayesian multi-scale model of photon counts. Signal Process. **93**, 2906–2914 (2012)
14. Willett, R.M., Nowak, R.D.: Platelets: A multiscale approach for recovering edges and surfaces in photon-limited medical imaging. IEEE Trans. Med. Imag. **22**, 332–350 (2003)

Chapter 11
Peak-Load Forecasting Using a Functional Semi-Parametric Approach

Frédéric Ferraty, Aldo Goia, Ernesto Salinelli, and Philippe Vieu

Abstract We consider the problem of short-term peak load forecasting in a district-heating system using past heating demand data. Taking advantage of the functional nature of the data, we introduce a forecasting methodology based on functional regression approach. To avoid the limitations due to the linear specification when one uses the linear model and to the well-known dimensionality effects when one uses the full nonparametric model, we adopt a flexible semi-parametric approach based on the Projection Pursuit Regression idea. It leads to an additive decomposition which exploits the most interesting projections of the prediction variable to explain the response. The terms of such decomposition are estimated with a procedure which combines a spline approximation and the one-dimensional Nadaraya–Watson approach.

Keywords Semiparametric functional data analysis • Projection pursuit regression • Forecasting • District heating

11.1 Introduction

District heating (or "teleheating") system consists in distributing the heat for residential and commercial requirements, via a network of insulated pipes. To guarantee an efficient generating capacity, maintaining the system stability, the short-term forecasting (made within the 24 h of the day after), and in particular the prevision of peaks of demand, plays a central role.

Many statistical methods have been introduced to make forecasting in a district-heating system: one can see, for example, Dotzauer [4] and Nielsen and

F. Ferraty • P. Vieu
Institut de Mathématiques de Toulouse Équipe LSP, Université Paul Sabatier,
118, route de Narbonne, 31062 Toulouse Cedex, France
e-mail: ferraty@math.univ-toulouse.fr; vieu@math.univ-toulouse.fr

A. Goia (✉) • E. Salinelli
Dipartimento di Studi per l'Economia e l'Impresa, Unversità del Piemonte
Orientale "A. Avogadro", Via Perrone, 18, 28100 Novara, Italy
e-mail: aldo.goia@eco.unipmn.it; ernesto.salinelli@eco.unipmn.it

© Springer Science+Business Media New York 2014   105
M.G. Akritas et al. (eds.), *Topics in Nonparametric Statistics*, Springer Proceedings
in Mathematics & Statistics 74, DOI 10.1007/978-1-4939-0569-0_11

Madsen [11] for some applications and references. The techniques employed in these works are based on regression or time series models; in practice, however, they skip the fact that the data used are discretization points of curves, are highly correlated, and exhibit some seasonality patterns. Hence, it seems natural in this context to use a functional approach (for a review on functional data analysis the reader can see, e.g., Bosq [1], Ferraty and Vieu [7], and Ramsay and Silverman [12]). Linear modelling for functional time series analysis have been proposed by Bosq [1] while the first nonparametric methodological advances have been given in Ferraty et al. [6]. From an applied point of view, in Goia et al. [9] and in Goia [10] some of such techniques dealing with functional linear modelling have been employed in predicting the peaks load and compared with the multivariate ones.

The linear specification, although it allows to interpret the estimated coefficients, appears quite restrictive and is very difficult to verify in the functional regression context. On the other hand, however, a full nonparametric approach, which should provide a flexible exploratory tool, would suffer some limitations: in particular it would not permit a direct interpretation of estimates and it would be subject to the so-called problem of the dimensionality effects.

In order to avoid the drawbacks due to linear and nonparametric approaches, in this work we propose to make the predictions of peak of heat demand by using the semi-parametric functional approach proposed in Ferraty et al. [5] in the context of the regression model with a functional predictor and a scalar response. It is based on the Projection Pursuit Regression principle (see, e.g., Friedman and Stuetzle [8]), and leads to an additive decomposition which exploits the most interesting projections of the variable. This additive structure provides a flexible alternative to the nonparametric approach, offering often a reasonable interpretation and, as shown in the cited paper, being also insensitive to dimensionality effects.

The paper is organized as follows. We first give a brief presentation of data and the prediction problem (Sect. 11.2). Then, in the Sect. 11.3 we illustrate the Functional Projection Pursuit Regression approach and the estimation technique. Finally the Sect. 11.4 is devoted to present the obtained results.

11.2 A Brief Presentation of the Forecasting Problem

The dataset analyzed has been provided by AEM Turin Group, a municipal utility of the northern Italy city of Turin, which produces heat by means of cogeneration technology and distributes it, guaranteeing the heating to over a quarter of the town. The data consist of measurements of heat consumption taken every hour, during four periods: from 15 October 2001 to 20 April 2002, and the same dates in 2002–2003, 2003–2004 and 2004–2005.

The hourly data for the heating demand in three selected weeks have been plotted in Fig. 11.1. We can clearly distinguish an intra-daily periodical pattern which

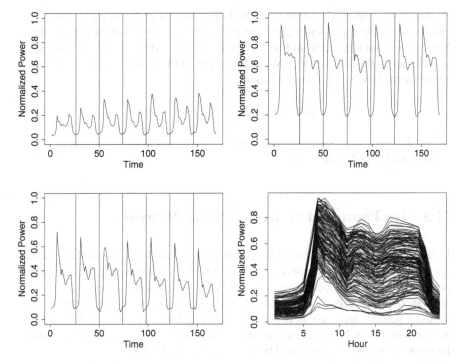

Fig. 11.1 Heating demand in three selected weeks (the panels contain data from November 2002, January and March 2003) and daily load curves of the whole first period (from October 2002 to April 2003).

reflects the aggregate behavior of consumers, and we can also note how it evolves over the seasons following the variations of the climatic conditions.

We take advantage of the functional nature of the data and we divide, in a natural way, the observed series of heat demand of each period in $n = 186$ functional observations, each one coinciding with a specific daily load curve.

We denote by $C_{y,d} = \{C_{y,d}(t), t \in [0, 24]\}$ the daily load curve of period y and day d, with $y = 1, \ldots, 4$ and $d = 1, 2, \ldots, 186$. Since each of these functional data is observed on a finite mesh of discrete times, $t_1 < t_2 < \cdots < t_{24}$, the resulting dataset consists in a matrix with 744 rows and 24 columns. This way for cutting time series into continuous pieces is commonly used when one wishes to apply functional data analysis techniques, as well linear ones (see Bosq [1]) as nonparametric ones (see Ferraty et al. [6]).

Let us consider now the forecasting problem. Define the daily peak load as

$$P_{y,d} = \max_{j=1,\ldots,24} C_{y,d}(t_j),$$

the aim is to predict $P_{y,d}$ on the basis of the load curve from the previous day $C_{y,d-1}$.

Due to the presence of a trend in the time series of peaks in each period, we consider the differenced series $\Delta P_{y,d} = P_{y,d} - P_{y,d-1}$, instead of the original data. Moreover, as we can see from Fig. 11.1, the load data exhibit a vertical shift, so it is more interesting to take the first derivative $DC_{y,d}$ of the load curves. Hence, we introduce the functional time series model:

$$\Delta P_{y,d+1} = \Phi\left[DC_{y,d}\right] + \varepsilon_{y,d}$$

where Φ is the regression operator and $\varepsilon_{y,d}$ is a centered real random error, uncorrelated with the predictor.

11.3 The Functional Projection Pursuit Regression

In this section we resume the most important aspects about the Functional Projection Pursuit Regression (FPPR) approach we use. Rather than developing the time series prediction problem directly, we will look at it as a special case of regression estimation problem from dependent observations. Thus, in the presentation, we follow a general approach as in Ferraty et al. [5]. Sections 11.3.1 and 11.3.2 are written with new self-contained regression notations. The link with our forecasting functional time series context will be made clear in Sect. 11.3.3.

11.3.1 The Additive Decomposition Principle

Let $\{(X_i, Y_i), i = 1, \ldots, n\}$ be n centered random variables (r.v.) identically distributed as (X, Y), where Y is a real r.v. and X is a functional r.v. having values in H, where $H = \{h : \int_I h^2(t)\,dt < +\infty\}$ (I interval of $\mathbb{R}$) is a separable Hilbert space equipped with the inner product $\langle g, f \rangle = \int_I g(t) f(t)\,dt$ and induced norm $\|g\|^2 = \langle g, g \rangle$. The general regression problem can be stated in a standard way as

$$Y = r[X] + \mathscr{E}$$

with $r[X] = \mathbb{E}[Y|X]$. As usual, we assume $\mathbb{E}[\mathscr{E}|X] = 0$ and $\mathbb{E}[\mathscr{E}^2|X] < \infty$.

The main principle of the FPPR is to approximate the unknown regression operator r by a finite sum of terms

$$r[X] \approx \sum_{j=1}^{m} g_j^*\left(\left\langle \theta_j^*, X \right\rangle\right) \tag{11.1}$$

where $\theta_j^* \in H$ with $\left\| \theta_j^* \right\|^2 = 1$, g_j^*, for $j = 1, \ldots, m$, are real univariate functions and m is a positive integer to determine. The aim is to find the *most predictive directions* θ_j^* along which to project X that are the most interesting for explaining Y and, at the same time, describing the relation with Y by using a sum of functions g_j^*, that we call *most predictive components*. We make this by looking at the pairs $\left(\theta_j^*, g_j^* \right)$ iteratively. At first step, we determine θ_1^* by solving

$$\min_{\|\theta_1\|^2 = 1} \mathbb{E}\left[(Y - \mathbb{E}\left[Y \mid \langle \theta_1, X \rangle \right])^2 \right].$$

Once θ_1^* is obtained, we have $g_1^*(u) = \mathbb{E}\left[Y \mid \langle \theta_1^*, X \rangle = u \right]$. If we set $\mathscr{E}_{1,\theta^*} = Y - g_1^* \left(\langle \theta_1^*, X \rangle \right)$, then $\mathscr{E}_{1,\theta_1^*}$ and $\langle \theta_1^*, X \rangle$ are uncorrelated. So, in an iterative way, we can define

$$\mathscr{E}_{j,\theta_j^*} = Y_i - \sum_{s=1}^{j} g_s^* \left(\langle \theta_s^*, X_i \rangle \right) \qquad j = 1, \ldots, m$$

with at each stage $\mathbb{E}\left[\mathscr{E}_{j,\theta_j^*} \mid \langle \theta_j^*, X \rangle \right] = 0$. Then, one can obtain for $j > 1$ the j-th direction θ_j^* by solving the minimization problem

$$\min_{\|\theta_j\|^2 = 1} \mathbb{E}\left[\left(\mathscr{E}_{j-1,\theta_{j-1}^*} - \mathbb{E}\left[\mathscr{E}_{j-1,\theta_{j-1}^*} \mid \langle \theta_j, X \rangle \right] \right)^2 \right]$$

and then define the j-th component as $g_j^*(u) = \mathbb{E}\left[\mathscr{E}_{j-1,\theta_{j-1}^*} \mid \langle \theta_j^*, X \rangle = u \right]$.

By this way, the directions θ_j^* entering in (11.1) are explicitly constructed and so, after the m-th step, one has the additive decomposition with $\mathbb{E}\left[\mathscr{E}_{m,\theta_m^*} \mid \langle \theta_m^*, X \rangle \right] = 0$:

$$Y = \sum_{j=1}^{m} g_j^* \left(\langle \theta_j^*, X \rangle \right) + \mathscr{E}_{m,\theta_m^*}.$$

We remark that two different pairs $\left(\theta_j^{1*}, g_j^{1*} \right)$ and $\left(\theta_j^{2*}, g_j^{2*} \right)$ may produce the same additive estimation (i.e., $g_j^{1*}(\langle \theta_j^{1*}, X \rangle) = g_j^{2*}(\langle \theta_j^{2*}, X \rangle)$). However, that lack of unicity is not a problem in a prediction perspective.

11.3.2 Estimation Strategy in FPPR

We illustrate now how to estimate the functions g_j^* and θ_j^*. The procedure is based on an alternating optimization strategy combining a spline approximation of directions and the Nadaraya–Watson kernel regression estimate of the additive components.

Denote by $\mathscr{S}_{d,N}$ the $(d + N)$-dimensional space of spline functions defined on I with degree d and with $N - 1$ interior equispaced knots (with $d > 2$ and $N > 1$, integers). Let $\{B_{d,N,s}\}$ be the normalized B-splines. For $j = 1, \ldots, m$, the spline approximation of θ_j is represented as $\boldsymbol{\gamma}_j^T \mathbf{B}_{d_j,N_j}(\cdot)$, where $\mathbf{B}_{d_j,N_j}(\cdot)$ is the vector of all the B-splines and $\boldsymbol{\gamma}_j$ is the vector of coefficients satisfying the normalization condition

$$\boldsymbol{\gamma}_j^T \int_I \mathbf{B}_{d_j,N_j}(t) \mathbf{B}_{d_j,N_j}(t)^T \, dt \, \boldsymbol{\gamma}_j = 1. \tag{11.2}$$

The estimation procedure is based on the following steps:

- **Step 1**: Initialize the algorithm by setting $m = 1$ and current residuals $\hat{\mathcal{E}}_{m-1,\hat{\boldsymbol{\gamma}}_{m-1},i} = Y_i, i = 1, \ldots, n$.
- **Step 2**: Choose the dimension $N_m + d_m$ of $\mathscr{S}_{d_m,N_m}$ and fix the initial direction setting the vector of initial coefficients $\boldsymbol{\gamma}_m^{(0)}$ satisfying (11.2).

  Find an estimate $\hat{g}_{m,\boldsymbol{\gamma}_m^{(0)}}^{-i}$ of g_m using the Nadaraya–Watson kernel regression approach excluding the i-th observation X_i:

$$\hat{g}_{m,\boldsymbol{\gamma}_m^{(0)}}^{-i}(z) = \sum_{l \neq i} \frac{K_m\left(\frac{z-(\boldsymbol{\gamma}_m^0)^T \mathbf{b}_{m,l}}{h_m}\right)}{\sum_{l \neq i} K_m\left(\frac{z-(\mathbf{O}_m^0)^T \mathbf{b}_{m,l}}{h_m}\right)} \hat{\mathcal{E}}_{m-1,\hat{\boldsymbol{\gamma}}_{m-1},l}$$

where $\mathbf{b}_{m,l} = \langle \mathbf{B}_{d_m,N_m}, X_l \rangle$.

  Then, compute an estimate $\hat{\boldsymbol{\gamma}}_m$ by minimizing

$$CV_m(\boldsymbol{\gamma}_m) = \frac{1}{n} \sum_{i=1}^n \left[\left(\hat{\mathcal{E}}_{m-1,\hat{\boldsymbol{\gamma}}_{m-1},i} - \hat{g}_{m,\boldsymbol{\gamma}_m^{(0)}}^{-i}\left(\boldsymbol{\gamma}_m^T \mathbf{b}_{m,i}\right) \right)^2 \right] I_{X_i \in \mathscr{G}}$$

over the set of vectors $\boldsymbol{\gamma}_m \in \mathbb{R}^{N_m + d_m}$ satisfying (11.2). Update $\boldsymbol{\gamma}_m^{(0)} = \hat{\boldsymbol{\gamma}}_m$, and repeat the cycle until the convergence: the algorithm terminates when the variation of CV_m passing from the previous to the current iteration (normalized dividing it by the variance of current residuals) is positive and less than a prespecified threshold.

- **Step 3**: Let u_n be a positive sequence tending to zero as n grows to infinity (for instance, one might take u_n proportional to $m \log n / n$). If the penalyzed criterion of fit

$$\text{PCV}(m) = \frac{1}{n} \sum_{i=1}^{n} \left[\left(\hat{\mathcal{E}}_{m-1,\hat{\gamma}_{m-1},i} - \sum_{j=1}^{m} \hat{g}_{m,\hat{\gamma}_m}^{-i} \left(\hat{\gamma}_m^T \mathbf{b}_{m,i} \right) \right)^2 \right] (1+u_n) I_{X_i \in \mathcal{G}}$$

does not decrease, then stop the algorithm. Otherwise, construct the next set of residuals

$$\hat{\mathcal{E}}_{m,\hat{\gamma}_m,i} = \hat{\mathcal{E}}_{m-1,\hat{\gamma}_{m-1},i} - \hat{g}_{m,\hat{\gamma}_m} \left(\hat{\gamma}_m^T \mathbf{b}_{m,i} \right),$$

update the term counter $m = m + 1$, and go to Step 2.

Once the m^* most predictive directions θ_j^* and functions g_j^* which approximate the link between the functional regressor and the scalar response are estimated, one can try to improve the performances of the model, by using a boosting procedure consisting in final full nonparametric step. In practice, we compute the residuals

$$Y_i - \sum_{j=1}^{m^*} \hat{g}_{j,\hat{\theta}_j} \left(\langle \hat{\theta}_j, X_i \rangle \right)$$

and after we estimate the regression function between these residuals and the functional regressors X_i by using the Nadaraya–Watson type estimator proposed in Ferraty and Vieu [7], deriving the following additive decomposition

$$Y_i = \sum_{j=1}^{m^*} \hat{g}_{j,\hat{\theta}_j} \left(\langle \hat{\theta}_j, X_i \rangle \right) + \hat{r}_{m^*+1}(X_i) + \eta_i$$

where η_i are centered independents random errors with finite variance, uncorrelated with X_i, and

$$\hat{r}_{m^*+1}(x) = \sum_{i=1}^{n} \frac{K_{m^*} \left(\frac{\varphi(X_i,x)}{h_{m^*+1}} \right)}{\sum_{i=1}^{n} K_{m^*} \left(\frac{\varphi(X_i,x)}{h_{m^*+1}} \right)} \left(Y_i - \sum_{j=1}^{m^*} \hat{g}_{j,\hat{\theta}_j} \left(\langle \hat{\theta}_j, X_i \rangle \right) \right),$$

where K_{m^*} is a standard kernel function, $h_{m^*+1} > 0$ is a suitable bandwidth, and $\varphi(\cdot, \cdot)$ is a relevant semi-metric.

Remark 1. In the situations when the pairs (X_i, Y_i) are independent, various theoretical properties have been stated for this method along two almost simultaneous papers (see Chen et al. [3], Ferraty et al. [5]), and its nice behavior on real curves datasets has also been highlighted. While theoretical advances for dependent data are still an open challenging problem, we will see in Sect. 11.4 how this methodology behaves nicely for real-time series forecasting problems.

11.3.3   Time Series Application

The general regression methodology described in Sects. 11.3.1 and 11.3.2 below applies directly to the forecasting time series problem described in Sect. 11.3, by putting:

- $X_i = DC_{y,d}$
- $Y_i = \Delta P_{y,d}$
- $I = [0, 24]$
- $n = 744$.

11.4   Application to the District Heating Data

Let us consider the forecasting problem of the daily peak load using the FPPR approach introduced above. We split the initial sample into two sub-samples: a learning sample corresponding to the data observed in the first three periods (2001–2002, 2002–2003, and 2003–2004) and a test sample containing the data of whole fourth period (2004–2005). The estimation procedure is based on cubic splines and the number of knots at each step has been fixed to 10. The initialized values of the vector $\gamma_m^{(0)}$ are random. The bandwidths are selected by a cross-validation procedure over the learning sample.

To evaluate the predictive abilities of the introduced approach, we use the Mean of the Absolute Percentage Errors $e_{4,d} = \left| P_{4,d} - \hat{P}_{4,d} \right| / P_{4,d}$ ($\hat{P}_{4,d}$ are the predicted values) and the classical Relative Mean Square Error of Prediction (RMSEP). The out-of-sample performances are compared with those obtained by using two functional competitors: the functional linear model and the nonparametric model. About the first, the estimation of the functional linear coefficient is based on the penalized B-spline approach (see Cardot et al. [2]) with 40 knots and the penalized coefficient chosen by cross-validation. Concerning the nonparametric model, we use the kernel estimator proposed in Ferraty and Vieu [7] with the semi-metric $\varphi_1(u, v) = \left(\int_I \left(u^{(1)}(t) - v^{(1)}(t) \right)^2 dt \right)^{1/2}$ and the K-nearest neighbors bandwidths, with K selected by a cross-validation over the training-set.

Stopping the FPPR algorithm at $\hat{m} = 1$ (any supplementary steps do not improve the performances), we obtain the results gathered in Table 11.1: we can note that our method fits well and its out-of-sample performances are equivalent to the competitors ones. To clarify what is the surplus of information in using FPPR, it is convenient to look at the estimates of the first most predictive direction θ_1 and the corresponding first additive component g_1 shown in Fig. 11.2. As we can see, the main feature revealed by the graphs is the linearity of the link between the predictor and the response, as we can deduce from the shape of $\hat{g}_1$.

Table 11.1 Out-of-sample performances of the model for the prediction of the peaks of consumption

	FPPR with $m = 1$	Linear model	Nonparametric model
RMSEP	0.9095	0.9151	0.8949
MAPE	0.0553	0.0558	0.0602
% of days in which $e_{4,d} \leq 10\%$	82.8	84.4	81.7
% of days in which $e_{4,d} \leq 5\%$	62.4	60.8	60.8
% of days in which $e_{4,d} \leq 1\%$	18.3	17.7	15.0

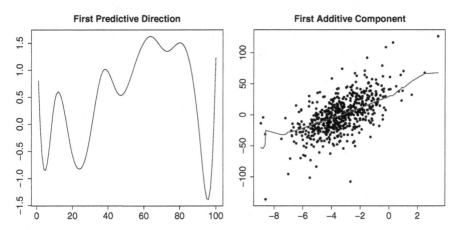

Fig. 11.2 Estimates of the first most predictive direction θ_1 (*left panel*) and the corresponding additive component g_1(*right panel*)

References

1. Bosq, D.: Linear Processes in Function Spaces: Thery and Applications. In: Lecture Notes in Statistics, vol. 149. Springer, New York (2000)
2. Cardot, H., Ferraty, F., Sarda, P.: Spline estimators for the functional linear model. Stat. Sinica **13**, 571–591 (2003)
3. Chen, D., Hall, P., Müller, H.G.: Single and multiple index functional regression models with nonparametric link. Ann. Stat. **39**, 1720–1747 (2011)
4. Dotzauer, E.: Simple model for prediction of loads in district-heating systems. Appl. Energy **73**, 277–284 (2002)
5. Ferraty, F., Goia, A., Salinelli, E., Vieu, P.: Functional projection pursuit regression. TEST **22**, 293–320 (2013)
6. Ferraty, F., Goia, A., Vieu, P.: Functional nonparametric model for time series: a fractal approach to dimension reduction. TEST **11**, 317–344 (2002)
7. Ferraty, F., Vieu, P.: Nonparametric Functional Data Analysis. Springer, New York (2006)
8. Friedman, J.H., Stuetzle, W.: Projection pursuit regression. J. Am. Stat. Assoc. **76**, 817–823 (1981)
9. Goia, A.: A functional linear model for time series prediction with exogenous variables. Stat. Probab. Lett. **82**, 1005–1011 (2012)

10. Goia, A., May, C., Fusai, G.: Functional clustering and linear regression for peak load forecasting. Int. J. Forecast. **26**, 700–711 (2010)
11. Nielsen, H.A., Madsen, H.: Modelling the heat consumption in district heating systems using a grey-box approach. Energy Build. **38**, 63–71 (2006)
12. Ramsay, J.O., Silverman, B.W.: Functional Data Analysis, 2nd edn. Springer, New York (2006)

Chapter 12
An Open Problem on Strongly Consistent Learning of the Best Prediction for Gaussian Processes

László Györfi and Alessio Sancetta

Abstract For Gaussian process, we present an open problem whether or not there is a data driven predictor of the conditional expectation of the current value given the past such that the difference between the predictor and the conditional expectation tends to zero almost surely for all stationary, ergodic, Gaussian processes. We show some related negative and positive findings.

Keywords Conditional expectation • Data driven predictor • Ergodic process • Gaussian time series • Linear regression • Strong consistency

12.1 Open Problem

Let $\{Y_n\}_{-\infty}^{\infty}$ be a stationary, ergodic, mean zero Gaussian process. The predictor is a sequence of functions $g = \{g_i\}_{i=1}^{\infty}$. It is an open problem whether it is possible to learn the best predictor from the past data in a strongly consistent way, i.e., whether there exists a prediction rule g such that

$$\lim_{n \to \infty} \left(\mathbf{E}\{Y_n \mid Y_1^{n-1}\} - g_n(Y_1^{n-1}) \right) = 0 \quad \text{almost surely} \qquad (12.1)$$

for all stationary and ergodic Gaussian processes. (Here Y_1^{n-1} denotes the string $Y_1, \ldots, Y_{n-1}$.)

Bailey [3] and Ryabko [29] proved that just stationarity and ergodicity is not enough, i.e., for any predictor g, there is a binary valued stationary ergodic process such that

L. Györfi (✉)
Department of Computer Science and Information Theory,
Budapest University of Technology and Economics, Stoczek u.2, 1521 Budapest, Hungary
e-mail: gyorfi@cs.bme.hu

A. Sancetta
Department of Economics, Royal Holloway University of London, London, UK
e-mail: asancetta@gmail.com

© Springer Science+Business Media New York 2014   115
M.G. Akritas et al. (eds.), *Topics in Nonparametric Statistics*, Springer Proceedings
in Mathematics & Statistics 74, DOI 10.1007/978-1-4939-0569-0_12

$$\mathbf{P}\left\{\limsup_{n\to\infty}|g_n(Y_1^{n-1}) - \mathbf{E}\{Y_n \mid Y_1^{n-1}\}| \geq 1/2\right\} \geq 1/8,$$

(cf. Győrfi, Morvai and Yakowitz [17]).

In this paper we try to collect some related results such that the main aim is to have as mild conditions on the stationary, ergodic, Gaussian process $\{Y_n\}_{-\infty}^{\infty}$ as possible.

Concerning ergodicity of stationary Gaussian processes, L_2 ergodicity means that

$$\mathbf{E}\left\{\left(\frac{1}{n}\sum_{i=1}^{n} Y_i\right)^2\right\} \to 0, \tag{12.2}$$

which is equivalent to

$$\frac{1}{n}\sum_{i=1}^{n} r(i) \to 0, \tag{12.3}$$

where

$$r(i) = \operatorname{cov}(Y_i, Y_0),$$

(cf. Karlin and Taylor [21]). Moreover, because of stationarity the ergodic theorem implies that

$$\frac{1}{n}\sum_{i=1}^{n} Y_i \to \mathbf{E}\{Y_1 \mid \mathscr{F}\} \tag{12.4}$$

a.s. such that $\mathscr{F}$ is the σ-algebra of invariant sets. From (12.2) and (12.4) we get that

$$\mathbf{E}\{Y_1 \mid \mathscr{F}\} = 0 \tag{12.5}$$

a.s. Thus, from (12.3) we get the strong law of large numbers, and so (12.3) is a necessary condition for ergodicity of a stationary Gaussian process. Maruyama [24] and Grenander [14] proved that the necessary and sufficient condition for ergodicity of a stationary Gaussian process is that the spectral distribution function F is everywhere continuous. Lindgren [23] showed that

$$\frac{1}{n}\sum_{i=1}^{n} r(i)^2 \to 0 \tag{12.6}$$

is a necessary condition for ergodicity, while

$$r(i) \to 0 \tag{12.7}$$

is a sufficient condition. Because of Jensen inequality, we get that

$$\left(\frac{1}{n}\sum_{i=1}^{n} r(i)\right)^2 \le \frac{1}{n}\sum_{i=1}^{n} r(i)^2,$$

therefore (12.6) implies (12.3). Cornfeld et al. [8] showed that for stationary Gaussian process with absolutely continuous spectral distribution, (12.7) is a necessary, too.

In the theory of prediction of stationary Gaussian process, the Wold decomposition plays an important role. It says that we have $Y_n = U_n + V_n$, where the stationary Gaussian processes $\{U_n\}_{-\infty}^{\infty}$ and $\{V_n\}_{-\infty}^{\infty}$ are independent, $\{U_n\}_{-\infty}^{\infty}$ has the MA(∞) representation

$$\sum_{j=0}^{\infty} a_j^* Z_{n-j}, \qquad (12.8)$$

with i.i.d. Gaussian innovations $\{Z_n\}$ and with

$$\sum_{i=1}^{\infty} |a_i^*|^2 < \infty, \qquad (12.9)$$

while the process $\{V_n\}_{-\infty}^{\infty}$ is deterministic: $V_n = \mathbf{E}\{V_n \mid V_{-\infty}^{n-1}\}$.

For a stationary, ergodic Gaussian process, we may get a similar decomposition, if we write the continuous spectral distribution function in the form $F(\lambda) = F^{(a)}(\lambda) + F^{(s)}(\lambda)$, where $F^{(a)}(\lambda)$ is an absolutely continuous distribution function with density function f and $F^{(s)}(\lambda)$ is singular continuous distribution function. Then we have the decomposition $Y_n = U_n' + V_n'$, where the stationary, ergodic Gaussian processes $\{U_n'\}_{-\infty}^{\infty}$ and $\{V_n'\}_{-\infty}^{\infty}$ are independent, $\{U_n'\}_{-\infty}^{\infty}$ has the spectral distribution function $F^{(a)}$, while $\{V_n'\}_{-\infty}^{\infty}$ has the spectral distribution function $F^{(s)}$. If $\int_{-\pi}^{\pi} \ln f(\lambda) d\lambda > -\infty$, then $U_n = U_n'$ and $V_n = V_n'$ (cf. Lindgren [23]).

In the analysis of stationary Gaussian processes one often assumes the MA(∞) or the AR(∞) representations such that these representations imply various type of mixing properties. The AR(∞) representation of the process $\{Y_n\}$ means that

$$Y_n = Z_n + \sum_{j=1}^{\infty} c_j^* Y_{n-j}, \qquad (12.10)$$

with the vector $c^* = (c_1^*, c_2^*, \dots)$. Bierens [4] introduced a non-invertible MA(1) process such that

$$Y_n = Z_n - Z_{n-1}, \qquad (12.11)$$

where the innovations $\{Z_n\}$ are i.i.d. standard Gaussian. Bierens [4] proved that this process has no AR(∞) representation.

The rest of the paper is organized as follows. In Sect. 12.2 we summarize the basic concepts of predicting Gaussian time series, while Sect. 12.3 contains some positive and negative findings concerning universally consistent prediction. The current machine learning techniques in Sect. 12.4 may result in universal consistency.

12.2 Prediction of Gaussian Processes

In this section we consider the classical problem of Gaussian time series prediction (cf. Brockwell and Davis [6]). In this context, parametric models based on distributional assumptions and structural conditions such as $AR(p)$, $MA(q)$, $ARMA(p,q)$, and $ARIMA(p,d,q)$ are usually fitted to the data (cf. Gerencsér and Rissanen [13], Gerencsér [11, 12]). However, in the spirit of modern nonparametric inference, we try to avoid such restrictions on the process structure. Thus, we only assume that we observe a string realization Y_1^{n-1} of a zero mean, stationary, and ergodic Gaussian process $\{Y_n\}_{-\infty}^{\infty}$, and try to predict Y_n, the value of the process at time n.

For Gaussian time series and for any integer $k > 0$, $\mathbf{E}\{Y_n \mid Y_{n-k}^{n-1}\}$ is a linear function of Y_{n-k}^{n-1}:

$$\mathbf{E}\{Y_n \mid Y_{n-k}^{n-1}\} = \sum_{j=1}^{k} c_j^{(k)} Y_{n-j}, \qquad (12.12)$$

where the coefficients $c_j^{(k)}$ minimize the risk

$$\mathbf{E}\left\{\left(\sum_{j=1}^{k} c_j Y_{-j} - Y_0\right)^2\right\},$$

therefore the main ingredient is the estimate of the coefficients $c_1^{(k)}, \ldots, c_k^{(k)}$ from the data Y_1^{n-1}. Such an estimate is called elementary predictor, it is denoted by $\tilde{h}^{(k)}$ generating a prediction of form

$$\tilde{h}^{(k)}(Y_1^{n-1}) = \sum_{j=1}^{k} C_{n,j}^{(k)} Y_{n-j}$$

such that the coefficients $C_{n,j}^{(k)}$ minimize the empirical risk

$$\sum_{i=k+1}^{n-1} \left(\sum_{j=1}^{k} c_j Y_{i-j} - Y_i\right)^2$$

if $n > k$, and the all-zero vector otherwise. Even though the minimum always exists, it is not unique in general, and therefore the minimum is not well defined. It is shown by Györfi [15] that there is a unique vector $C_n^{(k)} = (C_{n,1}^{(k)}, \ldots, C_{n,k}^{(k)})$ such that

$$\sum_{i=k+1}^{n-1} \left(\sum_{j=1}^{k} C_{n,j}^{(k)} Y_{i-j} - Y_i \right)^2 = \min_{(c_1,\ldots,c_k)} \sum_{i=k+1}^{n-1} \left(\sum_{j=1}^{k} c_j Y_{i-j} - Y_i \right)^2,$$

and it has the smallest Euclidean norm among the minimizer vectors.

For fixed k, an elementary predictor

$$\tilde{h}^{(k)}(Y_1^{n-1}) = \sum_{j=1}^{k} C_{n,j}^{(k)} Y_{n-j}$$

cannot be consistent. In order to get consistent predictions there are three main principles:

- k is a deterministic function of n,
- k depends on the data Y_1^{n-1},
- aggregate the elementary predictors $\{\tilde{h}^{(k)}(Y_1^{n-1}), k = 1, 2, \ldots, n-2\}$.

12.3   Deterministic k_n

Schäfer [30] investigated the following predictor: for $a > 0$, introduce the truncation function

$$T_a(z) = \begin{cases} a & \text{if } z > a; \\ z & \text{if } |z| < a; \\ -a & \text{if } z < -a. \end{cases}$$

Choose $L_n \uparrow \infty$, then his predictor is

$$\bar{g}_n(Y_1^{n-1}) = \sum_{j=1}^{k_n} C_{n,j}^{(k_n)} T_{L_n}(Y_{n-j}).$$

Schäfer [30] proved that, under some conditions on the Gaussian process, we have that

$$\lim_{n \to \infty} \left(\mathbf{E}\{Y_n \mid Y_{n-k_n}^{n-1}\} - \bar{g}_n(Y_1^{n-1}) \right) = 0 \quad \text{a.s.}$$

His conditions include that the process has the MA(∞) representation (12.8) such that

$$\sum_{i=1}^{\infty} |a_i^*| < \infty, \tag{12.13}$$

and therefore it is purely nondeterministic and the spectral density exists. Moreover, he assumed that

$$\mathbf{E}\{Y_n \mid Y_{-\infty}^{n-1}\} - \mathbf{E}\{Y_n | Y_{n-k_n}^{n-1}\} \to 0$$

a.s. For example, he proved the strong consistency with $k_n = n^{1/4}$ if the spectral density is bounded away from zero. The question left is how to avoid these conditions such that we pose conditions only on the covariances just slightly stronger than (12.7).

For a deterministic sequence $k_n, n = 1, 2, \ldots$, consider the predictor

$$\tilde{g}_n(Y_1^{n-1}) = \tilde{h}^{(k_n)}(Y_1^{n-1}) = \sum_{j=1}^{k_n} C_{n,j}^{(k_n)} Y_{n-j}.$$

For the prediction error $\mathbf{E}\{Y_n \mid Y_1^{n-1}\} - \tilde{g}_n(Y_1^{n-1})$ we have the decomposition

$$\mathbf{E}\{Y_n \mid Y_1^{n-1}\} - \tilde{g}_n(Y_1^{n-1}) = I_n + J_n,$$

where

$$I_n = \mathbf{E}\{Y_n \mid Y_1^{n-1}\} - \mathbf{E}\{Y_n | Y_{n-k_n}^{n-1}\}$$

is the approximation error, and

$$J_n = \mathbf{E}\{Y_n | Y_{n-k_n}^{n-1}\} - \tilde{g}_n(Y_1^{n-1}) = \sum_{j=1}^{k_n} (c_j^{(k_n)} - C_{n,j}^{(k_n)}) Y_{n-j}$$

is the estimation error. In order to have small approximation error, we need $k_n \to \infty$, while the control of the estimation error is possible if this convergence to ∞ is slow.

We guess that the following is true:

Conjecture 1. For any deterministic sequence k_n, there is a stationary, ergodic Gaussian process such that the prediction error $\mathbf{E}\{Y_n \mid Y_1^{n-1}\} - \sum_{j=1}^{k_n} C_{n,j}^{(k_n)} Y_{n-j}$ does not converge to 0 a.s.

Next we show that the approximation error tends to zero in L_2 without any condition:

Lemma 1. *For any sequence $k_n \to \infty$ and for any stationary process $\{Y_n\}_{-\infty}^{\infty}$,*

$$\lim_{n\to\infty} \mathbf{E}\{(I_n)^2\} = 0.$$

Proof. We follow the argument from Doob [9]. Because of stationarity,

$$\mathbf{E}\{Y_n \mid Y_1^{n-1}\} - \mathbf{E}\{Y_n | Y_{n-k_n}^{n-1}\}$$

and

$$\mathbf{E}\{Y_0 \mid Y_{-n+1}^{-1}\} - \mathbf{E}\{Y_0 | Y_{-k_n}^{-1}\}$$

have the same distribution. The sequence $\mathbf{E}\{Y_0 \mid Y_{-n+1}^{-1}\}$, $n = 1, 2, \ldots$ is a martingale such that $\mathbf{E}\{Y_0 \mid Y_{-n+1}^{-1}\} \to \mathbf{E}\{Y_0 \mid Y_{-\infty}^{-1}\}$ a.s. and in L_2, too. Similarly, if $k_n \to \infty$, then $\mathbf{E}\{Y_0 \mid Y_{-k_n}^{-1}\} \to \mathbf{E}\{Y_0 \mid Y_{-\infty}^{-1}\}$ a.s. and in L_2. These imply that

$$\mathbf{E}\{Y_0 \mid Y_{-n+1}^{-1}\} - \mathbf{E}\{Y_0 | Y_{-k_n}^{-1}\} \to 0$$

a.s. and in L_2, therefore for the variance of the approximation error, $k_n \to \infty$ implies that

$$\mathbf{Var}(I_n) = \mathbf{Var}(\mathbf{E}\{Y_n \mid Y_1^{n-1}\} - \mathbf{E}\{Y_n | Y_{n-k_n}^{n-1}\}) \to 0. \qquad (12.14)$$

$\square$

Next we consider the problem of strong convergence of the approximation error. First we show a negative finding:

Proposition 1. *Put $k_n = (\ln n)^{1-\delta}$ with $0 < \delta < 1$. Then for the MA(1) process defined in (12.11), the approximation error does not converge to zero a.s.*

Proof. For the MA(1) process defined in (12.11), we get that

$$\mathbf{E}\{Y_n | Y_1^{n-1}\} = \sum_{j=1}^{n-1} \left(\frac{j}{n} - 1\right) Y_{n-j},$$

(see (5) in Bierens [4]). Similarly,

$$\mathbf{E}\{Y_{k_n+1} | Y_1^{k_n}\} = \sum_{j=1}^{k_n} \left(\frac{j}{k_n + 1} - 1\right) Y_{k_n+1-j},$$

so stationarity implies that

$$\mathbf{E}\{Y_n | Y_{n-k_n}^{n-1}\} = \sum_{j=1}^{k_n} \left(\frac{j}{k_n + 1} - 1\right) Y_{n-j}.$$

On the one hand,

$$
\mathbf{E}\left\{Y_n | Y_1^{n-1}\right\} = \sum_{j=1}^{n-1} \left(\frac{j}{n} - 1\right) Y_{n-j}
$$

$$
= \sum_{j=1}^{n-1} \left(\frac{j}{n} - 1\right) (Z_{n-j} - Z_{n-j-1})
$$

$$
= \left(\frac{1}{n} - 1\right) Z_{n-1} + \frac{1}{n} \sum_{j=0}^{n-2} Z_j,
$$

and, on the other hand,

$$
\mathbf{E}\left\{Y_n | Y_{n-k_n}^{n-1}\right\} = \sum_{j=1}^{k_n} \left(\frac{j}{k_n + 1} - 1\right) Y_{n-j}
$$

$$
= \sum_{j=1}^{k_n} \left(\frac{j}{k_n + 1} - 1\right) (Z_{n-j} - Z_{n-j-1})
$$

$$
= \left(\frac{1}{k_n + 1} - 1\right) Z_{n-1} + \frac{1}{k_n + 1} \sum_{j=n-k_n-1}^{n-2} Z_j.
$$

Thus

$$
\mathbf{E}\left\{Y_n | Y_1^{n-1}\right\} - \mathbf{E}\left\{Y_n | Y_{n-k_n}^{n-1}\right\}
$$

$$
= \left(\frac{1}{n} - \frac{1}{k_n + 1}\right) Z_{n-1} + \frac{1}{n} \sum_{j=0}^{n-2} Z_j - \frac{1}{k_n + 1} \sum_{j=n-k_n-1}^{n-2} Z_j
$$

$$
= \frac{1}{n} \sum_{j=0}^{n-1} Z_j - \frac{1}{k_n + 1} \sum_{j=n-k_n-1}^{n-1} Z_j.
$$

The strong law of large numbers implies that $\frac{1}{n} \sum_{j=0}^{n-1} Z_j \to 0$ a.s., therefore we have to prove that

$$
\limsup_n \frac{1}{k_n + 1} \sum_{j=n-k_n-1}^{n-1} Z_j = \infty
$$

a.s. Let $n_m = \lfloor m \ln m \rfloor$ be a subsequence of the positive integers, then we show that

$$
\limsup_m \frac{1}{k_{n_m} + 1} \sum_{j=n_m-k_{n_m}-1}^{n_m-1} Z_j = \infty
$$

a.s. One can check that $n_m - k_{n_m} > n_{m-1}$, therefore the intervals $[n_m - k_{n_m} - 1, n_m - 1]$, $m = 1, 2, \ldots$ are disjoint, and so for $C > 0$ the error events

$$A_m := \left\{ \frac{1}{k_{n_m} + 1} \sum_{j = n_m - k_{n_m} - 1}^{n_m - 1} Z_j > C \right\}$$

$m = 1, 2, \ldots$ are independent. If φ and Φ denote the density and the distribution function of a standard normal distribution, then the tail probabilities of the standard Gaussian satisfy

$$\frac{\varphi(z)}{z} \left(1 - \frac{1}{z^2} \right) \leq \Phi(-z) \leq \frac{\varphi(z)}{z}$$

for $z > 0$, (cf. Feller [10, p. 179]). These imply that

$$\mathbf{P}\{A_m\} = \mathbf{P}\left\{ \frac{1}{k_{n_m} + 1} \sum_{j = n_m - k_{n_m} - 1}^{n_m - 1} Z_j > C \right\}$$

$$= \Phi(-C\sqrt{k_{n_m} + 1})$$

$$\geq \frac{\varphi\left(C\sqrt{k_{n_m} + 1}\right)}{C\sqrt{k_{n_m} + 1}} \left(1 - \frac{1}{C^2(k_{n_m} + 1)} \right).$$

Because of the choice of k_n, we get that

$$\sum_{m=1}^{\infty} \mathbf{P}\{A_m\} \geq \sum_{m=1}^{\infty} \frac{\varphi\left(C\sqrt{k_{n_m} + 1}\right)}{C\sqrt{k_{n_m} + 1}} \left(1 - \frac{1}{C^2(k_{n_m} + 1)} \right) = \infty,$$

so the (second) Borel–Cantelli Lemma for independent events implies that

$$\mathbf{P}\left\{ \limsup_m A_m \right\} = 1,$$

and the proof of the proposition is finished.   □

Proposition 2. *Assume that for all $n > k$,*

$$\sum_{j=k+1}^{n-1} c_j^{(n-1)} r(j) \leq C_1 k^{-\gamma}, \tag{12.15}$$

and

$$\sum_{j=1}^{k} (c_j^{(n-1)} - c_j^{(k)}) r(j) \leq C_2 k^{-\gamma}, \tag{12.16}$$

with $\gamma > 0$. If

$$k_n = (\ln n)^{(1+\delta)/\gamma} \tag{12.17}$$

($\delta > 0$), then for the approximation error, we have that $I_n = \mathbf{E}\{Y_n \mid Y_1^{n-1}\} - \mathbf{E}\{Y_n \mid Y_{n-k_n}^{n-1}\} \to 0$ a.s.

Proof. The approximation error I_n is a zero mean Gaussian random variable. Buldygin and Donchenko [7] proved that $I_n \to 0$ a.s. if and only if $\mathbf{Var}(I_n) \to 0$ and for any $\epsilon > 0$,

$$\mathbf{P}\left\{\limsup_{n \to \infty} I_n < \epsilon\right\} > 0. \tag{12.18}$$

Because of (12.14), we have to verify (12.18), which is equivalent to

$$\mathbf{P}\left\{\limsup_{n \to \infty} I_n \geq \epsilon\right\} < 1.$$

Next we show that under the conditions of the proposition we have that

$$\mathbf{Var}(I_n) \leq \frac{c}{(\ln n)^{1+\delta}} \tag{12.19}$$

with some constants $c > 0$ and $\delta > 0$. In order to show (12.19), consider the representations $\mathbf{E}\{Y_n \mid Y_1^{n-1}\} = \sum_{j=1}^{n-1} c_j^{(n-1)} Y_{n-j}$ and $\mathbf{E}\{Y_n \mid Y_{n-k_n}^{n-1}\} = \sum_{j=1}^{k_n} c_j^{(k_n)} Y_{n-j}$. Introduce the vectors $X_i^{(k)} = (Y_{i-k}, \ldots, Y_{i-1})^T$ (where the superscript T denotes transpose), and the empirical covariance matrix $R_n^{(k)} = \frac{1}{n-k-1} \sum_{i=k+1}^{n-1} X_i^{(k)} (X_i^{(k)})^T$, and the vector of empirical covariances $M_n^{(k)} = \frac{1}{n-k-1} \sum_{i=k+1}^{n-1} Y_i X_i^{(k)}$. If $r(n) \to 0$, then the covariance matrix $R^{(k)} = \mathbf{E}\{R_n^{(k)}\}$ is not singular, and the optimal mean squared error of the prediction is

$$\mathbf{E}\{(Y_0 - \mathbf{E}\{Y_0 \mid Y_{-k}^{-1}\})^2\} = \mathbf{E}\{Y_0^2\} - \mathbf{E}\{\mathbf{E}\{Y_0 \mid Y_{-k}^{-1}\}^2\}$$
$$= \mathbf{E}\{Y_0^2\} - (M^{(k)})^T (R^{(k)})^{-1} M^{(k)},$$

(cf. Proposition 5.1.1 in Brockwell and Davis [6]), where $M^{(k)} = \mathbf{E}\{M_n^{(k)}\}$. Thus,

$$\mathbf{Var}(I_n) = \mathbf{E}\{I_n^2\}$$
$$= \mathbf{E}\{\mathbf{E}\{Y_0 \mid Y_{-(n-1)}^{-1}\}^2\} - \mathbf{E}\{\mathbf{E}\{Y_0 \mid Y_{-k_n}^{-1}\}^2\}$$
$$= (M^{(n-1)})^T (R^{(n-1)})^{-1} M^{(n-1)} - (M^{(k_n)})^T (R^{(k_n)})^{-1} M^{(k_n)}.$$

Moreover, we have that $c^{(n-1)} = (R^{(n-1)})^{-1} M^{(n-1)}$ and $c^{(k_n)} = (R^{(k_n)})^{-1} M^{(k_n)}$. Applying the conditions of the proposition, we get that

$$
\begin{aligned}
\mathbf{Var}(I_n) &= \sum_{j=1}^{n-1} c_j^{(n-1)} M_j^{(n-1)} - \sum_{j=1}^{k_n} c_j^{(k_n)} M_j^{(k_n)} \\
&= \sum_{j=1}^{n-1} c_j^{(n-1)} r(j) - \sum_{j=1}^{k_n} c_j^{(k_n)} r(j) \\
&= \sum_{j=1}^{k_n} (c_j^{(n-1)} - c_j^{(k_n)}) r(j) + \sum_{j=k_n+1}^{n-1} c_j^{(n-1)} r(j) \\
&\le (C_1 + C_2) k_n^{-\gamma}
\end{aligned}
$$

with $\gamma > 0$, then for the choice (12.17), (12.19) is proved. Thus, (12.19) implies that

$$
\mathbf{P}\{I_n \ge \epsilon\} = \Phi\left(-\frac{\epsilon}{\sqrt{\mathbf{Var}(I_n)}}\right) \le e^{-\frac{\epsilon^2}{2\mathbf{Var}(I_n)}} \le e^{-\frac{\epsilon^2 (\ln n)^{1+\delta}}{2c}} = n^{-\frac{\epsilon^2 (\ln n)^\delta}{2c}}
$$

therefore

$$
\sum_{n=1}^{\infty} \mathbf{P}\{I_n \ge \epsilon\} < \infty,
$$

so the Borel–Cantelli Lemma implies that

$$
\limsup_{n\to\infty} I_n < \epsilon
$$

a.s.   □

The partial autocorrelation function of Y_n is $\alpha(j) := c_j^{(j)}$ where $c_j^{(j)}$ is as defined before, i.e. the jth coefficient from the AR(j) approximation of Y_n. It is possible to explicitly bound the approximation error I_n using $\alpha(j)$. The asymptotic behavior of $\alpha(j)$ has been studied extensively in the literature. For example, $|\alpha(j)| \le c/j$ for fractionally integrated ARIMA processes (e.g., Inoue [19]). This includes Gaussian processes such that $|r(i)| \le c(j+1)^{-\beta}$ under the sole condition that $\beta > 0$, as conjectured in Remark 1 below. It is unknown whether all stationary and ergodic purely nondeterministic Gaussian processes have partial correlation function satisfying $|\alpha(j)| \le c/j$.

Proposition 3. *Suppose that*

$$
\sum_{j=k+1}^{\infty} \alpha^2(j) \le ck^{-\gamma}
$$

with $\gamma > 0$, $c > 0$. For the choice (12.17), we have that $I_n = \mathbf{E}\{Y_n \mid Y_1^{n-1}\} - \mathbf{E}\left\{Y_n \mid Y_{n-k_n}^{n-1}\right\} \to 0$ a.s.

Proof. We follow the line of the proof of Proposition 2 to verify (12.19). To ease notation, let σ_k^2 be the optimal mean square prediction error of the AR(k) approximation: $\sigma_k^2 := \mathbf{E}\{(Y_0 - \mathbf{E}\{Y_0 \mid Y_{-k}^{-1}\})^2\}$. By the Durbin–Levinson Algorithm (cf. Brockwell and Davis [6], Proposition 5.2.1),

$$\sigma_k^2 = \sigma_{k-1}^2(1 - \alpha^2(k)) = r(0) \prod_{j=1}^k (1 - \alpha^2(j)),$$

iterating the recursion and noting that $\sigma_0^2 = r(0) = \mathbf{Var}(Y_0)$. Since, as in the proof of Proposition 2,

$$\mathbf{E}\{(I_n)^2\} = \mathbf{E}\{\mathbf{E}\{Y_0 \mid Y_{-(n-1)}^{-1}\}^2\} - \mathbf{E}\{\mathbf{E}\{Y_0 \mid Y_{-k_n}^{-1}\}^2\}$$

$$= \sigma_{k_n}^2 - \sigma_{n-1}^2$$

$$= r(0) \prod_{j=1}^{k_n} (1 - \alpha^2(j)) \left(1 - \prod_{j=k+1}^{n-1} (1 - \alpha^2(j))\right).$$

Without loss of generality assume that $\alpha(j)^2 \le C < 1$. For $0 < x \le C < 1$ apply the inequality $-\frac{\ln C}{C} x \le \ln(1 - x)$, then

$$\mathbf{E}\{(I_n)^2\} \le r(0) \left(1 - \prod_{j=k_n+1}^{\infty} (1 - \alpha^2(j))\right)$$

$$= r(0) \left(1 - e^{\sum_{j=k_n+1}^{\infty} \ln(1-\alpha^2(j))}\right)$$

$$\le r(0) \left(1 - e^{-\frac{\ln C}{C} \sum_{j=k_n+1}^{\infty} \alpha^2(j)}\right)$$

$$\le r(0) \frac{\ln C}{C} \sum_{j=k_n+1}^{\infty} \alpha^2(j)$$

$$\le r(0) \frac{\ln C}{C} c k_n^{-\gamma}.$$

Hence, with the choice (12.17), (12.19) is verified.   □

Remark 1. We conjecture that under the condition

$$|r(i)| \le c(|i| + 1)^{-\beta},$$

$c < \infty$, $\beta > 0$, the conditions of Propositions 2 and 3 are satisfied. Notice that for any MA(p), the conditions of Proposition 2 are met, while for any AR(q), the conditions of Proposition 3 are satisfied.

Remark 2. Notice that the MA(1) example in the proof of Proposition 1 satisfies the conditions of the Propositions 2 and 3 with $\gamma = 1$, since $\alpha(j) = c_j^{(j)} = j^{-1}$, and $r(0) = 2$, $r(1) = -1$ and $r(i) = 0$ if $i \geq 2$, from which one gets that

$$\mathbf{Var}(I_n) = \frac{1}{k_n + 1} - \frac{1}{n}.$$

Moreover, the choice $k_n = (\ln n)^{1+\delta}$ is just slightly larger than in the proof of Proposition 1.

Remark 3. Under the AR(∞) representation (12.10), the derivation and the conditions of Proposition 2 can be simplified. Multiplying both sides of (12.10) by Y_n and taking expectations,

$$\mathbf{E}Y_n^2 = \sum_{i=1}^{\infty} c_i^* \mathbf{E}\{Y_n Y_{n-i}\} + \mathbf{E}\{Y_n Z_n\}$$

$$= \sum_{i=1}^{\infty} c_i^* r(i) + \mathbf{Var}(Z_0)$$

$$= \mathbf{E}\{\mathbf{E}\{Y_0 \mid Y_{-\infty}^{-1}\}^2\} + \mathbf{Var}(Z_0).$$

It implies that $\sum_{i=1}^{\infty} c_i^* r(i) < \infty$ and

$$\mathbf{Var}(I_n) \leq \mathbf{E}\{\mathbf{E}\{Y_0 \mid Y_{-\infty}^{-1}\}^2\} - \mathbf{E}\{\mathbf{E}\{Y_0 \mid Y_{-k_n}^{-1}\}^2\}$$

$$\leq \sum_{j=1}^{k_n} (c_j^* - c_j^{(k_n)}) r(j) + \sum_{j=k_n+1}^{\infty} c_j^* r(j)$$

$$\leq (C_1 + C_2) k_n^{-\gamma},$$

if the conditions $\sum_{i=k}^{\infty} c_i^* r(i) \leq C_1 k^{-\gamma}$ and $\sum_{j=1}^{k} (c_j^* - c_j^{(k)}) r(j) \leq C_2 k^{-\gamma}$ are satisfied.

Remark 4. If the process has the MA(∞) representation (12.8), then $r(i) = \sum_{j=0}^{\infty} a_j^* a_{j+i}^*$ assuming the innovations have variance one. The Cauchy–Schwarz inequality implies that

$$|r(i)| \leq \sqrt{\sum_{j=0}^{\infty} (a_j^*)^2 \sum_{j=0}^{\infty} (a_{j+i}^*)^2} = \sqrt{\sum_{j=0}^{\infty} (a_j^*)^2 \sum_{j=i}^{\infty} (a_j^*)^2} \to 0.$$

We show that for $a_j^* > 0$, $\beta > 1$ implies (12.13). To see this, note that $r(i) > 0$ and

$$
\left(\sum_{j=0}^{\infty} a_j^* \right)^2 \geq \sum_{i=0}^{\infty} \sum_{j=0}^{\infty} a_j^* a_{j+i}^* = \sum_{i=1}^{\infty} r(i).
$$

Moreover, $\sum_{i=1}^{\infty} r(i) < \infty$ implies that $\sum_{j=0}^{\infty} a_j^* < \infty$. Notice that without any conditions on $\{a_j^*\}$, $\sum_{i=1}^{\infty} |r(i)| < \infty$ does not imply that $\sum_{j=0}^{\infty} |a_j^*| < \infty$.

Remark 5. Concerning the estimation error the main difficulty is the possible slow rate of convergence of averages. For an arbitrary ergodic process, the rate of convergence of an average can be arbitrary slow, which means that for any sequence $a_n \downarrow 0$, there is a zero mean, stationary, ergodic process such that

$$
\limsup_n \frac{\mathbf{E}\{(\frac{1}{n} \sum_{i=1}^{n} Y_i)^2\}}{a_n} > 0. \tag{12.20}
$$

The question here is whether or not for arbitrary sequence $a_n \downarrow 0$, there is a zero mean, stationary, ergodic Gaussian process with covariances $\{r(i)\}$ such that (12.20) is satisfied. To see this, let $\{Y_i\}$ have the MA(∞) representation (12.10) with Z_j standard normal and $a_0^* = 1$, $a_j^* = j^{-\alpha}$ for $j > 0$, $1 > \alpha > 1/2$. Then $a_j^* \downarrow$, therefore $r(j) \downarrow$ and so we get that

$$
\limsup_n \frac{\mathbf{E}\{(\frac{1}{n} \sum_{i=1}^{n} Y_i)^2\}}{a_n} \geq \limsup_n \frac{1}{n a_n} \sum_{i=0}^{n} r(i) \left(1 - \frac{|i|}{n} \right)
$$

$$
\geq \limsup_n \frac{1}{2 n a_n} \sum_{i=0}^{n} r(i)
$$

$$
\geq \limsup_n \frac{1}{2 a_n} r(n)
$$

$$
= \limsup_n \frac{1}{2 a_n} \left(\sum_{j=0}^{\infty} a_j^* a_{j+n}^* \right).
$$

Then $a_j^* \downarrow$ implies that

$$
\limsup_n \frac{\mathbf{E}\{(\frac{1}{n} \sum_{i=1}^{n} Y_i)^2\}}{a_n} \geq \limsup_n \frac{1}{2 a_n} \left(\sum_{j=0}^{\infty} (a_{j+n}^*)^2 \right)
$$

$$
\geq \limsup_n \frac{n^{(1-2\alpha)}}{2 a_n}.
$$

For $\alpha \to 1/2$ the sequence can be made to diverge for any $a_n \to 0$ polynomially. We can make it logarithmic using $a_j^* = \left(j \ln^{1+\epsilon}(j) \right)^{-1/2}$ for $j > 1$ and some $\epsilon > 0$, but it is a bit more complex. Similar slow rate results can be derived for empirical covariances.

If the process $\{Y_n\}_{-\infty}^\infty$ satisfies some mixing conditions, then Meir [26], Alquier and Wintenberger [2], and McDonald et al. [25] analyzed the predictor $\tilde{h}^{(k)}(Y_1^{n-1})$. If the process $\{Y_n\}_{-\infty}^\infty$ is stationary and ergodic, then Klimo and Nelson [22] proved that

$$C_n^{(k)} \to c^{(k)} \tag{12.21}$$

a.s. Unfortunately, the convergence (12.21) does not imply that $\sum_{j=1}^k (C_{n,j}^{(k)} - c_j^{(k)})Y_{n-j} \to 0$ a.s.

Conjecture 2. For any fixed k, there is a stationary, ergodic Gaussian process such that $\sum_{j=1}^k (C_{n,j}^{(k)} - c_j^{(k)})Y_{n-j}$ does not converge to 0 a.s.

Lemma 2. *Let*

$$r_n(i) := n^{-1} \sum_{j=1}^n Y_j Y_{j+i}$$

be the empirical autocovariance. Suppose that $|r(i)| \le c(|i|+1)^{-\beta}, c < \infty, \beta > 0$. Then, for the sequence $a_n = (n^\alpha / k_n)$ with $\alpha \in (0, \beta \wedge (1/2))$,

$$a_n \max_{i \le k_n} |r_n(i) - r(i)| \to 0$$

a.s.

Proof. At first, we show that

$$n^2 \mathbf{E}|r_n(k) - r(k)|^2 \le c^2 c_k n^{2-2\beta} \tag{12.22}$$

with $c_k < \infty$. Note that

$$\mathbf{E}|r_n(k) - r(k)|^2 = \frac{1}{n^2} \sum_{i=1}^n \sum_{j=1}^n \mathbf{E}\left\{ (Y_i Y_{i+k} - \mathbf{E}Y_i Y_{i+k})(Y_j Y_{j+k} - \mathbf{E}Y_j Y_{j+k}) \right\}.$$

To this end,

$$\mathbf{E}\left\{ (Y_i Y_{i+k} - \mathbf{E}Y_i Y_{i+k})(Y_j Y_{j+k} - \mathbf{E}Y_j Y_{j+k}) \right\}$$
$$= \mathbf{E}\left\{ Y_i Y_{i+k} Y_j Y_{j+k} \right\} - \mathbf{E}Y_i Y_{i+k} \mathbf{E}Y_j Y_{j+k}.$$

By Isserlis Theorem [20],

$$\mathbf{E}\left\{Y_i Y_{i+k} Y_j Y_{j+k}\right\}$$
$$= \mathbf{E}Y_i Y_{i+k} \mathbf{E}Y_j Y_{j+k} + \mathbf{E}Y_i Y_j \mathbf{E}Y_{i+k} Y_{j+k} + \mathbf{E}Y_{i+k} Y_j \mathbf{E}Y_i Y_{j+k}.$$

Therefore

$$\mathbf{E}\left\{(Y_i Y_{i+k} - \mathbf{E}Y_i Y_{i+k}) \left(Y_j Y_{j+k} - \mathbf{E}Y_j Y_{j+k}\right)\right\}$$
$$= \mathbf{E}Y_i Y_j \mathbf{E}Y_{i+k} Y_{j+k} + \mathbf{E}Y_{i+k} Y_j \mathbf{E}Y_i Y_{j+k}$$
$$= r^2(i-j) + r(i-j+k) r(i-j-k).$$

Hence,

$$\sum_{i=1}^{n}\sum_{j=1}^{n}\mathbf{E}\left\{(Y_i Y_{i+k} - \mathbf{E}Y_i Y_{i+k}) \left(Y_j Y_{j+k} - \mathbf{E}Y_j Y_{j+k}\right)\right\}$$

$$= \sum_{i=1}^{n}\sum_{j=1}^{n} r^2(i-j) + \sum_{i=1}^{n}\sum_{j=1}^{n} r(i-j+k)r(i-j-k)$$

$$= nr^2(0) + 2\sum_{i=1}^{n-1}(n-i)r^2(i) + nr^2(k) + 2\sum_{i=1}^{n-1}(n-i)r(i+k)r(i-k)$$

$$\leq n\left(2r^2(0) + 2\sum_{i=1}^{n-1}r^2(i) + \sum_{i=1}^{n-1}(r^2(i+k) + r^2(i-k))\right)$$

$$\leq nc^2\left(2 + 2\sum_{i=1}^{n-1}(|i|+1)^{-2\beta} + \sum_{i=1}^{n-1}\left((|i+k|+1)^{-2\beta} + (|i-k|+1)^{-2\beta}\right)\right)$$

$$\leq c^2 c_k n^{2-2\beta},$$

and so (12.22) is proved. Ninness [27] proved that if an arbitrary sequence of random variables $X_n, n = 1, 2, \ldots$ satisfies

$$\mathbf{E}\left\{\left(\frac{1}{n}\sum_{i=1}^{n} X_i\right)^2\right\} \leq Cn^{-2\beta}$$

with $C < \infty$, then

$$n^\alpha \left(\frac{1}{n}\sum_{i=1}^{n} X_i\right) \to 0$$

a.s., where $0 < \alpha < \beta \wedge (1/2)$. Thus, (12.22) satisfies the condition in Theorem 2.1 of Ninness [27], which implies $n^{\alpha}|r_n(k) - r(k)| \to 0$ a.s. for each k. Hence, by the union bound, the lemma is true for any $a_n \le n^{\alpha}/k_n$.   □

We slightly modify the coefficient vector as follows: introduce the notations $\tilde{R}_n^{(k)} = R_n^{(k)} + \frac{1}{\ln n}I$ and $\tilde{C}_n^{(k)} = (\tilde{R}_n^{(k)})^{-1}M_n^{(k)}$. Moreover, put $\tilde{c}^{(k)} = (\tilde{R}^{(k)})^{-1}M^{(k)}$, where $\tilde{R}^{(k)} = \mathbf{E}\{\tilde{R}_n^{(k)}\}$.

Proposition 4. *Under the conditions of Lemma 2 and for the choice $k_n = (\ln n)^{\gamma}$ with $\gamma > 0$,*

$$\sum_{j=1}^{k_n}(\tilde{C}_{n,j}^{(k_n)} - \tilde{c}_j^{(k_n)})Y_{n-j} \to 0$$

a.s.

Proof. Because of the Cauchy–Schwarz inequality

$$|J_n| = \left|\sum_{j=1}^{k_n}(\tilde{c}_j^{(k_n)} - \tilde{C}_{n,j}^{(k_n)})Y_{n-j}\right|$$

$$\le \sqrt{\sum_{j=1}^{k_n}(\tilde{c}_j^{(k_n)} - \tilde{C}_{n,j}^{(k_n)})^2 \sum_{j=1}^{k_n}Y_{n-j}^2}$$

$$\le \sqrt{\sum_{j=1}^{k_n}(\tilde{c}_j^{(k_n)} - \tilde{C}_{n,j}^{(k_n)})^2 k_n \max_{1 \le i \le n} Y_i^2}.$$

Pisier [28] proved the following: let $Z_1, \ldots, Z_n$ be zero mean Gaussian random variables with $\mathbf{E}\{Z_i^2\} = \sigma^2$, $i = 1, \ldots, n$. Then

$$\mathbf{E}\left\{\max_{i \le n}|Z_i|\right\} \le \sigma\sqrt{2\ln(2n)},$$

and for each $u > 0$,

$$\mathbf{P}\left\{\max_{i \le n}|Z_i| - \mathbf{E}\left\{\max_{i \le n}|Z_i|\right\} > u\right\} \le e^{-u^2/2\sigma^2}.$$

This implies, by taking $u = 2\sigma\sqrt{2\ln(2n)}$,

$$\mathbf{P}\left\{\max_{i \le n}|Y_i| > 3\sigma\sqrt{2\ln(2n)}\right\} \le \frac{1}{(2n)^4},$$

and therefore

$$\sum_{n=1}^{\infty} \mathbf{P} \left\{ \max_{1 \le i \le n} |Y_i| > 3\sigma \sqrt{2 \ln(2n)} \right\} < \infty,$$

and so the Borel–Cantelli Lemma implies that

$$\limsup_{n \to \infty} \frac{\max_{1 \le i \le n} |Y_i|}{\sqrt{\ln n}} \le \infty$$

a.s. Thus, we have to show that for the choice of $k_n = (\ln n)^{\gamma}$ we get that

$$k_n \ln n \sum_{j=1}^{k_n} (\tilde{c}_j^{(k_n)} - \tilde{C}_{n,j}^{(k_n)})^2 \to 0$$

a.s. Let $\| \cdot \|$ denote the Euclidean norm and the norm of a matrix. Then

$$\sum_{j=1}^{k_n} (\tilde{c}_j^{(k_n)} - \tilde{C}_{n,j}^{(k_n)})^2$$

$$= \| \tilde{c}^{(k_n)} - \tilde{C}_n^{(k_n)} \|^2$$

$$= \| (\tilde{R}^{(k)})^{-1} M^{(k)} - (\tilde{R}_n^{(k)})^{-1} M_n^{(k)} \|^2$$

$$\le 2\| (\tilde{R}_n^{(k)})^{-1} (M^{(k)} - M_n^{(k)}) \|^2 + 2\| ((\tilde{R}^{(k)})^{-1} - (\tilde{R}_n^{(k)})^{-1}) M^{(k)} \|^2$$

Concerning the first term of the right-hand side, we have that

$$\| (\tilde{R}_n^{(k)})^{-1} (M^{(k)} - M_n^{(k)}) \|^2 \le \| (\tilde{R}_n^{(k)})^{-1} \|^2 \| M^{(k)} - M_n^{(k)} \|^2$$

$$\le (\ln n)^2 \sum_{i=1}^{k_n} (r(i) - r_n(i))^2$$

$$\le (\ln n)^2 k_n \max_{1 \le i \le k_n} (r(i) - r_n(i))^2.$$

The derivation for the second term of the right-hand side is similar:

$$\| ((\tilde{R}^{(k)})^{-1} - (\tilde{R}_n^{(k)})^{-1}) M^{(k)} \|^2 \le \| (\tilde{R}^{(k)})^{-1} - (\tilde{R}_n^{(k)})^{-1} \|^2 \| M^{(k)} \|^2$$

$$\le \| (\tilde{R}^{(k)})^{-1} \|^2 \| (\tilde{R}_n^{(k)})^{-1} \|^2 \| \tilde{R}^{(k)} - \tilde{R}_n^{(k)} \|^2 \| M^{(k)} \|^2$$

$$\le (\ln n)^4 \| R^{(k)} - R_n^{(k)} \|^2 \sum_{i=1}^{k_n} r(i)^2$$

$$\leq (\ln n)^4 \sum_{i=1}^{k_n} \sum_{j=1}^{k_n} (r(i-j) - r_n(i-j))^2 \sum_{i=1}^{\infty} r(i)^2$$

$$\leq c_1 (\ln n)^4 2k_n \sum_{i=1}^{k_n} (r(i) - r_n(i))^2$$

$$\leq c_1 (\ln n)^4 2k_n^2 \max_{1 \leq i \leq k_n} (r(i) - r_n(i))^2.$$

For the choice $k_n = (\ln n)^\gamma$, summarizing these inequalities we get that

$$k_n \ln n \sum_{j=1}^{k_n} (\tilde{c}_j^{(k_n)} - \tilde{C}_{n,j}^{(k_n)})^2 \leq c_2 (k_n^2 (\ln n)^3 + k_n^3 (\ln n)^5) \max_{1 \leq i \leq k_n} (r(i) - r_n(i))^2)$$

$$\leq c_2 (\ln n)^{5+3\gamma} \max_{1 \leq i \leq k_n} (r(i) - r_n(i))^2)$$

$$\to 0$$

a.s., where we used Lemma 2 with $a_n = (\ln n)^{5+3\gamma}$. $\qquad\qquad\square$

Remark 6. In this section we considered deterministic choices of k_n. One can introduce data driven choices of K_n, for example, via complexity regularization or via boosting. In principle, it is possible that there is a data driven choice, for which the corresponding prediction is strongly consistent without any condition on the process. We conjecture the contrary: for any data driven sequence K_n, there is a stationary, ergodic Gaussian process such that the prediction error

$$\mathbf{E}\{Y_n \mid Y_1^{n-1}\} - \sum_{j=1}^{K_n} C_{n,j}^{(K_n)} Y_{n-j}$$

does not converge to 0 a.s.

12.4 Aggregation of Elementary Predictors

After n time instants, the *(normalized) cumulative squared prediction error* on the strings Y_1^n is

$$L_n(g) = \frac{1}{n} \sum_{i=1}^{n} (g_i(Y_1^{i-1}) - Y_i)^2.$$

There is a fundamental limit for the predictability of the sequence, which is determined by a result of Algoet [1]: for any prediction strategy g and stationary ergodic process $\{Y_n\}_{-\infty}^{\infty}$ with $\mathbf{E}\{Y_0^2\} < \infty$,

$$\liminf_{n\to\infty} L_n(g) \geq L^* \quad \text{almost surely}, \tag{12.23}$$

where

$$L^* = \mathbf{E}\left\{\left(Y_0 - \mathbf{E}\{Y_0|Y_{-\infty}^{-1}\}\right)^2\right\}$$

is the minimal mean squared error of any prediction for the value of Y_0 based on the infinite past observation sequences $Y_{-\infty}^{-1} = (\ldots, Y_{-2}, Y_{-1})$. A prediction strategy g is called *universally consistent* with respect to a class $\mathcal{C}$ of stationary and ergodic processes $\{Y_n\}_{-\infty}^{\infty}$ if for each process in the class,

$$\lim_{n\to\infty} L_n(g) = L^* \quad \text{almost surely}.$$

There are universally consistent prediction strategies for the class of stationary and ergodic processes with $\mathbf{E}\{Y^4\} < \infty$, (cf. Györfi and Ottucsák [18], and Bleakley et al. [5]).

With respect to the combination of elementary experts $\tilde{h}^{(k)}$, Györfi and Lugosi applied in [16] the so-called doubling-trick, which means that the time axis is segmented into exponentially increasing epochs and at the beginning of each epoch the forecaster is reset.

Bleakley et al. [5] proposed a much simpler procedure which avoids in particular the doubling-trick. Set

$$h_n^{(k)}(Y_1^{n-1}) = T_{\min\{n^\delta, k\}}\left(\tilde{h}_n^{(k)}(Y_1^{n-1})\right),$$

where the truncation function T_a was introduced in Sect. 12.3 and $0 < \delta < \frac{1}{8}$.

Combine these experts as follows. Let $\{q_k\}$ be an arbitrarily probability distribution over the positive integers such that for all k, $q_k > 0$, and define the weights

$$w_{k,n} = q_k e^{-(n-1)L_{n-1}(h_n^{(k)})/\sqrt{n}} = q_k e^{-\sum_{i=1}^{n-1}(h_i^{(k)}(Y_1^{i-1})-Y_i)^2/\sqrt{n}}$$

$(k = 1, \ldots, n-2)$ and their normalized values

$$p_{k,n} = \frac{w_{k,n}}{\sum_{i=1}^{n-2} w_{i,n}}.$$

The prediction strategy g at time n is defined by

$$g_n(Y_1^{n-1}) = \sum_{k=1}^{n-2} p_{k,n} h_n^{(k)}(Y_1^{n-1}), \qquad n = 1, 2, \ldots$$

Bleakley et al. [5] proved that the prediction strategy g defined above is universally consistent with respect to the class of all stationary and ergodic zero-mean Gaussian processes, i.e.,

$$\lim_{n \to \infty} L_n(g) = L^* \quad \text{almost surely,}$$

which implies that

$$\lim_{n \to \infty} \frac{1}{n} \sum_{i=1}^{n} \left(\mathbf{E}\{Y_i \mid Y_1^{i-1}\} - g_i(Y_1^{i-1}) \right)^2 = 0 \quad \text{almost surely.}$$

(cf. Györfi and Lugosi [16], and Györfi and Ottucsák [18]).

This later convergence is expressed in terms of an almost sure Cesáro consistency. We guess that even the almost sure consistency (12.1) holds. In order to support this conjecture mention that

$$g_n(Y_1^{n-1}) = \sum_{k=1}^{n-2} p_{k,n} h_n^{(k)}(Y_1^{n-1}) \approx \sum_{k=1}^{n-2} p_{k,n} \tilde{h}_n^{(k)}(Y_1^{n-1}) = \sum_{k=1}^{n-2} p_{k,n} \sum_{j=1}^{k} c_{n,j}^{(k)} Y_{n-j},$$

and so

$$g_n(Y_1^{n-1}) = \sum_{j=1}^{n-2} c_{n,j} Y_{n-j},$$

where

$$c_{n,j} = \sum_{k=j}^{n-2} p_{k,n} c_{n,j}^{(k)}.$$

Acknowledgements This work was partially supported by the European Union and the European Social Fund through project FuturICT.hu (grant no.: TAMOP-4.2.2.C-11/1/KONV-2012-0013).

References

1. Algoet, P.: The strong law of large numbers for sequential decisions under uncertainty. IEEE Trans. Inf. Theory **40**, 609–634 (1994)
2. Alquier, P., Wintenberger, O.: Model selection for weakly dependent time series forecasting. Bernoulli **18**, 883–913 (2012)
3. Bailey, D.H.: Sequential schemes for classifying and predicting ergodic processes. PhD thesis, Stanford University (1976)
4. Bierens, H.J.: The Wold decomposition. Manuscript (2012). http://econ.la.psu.edu/~hbierens/WOLD.PDF

5. Bleakley, K., Biau, G., Győrfi, L., Ottucs, G.: Nonparametric sequential prediction of time series. J. Nonparametr. Stat. **22**, 297–317 (2010)
6. Brockwell, P., Davis, R.A.: Time Series: Theory and Methods, 2nd edn. Springer, New York (1991)
7. Buldygin, V.V., Donchenko, V.S.: The convergence to zero of Gaussian sequences. Matematicheskie Zametki **21**, 531–538 (1977)
8. Cornfeld, I., Fomin, S., Sinai, Y.G.: Ergodic Theory. Springer, New York (1982)
9. Doob, J.L.: The elementary Gaussian processes. Ann. Math. Stat. **15**, 229–282 (1944)
10. Feller, W. An Introduction to Probability and Its Applications, vol. 1. Wiley, New York (1957)
11. Gerencsér, L.: $AR(\infty)$ estimation and nonparametric stochastic complexity. IEEE Trans. Inf. Theory **38**, 1768–1779 (1992)
12. Gerencsér, L.: On Rissanen's predictive stochastic complexity for stationary ARMA processes. J. Stat. Plan. Inference **41**, 303–325 (1994)
13. Gerencsér, L., Rissanen, J.: A prediction bound for Gaussian ARMA processes. In: Proc. of the 25th Conference on Decision and Control, pp. 1487–1490 (1986)
14. Grenander, U.: Stochastic processes and statistical inference. Arkiv Math. **1**, 195–277 (1950)
15. Győrfi, L.: Adaptive linear procedures under general conditions. IEEE Trans. Inf. Theory **30**, 262–267 (1984)
16. Győrfi, L. and Lugosi, G.: Strategies for sequential prediction of stationary time series. In: Dror, M., L'Ecuyer, P., Szidarovszky, F. (eds.) Modelling Uncertainty: An Examination of its Theory, Methods and Applications, pp. 225–248. Kluwer, Boston (2001)
17. Győrfi, L., Morvai, G., Yakowitz, S.: Limits to consistent on-line forecasting for ergodic time series. IEEE Trans. Inf. Theory **44**, 886–892 (1998)
18. Győrfi, L., Ottucsák, G.: Sequential prediction of unbounded time series. IEEE Trans. Inf. Theory **53**, 1866–1872 (2007)
19. Inoue, A.: AR and MA representations of partial autocorrelation functions, with applications. Probab. Theory Relat. Fields **140**, 523–551 (2008)
20. Isserlis, L.: On a formula for the product-moment coefficient of any order of a normal frequency distribution in any number of variables. Biometrika **12**, 134–39 (1918)
21. Karlin, S., Taylor, H.M.: A First Course in Stochastic Processes. Academic Press, New York (1975)
22. Klimo, L.A. Nelson, P.I.: On conditional least squares estimation for stochastic processes. Ann. Stat. **6**, 629–642 (1978)
23. Lindgren, G.: Lectures on Stationary Stochastic Processes. Lund University (2002). http://www.maths.lu.se/matstat/staff/georg/Publications/lecture2002.pdf
24. Maruyama, G.: The harmonic analysis of stationary stochastic processes. Mem. Fac. Sci. Kyusyu Univ. **A4**, 45–106 (1949)
25. McDonald, D.J., Shalizi, C.R., Schervish, M.: Generalization error bounds for stationary autoregressive models (2012). http://arxiv.org/abs/1103.0942
26. Meir, R.: Nonparametric time series prediction through adaptive model selection. Mach. Learn. **39**, 5–34 (2000)
27. Ninness, B.: Strong laws of large numbers under weak assumptions with application. IEEE Trans. Autom. Control **45**, 2117–2122 (2000)
28. Pisier, G.: Probabilistic methods in the geometry of Banach spaces. In: Probability and Analysis. Lecture Notes in Mathematics, vol. 1206, pp. 167–241. Springer, New York (1986)
29. Ryabko, B.Y.: Prediction of random sequences and universal coding. Probl. Inf. Transm. **24**, 87–96 (1988)
30. Schäfer, D.: Strongly consistent online forecasting of centered Gaussian processes. IEEE Trans. Inf. Theory **48**, 791–799 (2002)

Chapter 13
Testing for Equality of an Increasing Number of Spectral Density Functions

Javier Hidalgo and Pedro C.L. Souza

Abstract Nowadays it is very frequent that a practitioner faces the problem of modelling large data sets. Some relevant examples include spatio-temporal or panel data models with large N and T. In these cases deciding a particular dynamic model for each individual/population, which plays a crucial role in prediction and inferences, can be a very onerous and complex task.

The aim of this paper is thus to examine a nonparametric test for the equality of the linear dynamic models as the number of individuals increases without bound. The test has two main features: (a) there is no need to choose any bandwidth parameter and (b) the asymptotic distribution of the test is a normal random variable.

Keywords Spectral density function • Canonical decomposition • Testing • High dimensional data

13.1 Introduction

It is arguable that one of the ultimate goals of a practitioner is to predict the future or to obtain good inferences about some parameter of interest. To provide either of them, knowledge of the dynamic structure of the data plays a crucial role. Often this is done by choosing an $ARMA$ specification via algorithms such as AIC or BIC. However, when the practitioner faces a large dimensional data set, such as panel data models with large N and T or spatio-temporal data sets, the problem to identify a particular model for every element of the "population" can be very onerous and time-consuming. In the aforementioned cases, it might be convenient to decide, before to embark in such a cumbersome task, whether the dynamic structure is the same across the different populations. This type of scenarios/models can be regarded as an example of the interesting and growing field of high-dimensional data analysis. When the number of spectral density functions that we wish to compare is finite, there has been some work, see among others Coates and Diggle [6], Diggle and Fisher [9] or Detter and Paparoditis [8] and the references therein.

J. Hidalgo (✉) • P.C.L. Souza
London School of Economics, London, UK
e-mail: f.j.hidalgo@lse.ac.uk; p.souza@lse.ac.uk

© Springer Science+Business Media New York 2014   137
M.G. Akritas et al. (eds.), *Topics in Nonparametric Statistics*, Springer Proceedings
in Mathematics & Statistics 74, DOI 10.1007/978-1-4939-0569-0_13

In a spatio-temporal data set we can mention the work by Zhu et al. [19]. The above work does not assume any particular model for the dynamic structure of the data. In a parametric context, the test is just a standard problem of comparing the equality of a number of parameters. Finally, it is worth mentioning the work in a semiparametric framework by Härdle and Marron [14] or Pinkse and Robinson [16], who compare the equality of shapes up to a linear transformation.

In this paper we are interested on a nonparametric test for equality of the dynamic structure of an increasing number of time series. Two features of the tests are as follows. First, although we will not specify any parametric functional form for the dynamics of the data, our test does not require to choose any bandwidth/smoothing parameter for its implementation, and second, the asymptotic distribution of the test is a Gaussian random variable, so that inferences are readily available.

More specifically, let $\{x_{t,p}\}_{t \in \mathbb{Z}, p \in \mathbb{N}}$ be sequences of linear random variables, to be more specific in Condition $C1$ below. Denoting the spectral density function of the pth sequence $\{x_{t,p}\}_{t \in \mathbb{Z}}$ by $f_p(\lambda)$, we are interested in the null hypothesis

$$H_0: \quad f_p(\lambda) = f(\lambda) \text{ for all } p \geq 1 \tag{13.1}$$

a.e. in $[0, \pi]$, being the alternative hypothesis

$$H_a: 0 < \iota(\mathbf{P}) = |\mathscr{P}|/\mathbf{P} < 1, \tag{13.2}$$

where $\mathscr{P} = \{p: \mu(\Lambda_p) > 0\}$ with $\Lambda_p = \{\lambda \in [0, \pi]: f_p(\lambda) \neq f(\lambda)\}$ and "$|\mathscr{A}|$" denotes the cardinality of the set $\mathscr{A}$ being $\mu(\cdot)$ the Lebesgue measure. Herewith $\mathbf{P}$ denotes the number of individuals in the sample. Thus $\mathscr{P}$ denotes the set of individuals for which $f_p(\lambda)$ is different than the "common" spectral density $f(\lambda)$. In this sense $\iota =: \iota(\mathbf{P})$ represents the proportion of sequences $\{x_{t,p}\}_{t \in \mathbb{Z}}, p \geq 1$ for which $f_p(\lambda) \neq f(\lambda)$. One feature of (13.2) is that the proportion of sequences ι can be negligible. More specifically, as we show in the next section, the test has nontrivial power under local alternatives such that $\iota(\mathbf{P}) = O(\mathbf{P}^{1/2})$. The situation when $\iota = \iota(\mathbf{P})$ such that $\iota \searrow 0$ can be of interest for, say, classification purposes or when we want to decide if a new set of sequences share the same dynamic structure. Also, it could be interesting to relax the condition that $\mu(\Lambda_p) > 0$. This scenario is relevant if we were only concerned about the behaviour of the spectral density function in a neighbourhood of a frequency, say zero. An example of interest could be to test whether the so-called long range parameter is the same across the different sequences, which generalizes work on testing for unit roots in a panel data model with an increasing number of sequences, see, for instance, [15]. However, for the sake of brevity and space, we will examine this topic somewhere else. Finally it is worth mentioning that we envisage that the results given below can be used in other scenarios which are of interest at a theoretical as well as empirical level. Two of these scenarios are: (1) testing for a break in the covariance structure of a sequence of random variables, where the covariance structure of the data is not even known under the null hypothesis, and (2) as exploratory analysis on whether or not there is separability in a spatial-temporal data set, see [11] and the references there in, although contrary to the latter manuscript we will allow the number of locations to increase to infinity. However, the relevant technical details for the latter problems are beyond the scope of this paper.

We finish this section relating the results of the paper with the problem of classification with functional data sets, which is a topic of active research. The reason being that this paper tackles the problem of whether a set of curves, i.e. spectral density functions, are the same or not. Within the functional data analysis framework, this question translates on whether there is some common structure or if we can split the set into several classes or groups. See classical examples in [10], although their approach uses nonparametric techniques which we try to avoid so that the issue of how to choose a bandwidth parameter and/or the "metric" to decide closeness are avoided. With this in mind, we believe that our approach can be used for a classification scheme. For instance, in economics, are the dynamics across different industries the same? This, in the language of functional data analysis, is a problem of supervised classification which is nothing more than a modern name to one of the oldest statistical problems: namely to decide if an individual belongs to a particular population. The term supervised refers to the case where we have a "training" sample which has been classified without error. Moreover, we can envisage that our methodology can be extended to problems dealing with functional data in a framework similar to those examined by Chang and Ogden [5]. The latter is being under current investigation elsewhere, when one it is interested on testing in a partial linear model whether we have common trends across individuals/countries, see [18] or [7] among others for some related examples.

The remainder of the paper is as follows. Next section describes and examines a test for H_0 in (13.1) and we discuss the regularity conditions. Also we discuss the type of local alternatives for which the test has no trivial power. The proof of our main results is confined to Sects. 13.3 and 13.4. The paper finishes with a conclusion section.

13.2   The Test and Regularity Conditions

We begin describing our test. To that end, denote the periodogram of $\{x_{t,p}\}_{t=1}^{n}$, $p = 1, \ldots, \mathbf{P}$, by

$$I_p(j) = \frac{1}{n} \left| \sum_{t=1}^{n} x_{t,p} e^{-it\lambda_j} \right|^2,$$

where $\lambda_j = 2\pi j/n$, $j = 1, \ldots, [n/2] =: \tilde{n}$, are the Fourier frequencies.

Suppose that we were interested in H_0 but only at a particular frequency, say λ_j, for some integer $j = 1, \ldots, \tilde{n}$. Then, we might employ

$$\mathcal{T}_n(j) = \frac{1}{\mathbf{P}} \sum_{p=1}^{\mathbf{P}} \left(\frac{I_p(j)}{\mathbf{P}^{-1} \sum_{q=1}^{\mathbf{P}} I_q(j)} - 1 \right)^2$$

to decide whether or not $f_p(j) = f(j)$. The motivation is that as $\mathbf{P} \nearrow \infty$,

$$\frac{1}{\mathbf{P}} \sum_{q=1}^{\mathbf{P}} I_q(j) \xrightarrow{P} \lim_{\mathbf{P} \to \infty} \frac{1}{\mathbf{P}} \sum_{q=1}^{\mathbf{P}} f_q(j) =: \overline{f}(j), \tag{13.3}$$

under suitable regularity conditions. So, under H_0, we can expect that the sequence of random variables

$$\left(\frac{1}{\mathbf{P}} \sum_{q=1}^{\mathbf{P}} I_q(j) \right)^{-1} I_p(j) - 1$$

will have a "mean" equal to zero. On the other hand, under H_a, for all $p \in \mathscr{P}$ and $\lambda_j \in \Lambda_p$, we have that the last displayed expression develops a mean different than zero since $\mathscr{E}\left(I_p(j) / \overline{f}(j) \right) \neq 1$.

Now extending the above argument to every $j \geq 1$, we then test for H_0 using

$$\mathcal{T}_n = \frac{1}{\tilde{n}} \sum_{j=1}^{\tilde{n}} \left\{ \mathcal{T}_n(j) - \left(1 - \frac{2}{\mathbf{P}} \right) \right\}. \tag{13.4}$$

It is worth to discuss the technical reason for the inclusion of the term $(1 - 2/\mathbf{P})$ into the right side of (13.4). For that purpose, recall Barlett's decomposition, see [4], which implies that

$$\frac{I_p(j)}{\mathbf{P}^{-1} \sum_{q=1}^{\mathbf{P}} I_q(j)} - 1 \simeq \frac{I_{\varepsilon,p}(j)}{\mathbf{P}^{-1} \sum_{q=1}^{\mathbf{P}} I_{\varepsilon,q}(j)} - 1.$$

Now using standard linearization, see, for instance (13.15), the second moment of the right side of the last displayed expression is $1 - \frac{2}{\mathbf{P}} + o\left((n\mathbf{P})^{-1/2} \right)$, see the proof of Theorem 1 for some details. In fact, the reason to correct for the term $2/\mathbf{P}$ is due to the fact that the mean of $I_{\varepsilon,p}(j)$, which is 1, is estimated via $\mathbf{P}^{-1} \sum_{q=1}^{\mathbf{P}} I_{\varepsilon,q}(j)$. See, for instance, [15] for similar arguments.

We shall now introduce the regularity conditions.

Condition C1 $\{x_{t,p}\}_{t \in \mathbb{Z}}$, $p \in \mathbb{N}$, are mutually independent covariance stationary linear processes defined as

$$x_{t,p} = \sum_{j=0}^{\infty} b_{j,p} \varepsilon_{t-j,p}; \sum_{j=0}^{\infty} j \left| b_{j,p} \right| < \infty, \text{ with } b_{p,0} = 1,$$

where $\{\varepsilon_{t,p}\}_{t \in \mathbb{Z}}$, $p \in \mathbb{N}$, are iid sequences with $E\left(\varepsilon_{t,p}\right) = 0$, $E\left(\varepsilon_{t,p}^2\right) = \sigma_{\varepsilon,p}^2$, $E\left(\left|\varepsilon_{t,p}\right|^{\ell}\right) = \mu_{\ell,p} < \infty$ for some $\ell > 8$. Finally, we denote the fourth cumulant of $\{\varepsilon_{t,p}/\sigma_{\varepsilon,p}\}_{t \in \mathbb{Z}}$ by $\kappa_{4,p}$, $p \in \mathbb{N}$.

Condition C2   For all $p \in \mathbb{N}$, $f_p(\lambda)$ are bounded away from zero in $[0, \pi]$.

Condition C3 n and $\mathbf{P}$ satisfy $\frac{\mathbf{P}}{n} + \frac{n}{\mathbf{P}^2} \to 0$.

Condition $C1$ is standard and very mild. This condition implies that

$$f_p(\lambda) = \frac{\sigma_{\varepsilon,p}^2}{2\pi} \left| B_p(\lambda) \right|^2,$$

where $B_p(z) = \sum_{j=0}^{\infty} b_{j,p} e^{ijz}$. Thus, under H_0, we have that

$$B_p(z) =: B(z) = \sum_{j=0}^{\infty} b_j e^{ijz}, \text{ and } \sigma_{\varepsilon,p}^2 = \sigma_{\varepsilon}^2, \quad p \geq 1. \tag{13.5}$$

$C1$ together with $C2$ implies that the spectral density functions $\{f_p(\lambda)\}$, $p = 1, 2, \ldots$, are twice continuously differentiable. We could relax the condition to allow for strong dependent data at the expense of some strengthening of Condition $C3$. However, since the literature is full of scenarios where results for weakly dependent sequences follow for strong dependence, we have decided to keep $C1$ as it stands for the sake of clarity. Also it is worth emphasizing that we do not assume that the sequences are identically distributed, as we allow the fourth cumulant to vary among the sequences. It is worth signaling out that $C2$ implies that the sequences $\{x_{t,p}\}_{t \in \mathbb{Z}}$, $p \in \mathbb{N}$, have also a autoregression representation given by

$$x_{t,p} = \sum_{j=1}^{\infty} a_{j,p} x_{t-j,p} + \varepsilon_{t,p}; \sum_{j=0}^{\infty} j \left| a_{j,p} \right| < \infty$$

and $f_p(\lambda) = \frac{\sigma_{\varepsilon,p}^2}{2\pi} \left| A_p(\lambda) \right|^{-2}$, where $A_p(z) = 1 - \sum_{j=1}^{\infty} a_j e^{ijz}$. Finally a word regarding Condition $C1$ is worth considering. We have assumed that the sequences $\{x_{t,p}\}_{t \in \mathbb{Z}}$ and $\{x_{t,q}\}_{t \in \mathbb{Z}}$, for all $p, q = 1, 2, ..$ are mutually independent. It is true that this assumption in many settings can be difficult to justify and we can expect some "spatial" dependence among the sequences. An inspection of our proofs indicate that the results would follow provided some type of "weak" dependence. That is, if we denote the dependence between $\{\varepsilon_{t,p}\}_{t \in \mathbb{Z}}$ and $\{\varepsilon_{t,q}\}_{t \in \mathbb{Z}}$ by $\gamma_{p,q}(t)$, then we have that $\mathbf{P}^{-1} \sum_{p,q=1}^{\mathbf{P}} \left| \gamma_{p,q}(t) \right| \leq C < \infty$ uniformly in t. The only main noticeable difference when we compare the results that we shall have to those in Theorem 1 below is that the variance of the asymptotic distribution in Theorem 1 below would reflect this dependence among the sequences $\{\varepsilon_{t,p}\}_{p\in\mathbb{N}}$. However for simplicity and to follow more easily the arguments we have decided to keep $C1$ as it stands.

Denote

$$\kappa_4 = \lim_{P\to\infty} \frac{1}{P} \sum_{p=1}^{P} \kappa_{4,p}.$$

Theorem 1. *Under H_0 and assuming $C1 - C3$, we have that*

$$\tilde{n}^{1/2}P^{1/2}\mathscr{T}_n \to_d \mathscr{N}(0, 4 + \kappa_4).$$

Proof. The proof of this theorem is confined to Sect. 13.3 below.

The conclusion that we draw from Theorem 1 is that the asymptotic distribution is standard. However, its asymptotic variance depends on the "average" fourth cumulant κ_4. So, to be able to make inferences, we need to provide a consistent estimator of κ_4. One possibility comes from the well-known formula in [12]. However, in our context it can be a computational burden prospect, apart from the fact that all we need is not to obtain a consistent estimator of all $\kappa_{4,p}$ but for the average. Another potential problem to estimate κ_4 via [12] is that it would need a bandwidth or cut-off point to compute the estimator of $\kappa_{4,p}$, for all $p \geq 1$. This is the case as the computation of κ_4, i.e. $\kappa_{4,p}$, depends on the covariance structure of $x_{t,p}^2$ as well as that of $x_{t,p}$. This creates the problem of how to choose this bandwidth parameter for each individual sequence. Thus, we propose and examine an estimator of κ_4 which is easy to compute and in addition it will not require the choice of any bandwidth parameter in its computation.

To that end, denote the *discrete Fourier transform* of a generic sequence $\{u_t\}_{t=1}^{n}$ by

$$w_u(j) = \frac{1}{n^{1/2}} \sum_{t=1}^{n} u_t e^{-it\lambda_j}, \qquad j = 1, \ldots, \tilde{n}.$$

Also, $C1$ and H_0 suggest that the *discrete Fourier transform* of $\{\varepsilon_{t,p}\}_{t=1}^{n}$ is

$$w_{\varepsilon,p}(j) \approx A(-\lambda_j) w_{x,p}(j),$$

where $B^{-1}(\lambda_j) = A(\lambda_j)$ and using the inverse transform of $w_u(j)$, we conclude that

$$\varepsilon_{t,p} \approx \frac{1}{n^{1/2}} \sum_{j=1}^{n} e^{it\lambda_j} A(-\lambda_j) w_{x,p}(j) \tag{13.6}$$

where "$\approx$" should be read as "approximately". Notice that the last two expressions are valid under the alternative hypothesis if instead of $A(-\lambda_j)$ we write $A_p(-\lambda_j)$.

The latter indicates that the problem to obtain the residuals $\{\hat{\varepsilon}_{t,p}\}_{t=1}^n$ becomes a problem to compute an estimator of $A(-\lambda_j)$. To that end, denote by

$$\hat{f}(j) = \frac{1}{\mathbf{P}} \sum_{p=1}^{\mathbf{P}} I_p(j) \tag{13.7}$$

the estimator of $f(j)$ under H_0. Then, we compute $\{\hat{\varepsilon}_{t,p}\}_{t=1}^n$, $p = 1, \ldots, \mathbf{P}$, as

$$\hat{\varepsilon}_{t,p} = \frac{1}{n^{1/2}} \sum_{j=1}^{n-1} e^{it\lambda_j} \hat{A}(j) w_{x,p}(j), \quad \begin{cases} p = 1, \ldots, \mathbf{P}, \\ t = 1, \ldots, n, \end{cases}$$

where

$$\hat{A}(j) = \exp\left\{-\sum_{r=1}^n \hat{c}_r e^{-ir\lambda}\right\}, j = 1, \ldots, \tilde{n}$$

$$\hat{c}_r = \frac{1}{n} \sum_{\ell=1}^n \log \hat{f}(\lambda_\ell) \cos r\lambda_\ell, \quad r = 1, \ldots, n.$$

Note that $\hat{\sigma}_\varepsilon = 2\pi \exp(\hat{c}_0)$. The function $\exp\left\{\sum_{r=1}^n \hat{c}_r e^{-ir\lambda}\right\}$ is an estimator of $A(\lambda) = \exp\left\{\sum_{r=1}^\infty c_r e^{-ir\lambda}\right\}$ with

$$c_r = \frac{1}{\pi} \int_0^\pi \log f(\lambda) \cos(r\lambda) \, d\lambda. \tag{13.8}$$

Observe that $e^{c_0} |A(\lambda)|^{-2} = f(\lambda)$ and the motivation to estimate $A(\lambda)$ by $\hat{A}(j)$ comes from the canonical spectral decomposition of $f(\lambda)$, see, for instance, [2, pp. 78–79] or [13]. Moreover, denoting

$$\hat{a}_\ell = \frac{1}{n} \sum_{j=-\tilde{n}+1}^{\tilde{n}-1} \hat{A}(j) e^{i\ell\lambda_j}, \quad \ell = 1, \ldots, n,$$

we could also estimate $A(j)$ as $\hat{A}(j) = 1 + \hat{a}_1 e^{-i\lambda_j} + \cdots + \hat{a}_n e^{-in\lambda_j}$. This comes from the fact that $e^{c_0} \left|\exp\left\{\sum_{r=1}^\infty c_r e^{-ir\lambda}\right\}\right|^2 = f(\lambda)$ and that a_ℓ is the ℓth Fourier coefficient of $\exp\left\{\sum_{r=1}^\infty c_r e^{-ir\lambda}\right\}$. In fact, one of the implications of the canonical decomposition is that $\exp\left\{\sum_{r=1}^\infty c_r e^{-ir\lambda}\right\} = 1 - \sum_{j=1}^\infty a_j e^{ij\lambda}$. Also, we might consider $\hat{a}_\ell$ and $\hat{A}(j) = 1 + \hat{a}_1 e^{-i\lambda_j} + \cdots + \hat{a}_n e^{-in\lambda_j}$ as estimators of the average $\bar{a}_\ell = \mathbf{P}^{-1} \sum_{p=1}^{\mathbf{P}} a_{p,\ell}$ and $\bar{A}(j) = \mathbf{P}^{-1} \sum_{p=1}^{\mathbf{P}} A_p(j)$, respectively, as $\hat{f}(j)$ is an estimator of $\overline{f}(j)$ under the maintain hypothesis. In addition it is worth pointing out that from a computational and theoretically point of view, we may only estimate the first, say $Cn^{1/2}$, c_r coefficients as $C1$ implies that $c_r = O(r^{-2})$.

Then, we compute our estimator of κ_4 as

$$\hat{\kappa}_4 = \frac{1}{P} \sum_{p=1}^{P} \frac{1}{n} \sum_{t=1}^{n} \left(\frac{\hat{\varepsilon}_{t,p}^4}{\hat{\sigma}_\varepsilon^4} - 3 \right).$$

Nevertheless we have, as a by-product, that $\frac{1}{n} \sum_{t=1}^{n} \left(\frac{\hat{\varepsilon}_{t,p}^4}{\hat{\sigma}_\varepsilon^4} - 3 \right)$ is an estimator of the fourth cumulant $\kappa_{4,p}$, $p \geq 1$.

Theorem 2. *Under H_0 and assuming $C1 - C3$, we have that $\hat{\kappa}_4 \to_P \kappa_4$.*

Proof. The proof of this theorem is confined to Sect. 13.4 below.

We finish this section describing the local alternatives for which $\mathcal{T}_n$ does not have trivial power. In addition as a corollary to Proposition 1 below, we easily conclude that $\mathcal{T}_n$ will then provide a consistent test for H_0. To that end, we consider the local alternatives

$$H_l : \begin{cases} f_p(\lambda) = f(\lambda)\left(1 + \frac{1}{n^{1/2}} g_p(\lambda)\right) & \text{for all } p \leq C\mathbf{P}^{1/2} \\ f_p(\lambda) = f(\lambda) & \text{for all } C\mathbf{P}^{1/2} < p \leq \mathbf{P}, \end{cases}$$

where $g_p(\lambda)$ is different than zero in Λ_p. So, we hope that the test will have nontrivial power for alternatives converging to the null hypothesis at rate of order $O\left(n^{-1/2}\mathbf{P}^{-1/2}\right)$, which is the rate that one would expect in a parametric setting. Notice that, without loss of generality, we have ordered the sequences in such a way that the first $C\mathbf{P}^{1/2}$ sequences are those for which $f_p(\lambda) \neq f(\lambda)$. Assuming for notational simplicity $\Lambda_p = \Lambda$, we then have the following result.

Proposition 1. *Under H_l and assuming $C1 - C3$, we have that*

$$\tilde{n}^{1/2}\mathbf{P}^{1/2}\mathcal{T}_n \to_d \mathcal{N}(c, 4 + \kappa_4),$$

where $c = 2 \int_\Lambda g(\lambda) d\lambda$ with $g(\lambda) = \lim_{\mathbf{P} \nearrow \infty} \mathbf{P}^{-1/2} \sum_{p=1}^{\mathbf{P}^{1/2}} g_p(\lambda)$.

Proof. First, proceeding as in the proof of Theorem 1, it easily follows that $\mathcal{T}_n$ is governed by the behaviour of

$$\mathcal{T}_n^1 = \frac{1}{\tilde{n}} \sum_{j=1}^{\tilde{n}} \left\{ \frac{1}{\mathbf{P}} \sum_{p=1}^{\mathbf{P}} \left(\frac{f_p(j) \overset{\circ}{I}_{\varepsilon,p}(j) + f_p(j)}{\mathbf{P}^{-1} \sum_{q=1}^{\mathbf{P}} f_q(j) \overset{\circ}{I}_{\varepsilon,q}(j) + 1} - 1 \right)^2 - \left(1 - \frac{2}{\mathbf{P}}\right) \right\},$$

where $\mathring{I}_{\varepsilon,p}(j) = I_{\varepsilon,p}(j) - 1$ and where for notational simplicity we have abbreviated $f_p(j)/\overline{f}_{\mathbf{P}}(j)$ as $f_p(j)$ with $\overline{f}_{\mathbf{P}}(j) = \mathbf{P}^{-1}\sum_{q=1}^{\mathbf{P}} f_q(j)$. Also, to abbreviate the arguments and notation we assume that $\sigma_{\varepsilon,p}^2 = \sigma_\varepsilon^2$. Now, using (13.15) and the arguments in the proof of Theorem 1, we have that

$$\frac{1}{\left(\mathbf{P}^{-1}\sum_{q=1}^{\mathbf{P}} f_q(j)\,\mathring{I}_{\varepsilon,q}(j_j) + 1\right)^2} \overset{\mathrm{asym}}{\simeq} 1 - \left(b_n^2(j) + 2b_n(j)\right),$$

where $b_n(j) = \mathbf{P}^{-1}\sum_{q=1}^{\mathbf{P}} f_q(j)\,\mathring{I}_{\varepsilon,q}(j)$ and "$\overset{\mathrm{asym}}{\simeq}$" denotes that the left- and right-hand sides are asymptotically equivalent.

So, the asymptotic behaviour of $\mathscr{T}_n$ is that governed by

$$\frac{1}{\tilde{n}}\sum_{j=1}^{\tilde{n}}\left\{\frac{1}{\mathbf{P}}\sum_{p=1}^{\mathbf{P}}\left\{\left(f_p(j)\,\mathring{I}_{\varepsilon,p}(j) - b_n(j)\right) + \left(f_p(j) - 1\right)\right\}^2\right.$$

$$\left.\times\left(1 - \left(b_n^2(j) + 2b_n(j)\right)\right)\right\} - \left(1 - \frac{2}{\mathbf{P}}\right).$$

Now, except terms of smaller order of magnitude, the expectation of the last displayed expression is

$$\frac{1}{\tilde{n}}\sum_{j=1}^{\tilde{n}}\left\{\left(1 - \frac{2}{\mathbf{P}}\right)\frac{1}{\mathbf{P}}\sum_{q=1}^{\mathbf{P}} f_q^2(j) - 1 + \mathbf{P}^{-1}\sum_{q=1}^{\mathbf{P}}\left(f_p(j) - 1\right)^2\right\}$$

$$= \frac{2}{\tilde{n}}\sum_{j=1}^{\tilde{n}} n^{-1/2}\mathbf{P}^{-1/2}\sum_{q=1}^{\mathbf{P}^{1/2}} g_q(j)\left(1 + o\left(1\right)\right), \tag{13.9}$$

because, recalling the definition of H_l, we have that, for all $j = 1,\dots,\tilde{n}$,

$$\frac{1}{\mathbf{P}}\sum_{q=1}^{\mathbf{P}} f_q^2(j) - 1 = \frac{1}{\tilde{n}^{1/2}\mathbf{P}}\sum_{q=1}^{\mathbf{P}^{1/2}} g_q(j)\left(1 + o\left(1\right)\right).$$

So, standard arguments indicate that $\tilde{n}^{1/2}\mathbf{P}^{1/2}$ times the right side of (13.9) is

$$\frac{2^{1/2}}{\tilde{n}}\sum_{j=1}^{\tilde{n}}\left\{\frac{1}{\mathbf{P}^{1/2}}\sum_{q=1}^{\mathbf{P}^{1/2}} g_q(j)\right\} \underset{\mathbf{P},n\nearrow\infty}{\to} c$$

since $\overline{f}_{\mathbf{P}}(j) = 1 + n^{-1/2}\mathbf{P}^{-1}\sum_{q=1}^{\mathbf{P}^{1/2}} g_q(j)$ under H_l. From here the proof of the proposition proceeds as that of Theorem 1 and so it is omitted.   ∎

One immediate conclusion that we draw from Proposition 1 is that our test detect local alternatives shrinking to the null hypothesis at a parametric rate, even being our scenario a nonparametric one.

13.3 Proof of Theorem 1

In what follows we denote $\overset{\circ}{I}_{\varepsilon,p}(j) = I_{\varepsilon,p}(j) - 1$, where

$$I_{\varepsilon,p}(j) = \frac{1}{n}\sum_{t,s=1}^{n}\frac{\varepsilon_{t,p}}{\sigma_{\varepsilon}}\frac{\varepsilon_{s,p}}{\sigma_{\varepsilon}}e^{i(t-s)\lambda_j} \text{ and } R_p(j) = \frac{I_p(j)}{f(j)} - I_{\varepsilon,p}(j). \quad (13.10)$$

Using [3] Theorem 10.3.1 and then $C1$ we have that standard algebra implies that

$$\mathscr{E}\left(\frac{1}{\mathbf{P}}\sum_{p=1}^{\mathbf{P}}R_p^2(j)\right) = O\left(\frac{1}{n}\right); \quad \mathscr{E}(R_p(j)) = O\left(\frac{1}{n}\right) \quad (13.11)$$

$$\mathscr{E}\left(\frac{1}{\mathbf{P}}\sum_{p=1}^{\mathbf{P}}R_p(j)\right)^2 = O\left(n^{-2} + \mathbf{P}^{-1}n^{-1}\right). \quad (13.12)$$

In addition and denoting $\mathscr{I}(\cdot)$ as the indicator function, $C1$ also implies that

$$\mathscr{E}\left(\frac{1}{\mathbf{P}}\sum_{p=1}^{\mathbf{P}}\overset{\circ}{I}_{\varepsilon,p}(j)\right)^2 = O\left(\mathbf{P}^{-1}\right);$$

$$\mathscr{E}\left(\frac{1}{\mathbf{P}}\sum_{p=1}^{\mathbf{P}}\overset{\circ}{I}_{\varepsilon,p}(j)\frac{1}{\mathbf{P}}\sum_{p=1}^{\mathbf{P}}\overset{\circ}{I}_{\varepsilon,p}(k)\right) = \frac{\mathscr{I}(j=k)}{\mathbf{P}} + \frac{1}{n}\frac{1}{\mathbf{P}}\sum_{p=1}^{\mathbf{P}}\kappa_{4,p}. \quad (13.13)$$

Next, using (13.11) and (13.12), we obtain that

$$\left(\frac{1}{\mathbf{P}}\sum_{p=1}^{\mathbf{P}}\frac{I_p(j)}{f(j)}\right)^2 = \left(\frac{1}{\mathbf{P}}\sum_{p=1}^{\mathbf{P}}\left(R_p(j) + \overset{\circ}{I}_{\varepsilon,p}(j) + 1\right)\right)^2$$

$$= \left(\frac{1}{\mathbf{P}}\sum_{p=1}^{\mathbf{P}}\overset{\circ}{I}_{\varepsilon,p}(j)\right)^2 + \frac{2}{\mathbf{P}}\sum_{p=1}^{\mathbf{P}}\overset{\circ}{I}_{\varepsilon,p}(j) + 1 + \upsilon_n, \quad (13.14)$$

where υ_n is a sequence of random variables such that $\mathscr{E}\upsilon_n = O_p\left(n^{-1}\right)$. Thus, using the linearization

$$\frac{1}{a_n} = \frac{1}{a} - \frac{a_n - a}{a^2} + \frac{(a_n - a)^2}{a^3} + O\left(|a_n - a|^3\right), \tag{13.15}$$

with $a_n = \left(\mathbf{P}^{-1}\sum_{p=1}^{\mathbf{P}} I_p\left(j\right)/f\left(j\right)\right)^2$ and $a = 1$, it follows easily by $C3$ and then by (13.11) that

$$\mathscr{I}_n = \frac{1}{\tilde{n}}\sum_{j=1}^{\tilde{n}}\left\{\frac{1}{\mathbf{P}}\sum_{p=1}^{\mathbf{P}}\left[\left(R_p\left(j\right)+\mathring{I}_{\varepsilon,p}\left(j\right)-\frac{1}{\mathbf{P}}\sum_{q=1}^{\mathbf{P}}\left(R_q\left(j\right)+\mathring{I}_{\varepsilon,q}\left(j\right)\right)\right)^2\right.\right.$$

$$\times\left.\left.\left(1+3\left(\frac{1}{\mathbf{P}}\sum_{q=1}^{\mathbf{P}}\mathring{I}_{\varepsilon,q}\left(j\right)\right)^2 - \frac{2}{\mathbf{P}}\sum_{q=1}^{\mathbf{P}}\mathring{I}_{\varepsilon,q}\left(j\right)\right)\right] - \left(1-\frac{2}{\mathbf{P}}\right)\right\}$$

$$+ o_p\left((n\mathbf{P})^{-1/2}\right)$$

$$= \frac{1}{\tilde{n}}\sum_{j=1}^{\tilde{n}}\left\{\frac{1}{\mathbf{P}}\sum_{p=1}^{\mathbf{P}}\left[\left(\mathring{I}_{\varepsilon,p}^2\left(j\right)-\left(\frac{1}{\mathbf{P}}\sum_{q=1}^{\mathbf{P}}\mathring{I}_{\varepsilon,q}\left(j\right)\right)^2\right)\right.\right.$$

$$\times\left.\left.\left(1+3\left(\frac{1}{\mathbf{P}}\sum_{q=1}^{\mathbf{P}}\mathring{I}_{\varepsilon,q}\left(j\right)\right)^2 - \frac{2}{\mathbf{P}}\sum_{q=1}^{\mathbf{P}}\mathring{I}_{\varepsilon,q}\left(j\right)\right)\right] - \left(1-\frac{2}{\mathbf{P}}\right)\right\}$$

$$+ o_p\left((n\mathbf{P})^{-1/2}\right).$$

From here and observing that $\mathscr{E}\left(\sum_{p=1}^{\mathbf{P}}\left\{\mathring{I}_{\varepsilon,p}^2\left(j\right)-1\right\}\right)^2 = O\left(\mathbf{P}+n^{-1}\sum_{p=1}^{\mathbf{P}}\kappa_{4,p}\right)$, we obtain by (13.13) and standard arguments that

$$\mathscr{I}_n = \frac{1}{\tilde{n}}\sum_{j=1}^{\tilde{n}}\left\{\frac{1}{\mathbf{P}}\sum_{p=1}^{\mathbf{P}}\mathring{I}_{\varepsilon,p}^2\left(j\right)-2\mathring{I}_{\varepsilon,p}\left(j\right)+2\left(\frac{1}{\mathbf{P}}\sum_{q=1}^{\mathbf{P}}\mathring{I}_{\varepsilon,q}\left(j\right)\right)^2 - \frac{2}{\mathbf{P}^2}\sum_{q,p=1}^{\mathbf{P}}\mathring{I}_{\varepsilon,q}^2\left(j\right)\mathring{I}_{\varepsilon,p}\left(j\right)\right\}$$

$$- \left(1-\frac{2}{\mathbf{P}}\right) + o_p\left(n^{-1/2}\mathbf{P}^{-1/2}\right). \tag{13.16}$$

We can now see the motivation to include the term $(1-2/\mathbf{P})$ in (13.4) as standard manipulations indicate that

$$\mathcal{E}\left(\frac{1}{\mathbf{P}}\sum_{q=1}^{\mathbf{P}}\mathring{I}_{\varepsilon,q}(j)\right)^2 - \mathcal{E}\left(\frac{1}{\mathbf{P}^2}\sum_{q,p=1}^{\mathbf{P}}\mathring{I}_{\varepsilon,q}^2(j)\,\mathring{I}_{\varepsilon,p}(j)\right) = \frac{1}{\mathbf{P}} + o\left(\frac{1}{\mathbf{P}n}\right).$$

Next, $\mathcal{E}\left(\sum_{j=1}^{\tilde{n}}\left\{\left(\mathbf{P}^{-1/2}\sum_{p=1}^{\mathbf{P}}\mathring{I}_{\varepsilon,p}(j)\right)^2 - 1\right\}\right)^2$ is equal to

$$\sum_{j,k=1}^{\tilde{n}}\frac{1}{\mathbf{P}^2}\sum_{q_1,q_2,p_1,p_2=1}^{\mathbf{P}}\left\{\mathcal{E}\left(\mathring{I}_{\varepsilon,q_1}(j)\,\mathring{I}_{\varepsilon,q_2}(j)\,\mathring{I}_{\varepsilon,p_1}(k)\,\mathring{I}_{\varepsilon,p_2}(k)\right) - 1\right\}$$

$$= \sum_{j,k=1}^{\tilde{n}}\frac{1}{\mathbf{P}^2}\sum_{q,p=1}^{\mathbf{P}}\left[\mathcal{E}\mathring{I}_{\varepsilon,q}^2(j)\,\mathcal{E}\mathring{I}_{\varepsilon,p}^2(k) - 1\right] + 2\left(\frac{1}{\mathbf{P}}\sum_{p=1}^{\mathbf{P}}\mathcal{E}\left\{\mathring{I}_{\varepsilon,p}(j)\,\mathring{I}_{\varepsilon,p}(k)\right\}\right)^2$$

$$+ \frac{1}{\mathbf{P}^2}\sum_{p=1}^{\mathbf{P}}\mathcal{E}\left\{\mathring{I}_{\varepsilon,p}^2(j)\,\mathring{I}_{\varepsilon,p}^2(k)\right\} + O\left(\frac{n^2}{\mathbf{P}}\right)$$

because C1 implies that the expectation on the left of the last displayed equality is zero unless the subindexes (p_1, p_2, q_1, q_2) come by pairs. Now using [1] Theorems 2.3.2 and 4.3.1., and in particular expressions in (2.3.7) and (4.3.15), we obtain that $\mathrm{Cov}\left(I_{\varepsilon,p}(j), I_{\varepsilon,p}(k)\right) = \left(1 + O\left(n^{-1}\right)\right)\mathscr{I}(j = k) + O\left(n^{-1}\right)\mathscr{I}(j \neq k)$. The latter implies that the right side of the last displayed equation is $O\left(n^2/\mathbf{P}\right)$. Similarly

$$\mathcal{E}\left(\sum_{j=1}^{\tilde{n}}\left\{\frac{1}{\mathbf{P}}\sum_{q,p=1}^{\mathbf{P}}\mathring{I}_{\varepsilon,q}^2(j)\,\mathring{I}_{\varepsilon,p}(j) - 2\right\}\right)^2 = O\left(\frac{n^2}{\mathbf{P}}\right).$$

So, $(\tilde{n}\mathbf{P})^{1/2}\mathscr{T}_n = (\tilde{n}\mathbf{P})^{-1/2}\sum_{j=1}^{\tilde{n}}\sum_{p=1}^{\mathbf{P}}\left\{\mathring{I}_{\varepsilon,p}^2(j) - 1 - 2\mathring{I}_{\varepsilon,p}(j)\right\} + o_p(1)$ and thus the proof is completed if we show that

$$\frac{1}{(\tilde{n}\mathbf{P})^{1/2}}\sum_{j=1}^{\tilde{n}}\sum_{p=1}^{\mathbf{P}}\left\{\mathring{I}_{\varepsilon,p}^2(j) - 1 - 2\mathring{I}_{\varepsilon,p}(j)\right\} \xrightarrow{d} \mathcal{N}\left(0, 4 + \kappa_4\right).$$

To that end, we need to check the "generalized" Lindeberg's condition as in Theorem 2 of [15]. Indeed, a sufficient condition for that theorem to hold true is that

$$\mathcal{E}\left(\frac{1}{\mathbf{P}^{1/2}}\sum_{p=1}^{\mathbf{P}}\left\{\mathring{I}_{\varepsilon,p}^2(j) - 1 - 2\mathring{I}_{\varepsilon,p}(j)\right\}\right)^4 < \infty$$

uniformly in $\mathbf{P}$ and j. But this is the case by C1 and that $\mathcal{E}\mathring{I}_{\varepsilon,p}^8(j) < \infty$. This concludes the proof of the Theorem. ∎

13.4   Proof of Theorem 2

We first show that

$$
\hat{c}_j - c_j = w_j + \frac{1}{\tilde{n}} \sum_{\ell=1}^{\tilde{n}} \left(\frac{1}{P} \sum_{p=1}^{P} R_p(\ell) \right) \cos(j\lambda_\ell) + O_p\left(P^{-2}n^{-1/2}\right), \text{ for } j \geq 1,
$$

(13.17)

where $\mathscr{E}\left|w_j\right|^2 = O\left(n^{-1}P^{-1}\right)$ and $\mathscr{E}\left(w_j w_k\right) = 0$ for $j \neq k$, and the $O_p(\cdot)$ is uniformly in $j \geq 1$.

Indeed denoting $c_{j,n} = \tilde{n}^{-1} \sum_{\ell=1}^{\tilde{n}} \log f(\ell) \cos(j\lambda_\ell)$, by definition

$$
\hat{c}_j - c_j = \hat{c}_j - c_{j,n} + O\left(\min\left(n^{-1}, |c_j|\right)\right),
$$

because twice continuous differentiability of $\log f(\lambda)$ implies that

$$
\frac{1}{n} \sum_{\ell=1}^{n} \log f(\ell) \cos(j\lambda_\ell) - c_j = \frac{1}{n} \sum_{\ell=1}^{n} \left\{ \sum_{k=0}^{\infty} c_k e^{ik\lambda_\ell} \right\} \cos(j\lambda_\ell) - c_j
$$

$$
= O\left(\min\left(n^{-1}, |c_j|\right)\right).
$$

(13.18)

Recall that (13.8) indicates that c_k is the kth Fourier coefficient of $\log f(\lambda)$.

Next, using the inequality $\sup_{\ell=1,\dots,\tilde{n}} |a_p| \leq \left(\sum_\ell a_p^q\right)^{1/q}$ together with (13.12), we get that

$$
\mathscr{E} \sup_{\ell=1,\dots,\tilde{n}} \left|\hat{f}(\ell) - f(\ell)\right|^2 \leq \sum_{\ell=1}^{\tilde{n}} \mathscr{E} \left| \frac{1}{P} \sum_{p=1}^{P} R_p(\ell) \right|^2 + \left(\sum_{\ell=1}^{\tilde{n}} \mathscr{E} \left(\frac{1}{P} \sum_{p=1}^{P} \overset{\circ}{I}_{\varepsilon,p}(\ell) \right)^4 \right)^{1/2}
$$

$$
= O\left(n^{-1/2}P^{-1}\right).
$$

(13.19)

So, using that $\mathscr{E}\left|\hat{f}(\ell) - f(\ell)\right|^2 = O_p\left(P^{-1}\right)$ and (13.19), we obtain that Taylor's expansion of $\log f(\lambda)$ implies that

$$
\hat{c}_j - c_{j,n} = \frac{1}{\tilde{n}} \sum_{\ell=1}^{\tilde{n}} \left(\frac{\hat{f}(\ell)}{f(\ell)} - 1 \right) \cos(j\lambda_\ell)
$$

$$
+ \frac{1}{\tilde{n}} \sum_{\ell=1}^{\tilde{n}} \left(\frac{\hat{f}(\ell)}{f(\ell)} - 1 \right)^2 \cos(j\lambda_\ell) + O_p\left(P^{-2}n^{-1/2}\right),
$$

where the $O_p(\cdot)$ is uniformly in j. Using (13.10) and (13.12), we have that

$$
\hat{c}_j - c_{j,n} = \frac{1}{\tilde{n}} \sum_{\ell=1}^{\tilde{n}} \left(\frac{1}{\mathbf{P}} \sum_{p=1}^{\mathbf{P}} \overset{\circ}{I}_{\varepsilon,p}(\ell) \right) \cos(j\lambda_\ell)
$$

$$
+ \frac{1}{\tilde{n}} \sum_{\ell=1}^{\tilde{n}} \left(\frac{1}{\mathbf{P}} \sum_{p=1}^{\mathbf{P}} \overset{\circ}{I}_{\varepsilon,p}(\ell) \right)^2 \cos(j\lambda_\ell)
$$

$$
+ \frac{1}{\tilde{n}} \sum_{\ell=1}^{\tilde{n}} \left(\frac{1}{\mathbf{P}} \sum_{p=1}^{\mathbf{P}} R_p(\ell) \right) \cos(j\lambda_\ell) + O_p\left(\mathbf{P}^{-1} n^{-1} \right),
$$

where the $O_p(\cdot)$ is uniformly in $j \geq 1$. Because $\sum_{\ell=1}^{\tilde{n}} \cos(j\lambda_\ell) = 0$ if $j \geq 1$, we have that proceeding as in the proof of Theorem 1, the second moment of the first and second terms on the right of the last displayed expression is $O\left((n\mathbf{P})^{-1} + (n\mathbf{P}^2)^{-1} \right)$. From here it is standard to conclude (13.17). Finally, we have that

$$
\mathscr{E}\left(w_j w_k \right) = 0 \text{ if } j \neq k,
$$

which follows easily from the fact that the moments of either $\mathbf{P}^{-1} \sum_{p=1}^{\mathbf{P}} \overset{\circ}{I}_{\varepsilon,p}(\ell)$ or $\left(\mathbf{P}^{-1} \sum_{p=1}^{\mathbf{P}} \overset{\circ}{I}_{\varepsilon,p}(\ell) \right)^2$ are independent of ℓ. Recall that $\{\varepsilon_{t,p}\}_{t \in \mathbb{Z}}$ are independent sequences of random variables and $\sum_{\ell=1}^{\tilde{n}} \cos(j\lambda_\ell) = 0$ if $j \geq 1$. It is worth observing that in particular we have shown that

$$
\hat{\sigma}_\varepsilon^2 - \sigma_\varepsilon^2 = (\hat{c}_0 - c_0)(2\pi) e^{c_0} + o_p(1).
$$

Let $A_n(j) =: \exp\left\{ \sum_{u=1}^{\tilde{n}} c_u e^{-iu\lambda_j} \right\}$. Then uniformly in j, we have that

$$
\log\left(\hat{A}(j) / A_n(j) \right) = \sum_{u=1}^{\tilde{n}} (\hat{c}_u - c_{u,n}) e^{-iu\lambda_j}
$$

$$
= \sum_{u=1}^{\tilde{n}} \left\{ w_u + \frac{1}{\tilde{n}} \sum_{\ell=1}^{\tilde{n}} \cos(u\lambda_\ell) \frac{1}{\mathbf{P}} \sum_{p=1}^{\mathbf{P}} R_p(\ell) \right\} e^{-iu\lambda_j}
$$

$$
+ O_p\left(\mathbf{P}^{-2} n^{1/2} + \mathbf{P}^{-1} \right).
$$

Next,

$$\mathcal{E}\left(\sum_{u=1}^{\tilde{n}}\frac{1}{\tilde{n}}\sum_{\ell=1}^{\tilde{n}}\left(\frac{1}{P}\sum_{p=1}^{P}R_p\left(\ell\right)\right)\cos\left(u\lambda_\ell\right)e^{-iu\lambda_j}\right)^2$$

$$=\mathcal{E}\left(\sum_{\ell=1}^{\tilde{n}}\left(\frac{1}{P}\sum_{p=1}^{P}R_p\left(\ell\right)\right)\frac{1}{\tilde{n}}\sum_{u=1}^{\tilde{n}}\cos\left(u\lambda_\ell\right)e^{-iu\lambda_j}\right)^2$$

$$=\mathcal{E}\left(\sum_{\ell=1}^{\tilde{n}}\left(\frac{1}{|\ell-j|_+}+\frac{1}{|\ell+j|}\right)\frac{1}{P}\sum_{p=1}^{P}R_p\left(\ell\right)\right)^2=O\left(\mathbf{P}^{-1}n^{-1}\right),$$

where $|a|_+ = \max(1,|a|)$, using (13.12). Finally, proceeding as with the last expression, it is easy to observe that

$$\mathcal{E}\left(\sum_{u=1}^{\tilde{n}}w_u\cos\left(u\lambda_\ell\right)e^{-iu\lambda_j}\right)^2=O\left(\mathbf{P}^{-1}\right)$$

Next twice continuous differentiability of $f(\lambda)$ implies that $A_n(j)-A(j)=A(j)^{-1}\exp\{-\sum_{u=n}^{\infty}c_u e^{-iu\lambda_j}\}=O(n^{-2})$. So that, Taylor expansion of $A^{-1}(j)\hat{A}(j)-1$ indicates that

$$A^{-1}(j)\hat{A}(j)-1=b(j)+O_p\left(\mathbf{P}^{-2}n^{1/2}+\mathbf{P}^{-1}\right),$$

$$\mathcal{E}\left(b(j)^2\right)=O\left(\mathbf{P}^{-1}\right).$$

Now, by definition

$$\hat{\varepsilon}_{t,p}-\varepsilon_{t,p}=\frac{1}{n}\sum_{j=1}^{n}e^{it\lambda_j}\left\{\hat{A}(j)\sum_{s=1}^{n}x_{s,p}e^{-is\lambda_j}-\sum_{s=1}^{n}\varepsilon_{s,p}e^{-is\lambda_j}\right\}$$

$$=\frac{1}{n}\sum_{j=1}^{n}e^{it\lambda_j}\left(\hat{A}(j)A^{-1}(j)-1\right)\left\{A(j)\sum_{s=1}^{n}x_{s,p}e^{-is\lambda_j}-\sum_{s=1}^{n}\varepsilon_{s,p}e^{-is\lambda_j}\right\}$$

$$+\frac{1}{n}\sum_{j=1}^{n}e^{it\lambda_j}\left(\hat{A}(j)A^{-1}(j)-1\right)\sum_{s=1}^{n}\varepsilon_{s,p}e^{-is\lambda_j}\qquad(13.20)$$

$$+\frac{1}{n}\sum_{j=1}^{n}e^{it\lambda_j}\left\{A(j)\sum_{s=1}^{n}x_{s,p}e^{-is\lambda_j}-\sum_{s=1}^{n}\varepsilon_{s,p}e^{-is\lambda_j}\right\}.$$

By Barlett's decomposition, see Brockwell and Davis (2000), it is clear that the first term on the right is $O_p\left(\mathbf{P}^{-1}\right)$ uniformly in $t\geq1$. The second term on the right of (13.20) is

$$\frac{1}{n^{1/2}} \sum_{j=1}^{n} e^{it\lambda_j} \left(\frac{1}{\tilde{n}} \sum_{\ell=1}^{\tilde{n}} \cos{(j\lambda_\ell)} \frac{1}{\mathbf{P}} \sum_{p=1}^{\mathbf{P}} \overset{\circ}{I}_{\varepsilon,p}(\ell) \right) \frac{1}{n^{1/2}} \sum_{s=1}^{n} \varepsilon_{s,p} e^{-is\lambda_j}$$

$$+ O_p\left(n^{1/2}/\mathbf{P}\right)$$

uniformly in $t = 1, \ldots, n$. It is clear that using the inequality $\sup_{\ell=1,\ldots,\tilde{n}} |a_p| \le \left(\sum_\ell a_p^q\right)^{1/q}$, the last displayed expression is $O_p\left(\mathbf{P}^{-1/2} + n^{1/2}/\mathbf{P}\right)$ uniformly in $t \ge 1$. So, it remains to examine the third term on the right of (13.20), which is

$$\frac{1}{n^{1/2}} \sum_{j=1}^{n} e^{it\lambda_j} \left\{ A(j) \frac{1}{n^{1/2}} \left\{ \sum_{\ell=n+1}^{\infty} + \sum_{\ell=0}^{n} \right\} b_\ell e^{-i\ell\lambda_j} \left[\sum_{s=1-\ell}^{n-\ell} - \sum_{s=1}^{n} \right] \varepsilon_{s,p} e^{-is\lambda_j} \right\}$$

using formulae (10.3.12) of [3]. Using the standard result $\mathscr{E}\left(w_{\varepsilon,p}(j) w_{\varepsilon,p}(-k)\right) = \mathscr{I}(j = k)$, we have that

$$\frac{1}{n^{1/2}} \sum_{j=1}^{n-1} e^{it\lambda_j} \left\{ A(j) \frac{1}{n^{1/2}} \sum_{\ell=n+1}^{\infty} b_\ell e^{-i\ell\lambda_j} \left[\sum_{s=1-\ell}^{n-\ell} - \sum_{s=1}^{n} \right] \varepsilon_{s,p} e^{-is\lambda_j} \right\} = O_p\left(n^{-1}\right)$$

because by $C1$, $b_\ell = o\left(\ell^{-2}\right)$. Also, it is clear using the previous arguments that the last displayed expression is $o_p\left(n^{-1/2}\right)$ uniformly in $t \ge 1$. So, it remains to examine the contribution due to $\sum_{\ell=0}^{n}$, which using the change of subindex $s - n = r$, it is straightforward to notice that it suffices to examine

$$\frac{1}{n^{1/2}} \sum_{j=1}^{n} e^{it\lambda_j} \left\{ A(j) \frac{1}{n^{1/2}} \sum_{\ell=0}^{n} b_\ell e^{-i\ell\lambda_j} \sum_{s=1-\ell}^{n-\ell} \varepsilon_{s,p} e^{-is\lambda_j} \right\}$$

$$= \frac{1}{n^{1/2}} \sum_{j=1}^{n} e^{it\lambda_j} \left\{ \sum_{q=0}^{\infty} a_q e^{-iq\lambda_j} \frac{1}{n^{1/2}} \sum_{\ell=0}^{n} b_\ell e^{-i\ell\lambda_j} \sum_{s=1-\ell}^{n-\ell} \varepsilon_{s,p} e^{-is\lambda_j} \right\}$$

$$= \left\{ \sum_{q=0}^{\infty} a_q e^{-iq\lambda_j} \frac{1}{n^{1/2}} \sum_{\ell=0}^{n} b_\ell e^{-i\ell\lambda_j} \sum_{s=1-\ell}^{n-\ell} \varepsilon_{s,p} e^{-is\lambda_j} \frac{1}{n} \sum_{j=1}^{n} e^{-i(q+\ell+s-t)\lambda_j} \right\}$$

$$= \sum_{\ell=0}^{t} b_\ell \sum_{s=1-\ell}^{n-\ell} a_{t-(\ell+s)} \varepsilon_{s,p} + \sum_{\ell=t}^{n} b_\ell \left\{ \sum_{s=1-\ell}^{-t} + \sum_{s=-t}^{0} \right\} a_{t-(\ell+s)} \varepsilon_{s,p}$$

because $\sum_{j=1}^{n} e^{-i(q+\ell+s-t)\lambda_j} = n\mathscr{I}(q + \ell + s = t)$. The second and third terms are $o_p\left(t^{-1}\right)$ by summability of a_q and $C1$. Finally, the first absolute moment of the first term is easily seen to be $o\left(t^{-1}\right)$ uniformly in $t = 1, \ldots, n$.

So, we conclude that uniformly in $t = 1, \ldots, n$, $\hat{\varepsilon}_{t,p} - \varepsilon_{t,p} = o_p\left(t^{-1}\right)$. From here it is standard to conclude the proof of the theorem. ∎

13.5 Conclusion

In this paper we have described and examined a simple test for equality of an increasing number of spectral density functions of unspecified functional form. One interesting aspect of the test is that, even without knowledge of the spectral density function under the null, there was no need to choose any bandwidth or smoothing parameter for its implementation. This is possible by using all the information given in data, and in particular that the number of sequences **P** also increases without limit. The implementation of the test is straightforward after we make use of the Fast Fourier Transform. A second interesting aspect of the test is that its asymptotic distribution is a normal random variable, although its asymptotic variance depends on the "average" fourth cumulants of the sequences. So the implementation of the test might be thought to be challenging as to provide an estimator of this average of the fourth cumulant of the innovation sequences $\{\varepsilon_{t,p}\}_{t \in \mathbb{Z}, p \in \mathbb{N}}$. This is the case as we have not made any specification on the dynamic structure of the sequences $\{x_{t,p}\}_{t \in \mathbb{Z}, p \in \mathbb{N}}$. However, after realizing that all we need is to provide a consistent estimator of the average fourth cumulant instead of the fourth cumulants for each of the individual sequences, we suggest a very simple estimator based on the canonical decomposition of the spectral density function as given in [17], see also [13] or [2] for more details.

There are several interesting issues worth examining as those already mentioned in the introduction. One of them is how we can extend this methodology to the situation where the sample sizes for the different sequences are not necessarily the same. We believe that the methodology in the paper can be implemented after some smoothing has been put in place, for instance, via splines. A second relevant extension is what happens when there exists dependence among the sequences. This scenario might be the rule rather than the exception with, say, spatio-temporal data or with large panel data sets with cross-section dependence across individuals. That is, if we denote $\gamma_{p,q}(t)$ the dependence between $\{\varepsilon_{t,p}\}_{t \in \mathbb{Z}}$ and $\{\varepsilon_{t,q}\}_{t \in \mathbb{Z}}$, $p, q \in \mathbb{N}$, the question is how are our results going to change in this scenario? We conjecture that, after inspection of our proofs, a condition such as $\mathbf{P}^{-1} \sum_{p,q=1}^{\mathbf{P}} |\gamma_{p,q}(t)| \leq C < \infty$ uniformly in t will suffice for the main conclusions in Theorem 1 to go through. However, the technical details to accomplish this and in particular those to obtain a consistent estimator of the asymptotic variance might be cumbersome and lengthy. We envisage, though, that this is possible via bootstrap methods using results given in Sect. 13.2 together with those obtained by Chang and Ogden [5] to be able to obtain a simple computational estimator of the long run variance of the test. However the details are beyond the scope of this paper. Finally, there are a couple of problems, mentioned in the introduction, where the methods developed in the paper can be used. One of them was on testing for stationarity of the covariance structure of a sequence of random variables, being the second one on testing for separability of the covariance function in a spatio-temporal data set.

Acknowledgements Javier Hidalgo gratefully acknowledges the research support by a Catedra of Excellence by the Bank of Santander.

We are very grateful to the comments of a referee which have led to a much improve version of the paper. Of course, all the usual caveats are in placed for any remaining errors in the manuscript.

References

1. Brillinger, D.R.: Time series, Data Analysis and Theory. Holden-Day, San Francisco (1980)
2. Brillinger, D.R.: Time Series, Data Analysis and Theory. Holden-Day, San Francisco (1981)
3. Brockwell, P.J., Davis, R.A.: Time Series: Theory and Methods. Springer, New York (1991)
4. Brockwell, P.J., Davis, R.A.: Time series: Theory and Methods. Springer, New York (2000)
5. Chang, C., Ogden, R.T.: Bootstrapping sums of independent but not identically distributed continuous processes with applications to functional data. J. Multivariate Anal. **100**, 1291–1303 (2009)
6. Coates, D.S., Diggle, P.J.: Test for computing two estimated spectral densities. J. Time Ser. Anal. **7**, 7–20 (1986)
7. Degras, D., Xu, Z., Zhang, T., Wu, W.B.: Testing parallelism between trends in multiple time series. IEEE Trans. Signal Process. **60**, 1087–1097 (2012)
8. Detter, H., Paparoditis, E.: Bootstrapping frequency domain tests in multivariate time series with an application to testing equality of spectral densities. J. Roy. Stat. Soc. Ser. B **71**, 831–857 (2009)
9. Diggle, P.J., Fisher, N.I.: Nonparametric comparison of cumulative periodograms. Appl. Stat. **40**, 423–434 (1991)
10. Ferraty, F., Vieu, P.: Nonparametric Functional Data Analysis: Theory and Practice. Springer, Berlin (2006)
11. Fuentes, M.: Testing for separability of spatial–temporal covariance functions. J. Stat. Plan. Inference **136**, 447–466 (2006)
12. Grenander, U., Rosenblatt, M.: Statistical Analysis of Stationary Time Series. Wiley, New York (1957)
13. Hannan, E.J.: Multiple Time Series. Wiley, New York (1970)
14. Härdle, W., Marron, T.S.: Semiparametric comparison of regression curves. Ann. Stat. **18**, 63–89 (1990)
15. Phillips, P.C.B., Moon, R.: Linear regression limit theory for nonstationary panel data. Econometrica **67**, 1057–1111 (1999)
16. Pinkse, J., Robinson, P.M.: Pooling nonparametric estimates of regression functions with a similar shape. In: Maddala, G.S., Phillips, P.C.B., Srinivisan, T.N. (eds.) Statistical Methods in Econometrics and Quantitative Economics: A Volume in Honour of C.R. Rao, pp. 172–197. Wiley, Cambridge (1995)
17. Whittle, P.: On stationary processes in the plane. Biometrika **41**, 434–449 (1954)
18. Zhang, Y., Su, L., Phillips, P.C.B.: Testing for common trends in semi-parametric panel data models with fixed effects. Econometrics J. **15**, 56–100 (2012)
19. Zhu, J., Lahiri, S.N., Cressie, N.: Asymptotic inference for spatial CDFS over time. Statistica Sinica **12**, 843–861 (2002)

Chapter 14
Multiscale Local Polynomial Models for Estimation and Testing

Maarten Jansen

Abstract We present a wavelet-like multiscale decomposition based on iterated local polynomial smoothing with scale dependent bandwidths. For reasons of continuity and smoothness, a multiscale smoothing decomposition must be slightly overcomplete, but the redundancy is less than in the nondecimated wavelet transform. Unlike decimated wavelet transforms, multiscale local polynomial decompositions remain numerically stable and the reconstructions are still smooth when the decomposition is applied to time series data on irregular time points. In image denoising, local polynomials outperform nondecimated wavelet transforms, even though the latter have a higher degree of redundancy allowing additional smoothing upon reconstruction. Another benefit from the presented scheme is its ability to construct multiscale decompositions for derivatives of functions. The transform can also be extended towards nonlinear and observation-adaptive data decompositions.

Keywords Kernel • Local polynomial • Wavelet • Multiscale

14.1 Introduction

The success of wavelet methods in data smoothing hinges on the locality of the wavelet basis functions. Locality means that the basis function is concentrated around a given location in time or space, but also in frequency or scale. The property of locality makes wavelets well suited for the analysis of data with intermittent behavior. Examples of such data include the observations of functions that are piecewise smooth, with isolated jumps. As a consequence of locality, wavelet decompositions of such functions are sparse, since most basis functions correspond to locations and scales that do not appear in the data. Sparsity in its turn is the key to nonlinear processing, typically thresholding. Next to sparsity, wavelets are characterized by multiresolution, meaning that contributions at different scales can be linked to each other in a way so that the decomposition as a whole can be

M. Jansen (✉)

Department of Mathematics and Computer Science, Université Libre de Bruxelles (ULB),
Brussel, Belgium

e-mail: maarten.jansen@ulb.ac.be

© Springer Science+Business Media New York 2014   155

M.G. Akritas et al. (eds.), *Topics in Nonparametric Statistics*, Springer Proceedings
in Mathematics & Statistics 74, DOI 10.1007/978-1-4939-0569-0__14

seen as a tree structured analysis of the data. This structure can then be exploited in the processing of the data. Processing data, such as smoothing, with wavelets is thus based on nonlinearity and multiresolution. In accordance with the sparsity, the processing includes the selection of significant wavelet coefficients. This selection and the further estimation may proceed along the multiresolution structure.

In spite of the success of wavelet methods, criticisms quote the sometimes unpleasant visual artifacts present in the reconstruction from a wavelet decomposition. These artifacts come in two kinds. The first category is due to false positives among the selected wavelet coefficients. The false positives appear as spurious spikes in an interval of smooth behavior. False positives can be associated with a large variance in the observations. The second category are features missed by the selection. They appear as Gibbs-like fluctuations or blur near discontinuities. False negatives are due to strict selection criteria that introduce bias.

Another point of criticism is that wavelet methods are not easy to extend towards settings beyond that of equidistant observational points. The extensions exist [1, 3, 6, 9, 10], but are either computationally intensive, or have difficulties in combining smoothness with good numerical properties. It seems that in this respect, the classical fast wavelet transform, defined on equidistant point sets, is a lucky coincidence [4].

The new scheme is implemented as a repeated local polynomial smoothing operation, where bandwidths changing over the iterations define the subsequent levels of the multiscale decomposition. Bandwidths can be chosen in a dyadic way (as integer powers of two, that is), but other sequences of scales are equally possible. As a consequence, the bandwidth is not a smoothing parameter, but a design parameter for the decomposition. In general, the local polynomial smoothing is used as an intermediate step in a data transform procedure, and not for proper smoothing. Smoothing can take place after the decomposition. As usual with wavelets, the presented transform is particularly interesting for data that are piecewise smooth, that is, data that are smooth on subintervals, separated by singular points of jumps, cusps, or other forms of discontinuity. For such data, the multiscale local polynomial approach can be combined with nonlinear processing on the transformed data (typically thresholding), which would not be possible in a direct local polynomial smoothing approach. The reason behind this observation is that locating an unknown number of change points is in fact a naturally multiscale problem: indeed, the discovery of each change point involves the identification of the lengths or scales of the adjacent intervals of smooth behavior, which requires checking all possible scales. On the other hand, the multiscale local polynomial decomposition has the benefit over classical wavelet transforms, that it is not limited to dyadic scales or dyadic point sets, nor is it computationally or conceptually difficult to construct a multiscale local polynomial transform for data on nonequidistant point sets.

This paper discusses applications and extensions of a newly introduced multiresolution scheme [4]. The new scheme is slightly redundant, meaning that it produces more transform coefficients than observations. The redundancy is necessary in order to combine smoothness of the reconstruction with good numerical condition. The number of coefficients is twice the number of observations. This redundancy factor is lower than the $\log(n)$ for the nondecimated wavelet transform. Yet, as illustrated in Sect. 14.3, the new decomposition seems to be superior in image denoising than the nondecimated wavelet transform. Secondly, the new decomposition can easily be extended to be data-adaptive, as discussed in Sect. 14.5. Thirdly, the new decomposition can be used to construct a stable multiscale representation for the derivative of function, as shown in Sect. 14.4.

14.2 Multiscale Local Polynomial Transforms

The general scheme for a multiscale local polynomial transform starts off by associating the observations $Y_i = f(x_i) + \varepsilon_i$ for $i = 1, \ldots, n$ to the finest scale J, that is, the finest scaling coefficients are defined as $s_{J,k} = Y_k$ while the finest scale grid is $x_{J,k} = x_k$. By $\mathscr{J}_J = \{1, \ldots, n\}$, we denote the index set of the grid points. Then the proposed multiscale local polynomial decomposition is implemented as a loop over resolution level j, for starting at $j = J - 1$ down to the coarsest or lowest scale $j = L$. At each scale j, we perform the following actions.

1. We set the index set of scale j to the evens in the set of scale $j + 1$, that is, $\mathscr{J}_j = e = \{2k | k \in \mathscr{J}_{j+1}\}$. We coarsen the observational grid accordingly, that is, $\mathbf{x}_j = \mathbf{x}_{j+1,e}$. This subsampling operation is denoted by $\mathbf{x}_j = (\downarrow 2)\mathbf{x}_{j+1}$, where $(\downarrow 2)$ represents a rectangular matrix whose columns are the subsampled columns of the identity matrix.
2. We filter the coarse scaling coefficients $\mathbf{s}_j = (\downarrow 2)\tilde{\mathbf{H}}_j \cdot \mathbf{s}_{j+1}$. The square filter matrix can be the identity matrix. In that case, the even scaling coefficients will proceed unchanged to the next scale. As the routine is repeated scale after scale, a coarse scale coefficient would then follow from a single observation. The variance of the coarse scale coefficient would be reduced by a nontrivial smoothing filter $\tilde{\mathbf{H}}_j$ at each intermediate scale j, leading to coarse scaling coefficients that are weighted averages of the observations. In signal processing terms, such a filter can be described as anti-aliasing.
3. We define $\mathbf{w}_j$, the wavelet or detail coefficients at scale j as the offsets between the fine scaling coefficients and a prediction based on the coarse scale coefficients, that is $\mathbf{w}_j = \mathbf{s}_{j+1} - \mathbf{P}_j(\uparrow 2)\mathbf{s}_j$. The rectangular matrix $(\uparrow 2) = (\downarrow 2)^T$ maps the subsampled vector $\mathbf{s}_j$ onto a vector of the original size by inserting zeros. In practice, this upsampling is incorporated into the prediction matrix $\mathbf{P}_j$.
4. For reasons of numerical stability, the prediction can be followed by another update of the scaling coefficients, $\mathbf{s}_j^U = \mathbf{s}_j + \mathbf{U}_j(\downarrow 2)\mathbf{w}_j$. Such an update may serve to make the underlying basis or frame functions satisfy the vanishing integral condition. Basis functions without zero integral are known as hierarchical basis functions. They are perfectly fine with nonlinear processing in some smoothness spaces, including Sobolev spaces, but zero integrals are necessary in function spaces that allow discontinuities. Indeed, it is easy, for instance, to construct nontrivial decompositions of the zero function in a hierarchical basis

that converge in L_2 [5]. In signal processing terms, the update step can also be seen as an anti-aliasing step. We refer to [4] for further details about updating of the scaling coefficients.

The algorithm presented here can be seen as an extension of both the lifting scheme [7, 8] and the Laplacian pyramid [2]. It thus constructs a linear but overcomplete transform $\mathbf{w} = \tilde{\mathbf{W}}\mathbf{Y}$ of the observations $\mathbf{Y}$ onto a vector of multiscale local polynomial coefficients $\mathbf{w}$. The vector $\mathbf{w}$ is assembled as $\mathbf{w} = [\mathbf{s}_L, \mathbf{w}_L, \mathbf{w}_{L+1}, \dots, \mathbf{w}_{J-1}]$.

The redundancy of the transform occurs in the definition of $\mathbf{w}_j$, which has the same size as the scaling coefficients $\mathbf{s}_{j+1}$. In a wavelet transform using the lifting scheme, the detail coefficients $\mathbf{w}_j$ would have the same size as the odd subsamples of $\mathbf{s}_{j+1}$.

Because of the redundancy, the inverse transform is not unique. The most straightforward reconstruction is to use $\mathbf{w}_j$ and $\mathbf{s}_j$ in the expression $\mathbf{s}_{j+1} = \mathbf{w}_j + \mathbf{P}_j(\uparrow 2)\mathbf{s}_j$.

The redundancy is needed because the operation $\mathbf{P}_j$ is filled in by a local polynomial smoothing [4]. If the operation were applied on the odd subsamples of $\mathbf{s}_{j+1}$, as in a wavelet transform, then the subsampling would undo the smoothness of the operation in $\mathbf{P}_j$, leading to fractal-like reconstruction. In order to avoid this effect, subsampling requires $\mathbf{P}_j$ to be interpolating, but interpolation is unstable on irregular point sets. Therefore, smoothness and numerical stability seem to be somehow contradictory.

Both operations $\mathbf{P}_j$ and $\tilde{\mathbf{H}}_j$ can thus be implemented by local polynomial smoothing, see (13) in [4]. The designs of both operations together determine the properties of the transform. It is best to design the prefilter $\tilde{\mathbf{H}}_j$ in function of the choice of $\mathbf{P}_j$. For a full discussion, we refer to [4]. A good choice is to take $\tilde{\mathbf{H}}_j$ from the same family as $\mathbf{P}_j$. For instance, if $\mathbf{P}_j$ implements a local quadratic smoothing, then $\tilde{\mathbf{H}}_j$ should also be invariant under quadratic polynomials. On irregular point sets, it is important that the smoothing operation can at least reproduce the identical function on that set. Therefore, on irregular point sets, a local polynomial of degree at least one is recommended. Local constant smoothing, plain Nadaraya–Watson that is, kernel estimation is not sufficient: it will lead to reconstructions that reflect the structure of the observational grid.

It should be noted that the bandwidth in the multiscale local polynomial decomposition is not primarily a smoothing parameter, but it rather controls the scale. That means that the bandwidth should be scale dependent and that the transform can be generalized beyond the dyadic scales that are common in the classical discrete wavelet transform. For a more elaborated discussion of the bandwidth, we refer to [4].

Figure 14.1 illustrates the method with a test signal showing a bump, jump, and kink. The function is defined on $[0, 1]$ as

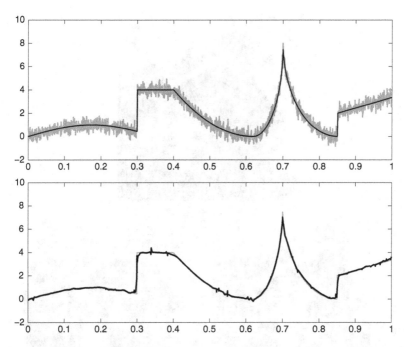

Fig. 14.1 Test signal with 2,049 observations, subject to independent, homoscedastic, additive normal errors and a reconstruction using multiscale local linear smoothing and scale dependent minimum prediction error soft thresholding

$$f(x) = \begin{cases} \sin(9x) \\ 4 \\ 4(0.62 - x)^2/(0.62 - 0.4)^2 \\ 8(1 - \sqrt{1 - (x - 0.62)^2/(0.7 - 0.62)^2}) \\ 8(1 - \sqrt{1 - (0.85 - x)^2/(0.85 - 0.7)^2}) \\ 2 + 9(x - 0.85) \end{cases}$$

for $x \in [0, 1]$ with transition points 0.3; 0.4; 0.62; 0.7; 0.85. The function has been observed in $n = 2,049$ points x_i uniformly distributed on $[0, 1]$, according to the additive model $Y_i = f(x_i) + \varepsilon_i$, where ε_i is independent, homoscedastic normal noise with $\sigma_\varepsilon = 1/3$. The simulation study then computes the multiscale local polynomial decomposition of the observations $\mathbf{w} = \tilde{\mathbf{W}}\mathbf{Y}$ along with its noise-free version $\mathbf{v} = \tilde{\mathbf{W}}\mathbf{f}$, where $\mathbf{f} = (f(x_1), \ldots, f(x_n))$. As before, denote $\mathbf{w} = [\mathbf{s}_L, \mathbf{w}_L, \mathbf{w}_{L+1}, \ldots, \mathbf{w}_{J-1}]$ and $\mathbf{v} = [\mathbf{r}_L, \mathbf{v}_L, \mathbf{v}_{L+1}, \ldots, \mathbf{v}_{J-1}]$, then the estimator of these noise-free coefficients used here is a level-dependent thresholding routine, which is $\hat{\mathbf{v}} = [\mathbf{s}_L, \hat{\mathbf{v}}_L, \hat{\mathbf{v}}_{L+1}, \ldots, \hat{\mathbf{v}}_{J-1}]$, where $\hat{\mathbf{v}}_j = \mathrm{ST}(\mathbf{w}_j, \lambda_j)$. The function $\mathrm{ST}(\mathbf{x}, \lambda)$ is the soft-threshold operation on vector $\mathbf{x}$ with threshold λ, which is defined componentwise by $\mathrm{ST}(\mathbf{x}, \lambda) = (\mathrm{ST}(x_i, \lambda))_{i=1,\ldots,n}$ and $\mathrm{ST}(x_i, \lambda) = \mathrm{sign}(x_i)(|x_i| - \lambda)\mathbf{1}(|x_i| \geq \lambda)$. In this expression, the function $\mathbf{1}(A)$ stands for

Fig. 14.2 Original noisy image (SNR = 3 dB); Denoising by simple thresholding the three finest scales in two decompositions: first using nondecimated Cohen–Daubechies–Feauveau wavelet family with less dissimilar lengths, four primal and four dual vanishing moments (SNR = 13.67 dB); second using multiscale local linear smoothing (SNR = 13.84 dB)

the indicator function or characteristic function for a set A. The threshold λ_j is chosen to minimize the prediction error $\text{PE}(\hat{\mathbf{v}}_j) = \frac{1}{n}\|\hat{\mathbf{v}}_j - \mathbf{v}_j\|^2$ over all soft-threshold estimators. Finally the vector of observations is estimated by an inverse transform $\hat{\mathbf{f}} = \mathbf{W}\hat{\mathbf{v}}$, where $\mathbf{W}$ is a reconstruction from a multiscale local polynomial decomposition.

14.3 Application in Image Denoising

The decomposition of Sect. 14.2 can be extended towards two-dimensional data, such as images, but also irregularly scattered data. On regular grids of pixels, a two-dimensional square wavelet transform can be constructed by using for $\mathbf{P}_j$ a local linear prediction with a cosine kernel and bandwidth equal to three pixels. The operation $\tilde{\mathbf{H}}_j$ is the same local linear prediction. The transform is completed by an update operation $\mathbf{U}_j$, ensuring that the frame functions of the reconstruction all have two vanishing moments. Figure 14.2 compares the denoising capacity of such a slightly overcomplete transform with a fully nondecimated wavelet transform using the filters of the Cohen–Daubechies–Feauveau wavelet family with less dissimilar lengths and four primal and four dual vanishing moments (CDF-LDL 4,4). These filters are quite popular in the image processing world. Somehow surprisingly, the local linear prediction transform, which relies just on a factor 2 redundancy, slightly outperforms the CDF-LDL 4,4 nondecimated transform, which can use a factor $\log(n)$ redundancy for additional smoothing upon reconstruction.

14.4   A Multiscale Estimation for the Derivative of a Function

In a local polynomial of degree one or higher, the coefficient of the linear term can be used to estimate the local value derivative in a numerically stable way. This procedure can be applied to the integral of the observations, which is numerically well-conditioned problem. The result is a differentiated multiscale local polynomial analysis of the integral.

More precisely, let $\mathbf{P}'_j$ denote the matrix that maps data $(\uparrow 2)\mathbf{s}_j$ onto the linear terms in the local weighted least squares polynomials that smooth $\mathbf{s}_j$ on the observational grid $\mathbf{x}_j$, with weights given by the moving kernel function. Furthermore, let $\mathbf{I}_j$ a numerical integration matrix, then the wavelet coefficients at scale j are defined as $\mathbf{w}'_j = \mathbf{s}_{j+1} - \mathbf{P}'_j(\uparrow 2)\mathbf{I}_j\mathbf{s}_j$. The prefilter can be constructed in a similar way, $\mathbf{s}_j = (\downarrow 2)\tilde{\mathbf{H}}'_j\mathbf{I}_{j+1}\mathbf{s}_{j+1}$, where $\tilde{\mathbf{H}}'_j$ maps a vector $\mathbf{s}_{j+1}$ onto the linear terms in the local polynomials for $\mathbf{s}_{j+1}$ on $\mathbf{x}_{j+1}$.

If we replace the finest scale $(J-1)$ prefilter by an operation without numerical integration, i.e., $\mathbf{s}_{J-1} = (\downarrow 2)\tilde{\mathbf{H}}'_{J-1}\mathbf{s}_J$, then the decomposition forms a multiscale analysis of the derivative. In particular assuming that $\mathbf{I}_J$ is invertible, and $\mathbf{I}_J\mathbf{D}_J = I$, with I the identity matrix, we can write $\tilde{\mathbf{H}}'_{J-1}\mathbf{s}_J = \tilde{\mathbf{H}}'_{J-1}\mathbf{I}_J(\mathbf{D}_J\mathbf{s}_J)$.

The choice of $\mathbf{I}_j$ requires some care. Complications include forward and backward shifts due to discretization, resulting in unsharp reconstruction of singularities. Another point of attention is numerical stability, especially when the differentiated integral prediction is followed by an update step.

The multiscale differentiated integral local linear smoothing is illustrated in Fig. 14.3.

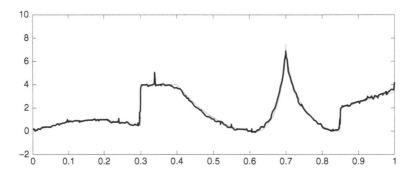

Fig. 14.3 Reconstruction using scale dependent minimum prediction error soft thresholding in a multiscale differentiated integral local linear smoothing for the observations in Fig. 14.1

14.5 Adaptive Multiscale Local Polynomial Smoothing

14.5.1 A Free Adaptive Prefilter

The reconstruction from the multiscale local polynomial decomposition is not unique. The most straightforward routine is based on repetitive application of the expression $\mathbf{s}_{j+1} = \mathbf{w}_j + \mathbf{P}_j(\uparrow 2)\mathbf{s}_j$. This expression does not involve the prefilter $\tilde{\mathbf{H}}_j$. As a consequence, the prefilter can be taken to be nonlinear or data-adaptive at no immediate cost. Nevertheless, as $\tilde{\mathbf{H}}_j$ and $\mathbf{P}_j$ are best designed together, both can be made adaptive if the analysis keeps track of the data-dependent choices for use at the reconstruction.

14.5.2 The Goal of Adaptivity

Multiscale transforms using local operations at each scale lead to sparse decompositions. These sparse data representations allow nonlinear processing with focus on the coefficients in the representation that correspond to the location of discontinuities. In this scheme, the multiscale transform itself plays a passive role: it is a linear decomposition into a representation from which discontinuities can be detected by a different routine.

The adaptive version that we propose here is a nonlinear, data-dependent transform, which, by itself, actively searches for possible discontinuities. The transform adapts itself to these possible discontinuities in a way that discontinuities detected during the transform will not give rise to large coefficients. More precisely, the set of input data at the current scale will be partitioned around the locations of the detected discontinuities, so that no subset of input data contains a discontinuity other than in its endpoints. As a result, the representation will be even sparser than before, leading to less false negatives and less false positives in the coefficient selection phase.

Fig. 14.4 Full and truncated
kernels in adaptive kernel
smoothing

14.5.3   Adaptivity Driven by Statistical Testing

The adaptivity is realized by choosing between alternative predictions. The standard
prediction, as defined in Sect. 14.2 is considered as default choice, unless significant
improvement is obtained by using a different prediction. More precisely, let $\mathbf{P}_j$ be a
local polynomial smoothing operation using a two-sided kernel $K(u)$ as depicted in
the middle of Fig. 14.4. Let $\mathbf{P}_j^L$ and $\mathbf{P}_j^R$ stand for the same operations, but this time
using, respectively, the right- and left truncated kernels $K_L(u) = K(u) \cdot I(u < 0)$
and $K_R(u) = K(u) \cdot 1(u > 0)$, where, as before, $1(\mathscr{A})$ is the indicator function
on a set $\mathscr{A}$. The use of a right- or left-truncated kernel suggests the presence of
a singularity, which would make smoothing across its location more difficult and
less meaningful. Denote k the index in $\mathbf{w}_j$ corresponding to the location of the
singularity, then smoothing across k being more difficult than truncated smoothing
would result in a large offset $w_{j,k}$, compared to one of the truncated versions, $w_{j,k}^L$
or $w_{j,k}^R$.

14.5.4   The Statistical Test

Let $\mu_{j,k}^s = E(w_{jk}^s)$ for $s \in \{L, C, R\}$, then we test if $H_0 : \mu_{j,k}^s = 0$ against
$H_1 : \mu_{j,k}^C \neq 0$ and $\min(|\mu_{j,k}^L|, |\mu_{j,k}^R|) = 0$. The test can be repeated for every
k separately, based on the test statistic $T_k = \max(|w_{jk}^C|/|w_{jk}^L|, |w_{jk}^C|/|w_{jk}^R|)$, or
any equivalent value. This test statistic is independent from the variance of the
coefficients. Its null distribution, however, is that of the maximum of two ratios,
which is typically a heavy tailed variable. The heavy tails lead to tests with little
power, many false positives or both.

  In order to increase the power of the statistical tests, we estimate the variance of
each coefficient based on all coefficients at scale j. This estimation proceeds in two
steps. First, assuming that the observations are independent, identically distributed,
we have for the covariance matrix $\Sigma_\mathbf{y} = \sigma^2 I$. From there we can, up to the unknown
constant σ^2, find the structure of the covariance matrix of the coefficients using the
recursion

$$\mathbf{s}_j = \tilde{V}_j \mathbf{s}_{j+1}$$
$$\mathbf{w}_j = \tilde{W}_j \mathbf{s}_{j+1}$$

Then

$$\Sigma_{\mathbf{s}_j} = \tilde{V}_j \, \Sigma_{\mathbf{s}_{j+1}} \, \tilde{V}_j^T$$
$$\Sigma_{\mathbf{w}_j} = \tilde{W}_j \, \Sigma_{\mathbf{s}_{j+1}} \, \tilde{W}_j^T,$$

which allows to standardize the detail coefficients as

$$w'_{j,k} = w_{j,k} / \sqrt{(\Sigma_{\mathbf{w}_j})_{kk}},$$

for which we know that $\mathrm{var}(w'_{j,k}) = \sigma^2$. In a second step, the parameter σ can be estimated, using the sparsity of the decompositions. For Gaussian data, this could be, for instance, the median absolute deviation (MAD) based estimator

$$\hat{\sigma}_j = \mathrm{median}(|\mathbf{w}'_j|)/\Phi^{-1}(3/4),$$

where $\Phi(x)$ is the cumulative Gaussian distribution, and $\Phi^{-1}(3/4) \approx 0.6745$. Based on this estimator, we can impose a threshold λ on $|\mathbf{w}'_j|/\hat{\sigma}_j$ and select coefficients that are above the threshold, while at least one of its truncated equivalents is below the threshold.

The presence of a single change point may lead to significant differences between full and truncated offsets at several locations. In order to eliminate multiple discoveries of a single change, all adjacent offsets are recomputed after the discovery of a change point, in a way that the newly discovered change point is excluded from the computations. Further details can be found in forthcoming publications, currently under submission.

14.5.5 Adaptivity and Further Update Steps

The adaptivity actively locates the singularities. It thus reduces the number of false negatives. It also avoids smoothing across singularities, thereby offering an alternative to decompositions with update operations $\mathbf{U}_j$ for use in function spaces that allow jumps. It is also an alternative for updated scaling coefficients in the reduction of aliasing. Updating the scaling coefficients after an adaptive prediction step is a nontrivial thing to do. Uncareful updates will easily introduce new numerical problems. This is because the adaptivity has lined up several basis functions along a singularity. As a consequence, these basis functions show a large degree of overlap, and are thus far from orthogonal. The impact of the overlap can be reduced by grouped processing, in particular by block or tree structured coefficient selection. This is subject of ongoing work.

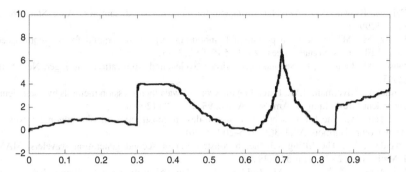

Fig. 14.5 Reconstruction using adaptive multiscale kernel smoothing the observations in Fig. 14.1

Figure 14.5 shows a reconstruction from simple thresholding within an adaptive multiscale kernel smoothing scheme without update step. The edges are much sharper than in the non-adaptive decompositions.

14.6 Conclusions

This paper has discussed multiscale local polynomial smoothing. Initially, this scheme has been proposed to reconcile the benefits from nonlinear processing within a sparse multiscale decompositions and smooth reconstructions using kernel based estimators. Thanks to a slight overcompleteness, the multiscale local polynomial decomposition is able to combine smoothness and numerical stability on an arbitrary irregular point set. This offers an elegant solution to the numerical problems encountered in the so-called second generation wavelet decompositions on irregular point sets, using the lifting scheme. The proposed transform performs surprisingly well on other settings, such as image denoising, which is obviously far from a problem on a nonequidistant point set. The transform can also be used to estimate derivatives in multiscale fashion. Data-adaptive versions are another extension of the presented method. Finally, it should be noted that the scale in the multiscale decomposition is not automatic and dyadic, as in a wavelet transform, but rather steered by the level dependent choice of the bandwidth. This leaves the user with a richer and yet very natural range of multiscale decomposition to choose from.

References

1. Antoniadis, A., Fan, J.: Regularized wavelet approximations. J. Am. Stat. Assoc. **96**(455), 939–955 (2001)
2. Burt, P.J., Adelson, E.H.: Laplacian pyramid as a compact image code. IEEE Trans. Commun. **31**(4), 532–540 (1983)

3. Cai, T., Brown, L.D.: Wavelet shrinkage for nonequispaced samples. Ann. Stat. **26**(5), 1783–1799 (1998)
4. Jansen, M.: Multiscale local polynomial smoothing in a lifted pyramid for non-equispaced data. IEEE Trans. Signal Process. **61**(3), 545–555 (2013)
5. Jansen, M., Oonincx, P.: Second Generation Wavelets and Applications. Springer, New York (2005)
6. Kovac, A., Silverman, B.W.: Extending the scope of wavelet regression methods by coefficient-dependent thresholding. J. Am. Stat. Assoc. **95**, 172–183 (2000)
7. Sweldens, W.: The lifting scheme: a custom-design construction of biorthogonal wavelets. Appl. Comp. Harmon. Anal. **3**(2), 186–200 (1996)
8. Sweldens, W.: The lifting scheme: a construction of second generation wavelets. SIAM J. Math. Anal. **29**(2), 511–546 (1998)
9. Van Aerschot, W., Jansen, M., Bultheel, A.: Adaptive splitting for stabilizing 1-d wavelet decompositions. Signal Process. **86**(9), 2447–2463 (2006)
10. Vanraes, E., Jansen, M., Bultheel, A.: Stabilizing wavelet transforms for non-equispaced data smoothing. Signal Process. **82**(12), 1979–1990 (2002)

Chapter 15
Distributions of Clusters of Exceedances and Their Applications in Telecommunication Networks

Natalia Markovich

Abstract In many applications it is important to evaluate the impact of clusters of observations caused by the dependence and heaviness of tails in time series. We consider a stationary sequence of random variables $\{R_n\}_{n\geq 1}$ with marginal cumulative distribution function $F(x)$ and the extremal index $\theta \in [0, 1]$. The clusters contain consecutive exceedances of the time series over a threshold u separated by return intervals with consecutive non-exceedances. We derive geometric forms of asymptotically equal distributions of the normalized cluster and inter-cluster sizes that depend on θ. The inter-cluster size determines the number $T_1(u)$ of inter-arrival times between observations of the process R_t arising between two consecutive clusters. The cluster size is equal to the number $T_2(u)$ of inter-arrival times within clusters. The inferences are valid when u is taken as a sufficiently high quantile of the process $\{R_n\}$. The derived geometric models allow us to obtain the asymptotically equal mean of $T_2(u)$ and other indices of clusters. Applications in telecommunication networks are discussed.

Keywords Clusters of exceedances • Extremal index • Geometric distribution Cluster size • Inter-cluster size

15.1 Introduction

In many applications it is important to evaluate the impact of clusters of observations caused by the dependence and heaviness of tails in time series. Clusters of extremal events, i.e. conglomerates of exceedances over a threshold, impact the risk of hazardous events like climate catastrophes, huge insurance claims, the loss and delay in telecommunication networks due to overloading. There are different definitions of a cluster. Clusters may be blocks of data with at least one exceedance over a threshold, or clusters are data blocks separated by a fixed number of non-exceedances over a threshold [2].

N. Markovich (✉)

Institute of Control Sciences, Russian Academy of Sciences, Moscow 117997, Russia
e-mail: nat.markovich@gmail.com

© Springer Science+Business Media New York 2014

M.G. Akritas et al. (eds.), *Topics in Nonparametric Statistics*, Springer Proceedings in Mathematics & Statistics 74, DOI 10.1007/978-1-4939-0569-0_15

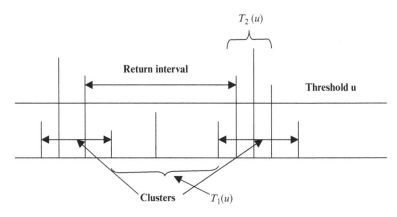

Fig. 15.1 Clusters of exceedances of the process R_t over the threshold u

We consider a stationary sequence of random variables (rvs) $\{R_n\}_{n\geq 1}$ with cumulative distribution function (cdf) $F(x)$, $M_n = \max\{R_1, \ldots, R_n\}$ and the extremal index $\theta \in [0, 1]$. The clusters contain consecutive exceedances of the $\{R_n\}$ over a threshold u separated by return intervals with consecutive non-exceedances, see Fig. 15.1.

We derive geometric-like asymptotically equal distributions of the cluster and inter-cluster sizes that depend on θ. The inter-cluster size determines the number $T_1(u)$ of inter-arrival times between non-exceedances of $\{R_n\}$ arising between two consecutive clusters, i.e. between two consecutive exceedances of the process $\{R_n\}$ over u. The cluster size is equal to the number $T_2(u)$ of inter-arrival times between exceedances within clusters, i.e. between two consecutive non-exceedances. The u is taken as a sufficiently high quantile of the process $\{R_n\}$. We denote

$$T_1(u) = \min\{j \geq 1 : M_{1,j} \leq u, R_{j+1} > u | R_1 > u\},$$

$$T_2(u) = \min\{j \geq 1 : L_{1,j} > u, R_{j+1} \leq u | R_1 \leq u\},$$

where $M_{1,j} = \max\{R_2, \ldots, R_j\}$, $M_{1,1} = -\infty$, $L_{1,j} = \min\{R_2, \ldots, R_j\}$, $L_{1,1} = +\infty$.

The $(1 - F(u))T_1(u)$ is derived to be exponentially distributed [4]. In telecommunication systems [10] one has to exclude the cases $T_1(u) = 1$ and $T_2(u) = 1$ since they correspond to inter-arrival times between consecutive events $\{R_i\}$ and to use

$$T_1^*(u) = \min\{j > 1 : M_{1,j} \leq u, R_{j+1} > u | R_1 > u\},$$

$$T_2^*(u) = \min\{j > 1 : L_{1,j} > u, R_{j+1} \leq u | R_1 \leq u\}.$$

Then $T_2^*(u) - 1$ is the number of exceedances of $\{R_n\}$ over u in the cluster (the size of the cluster) and $T_1^*(u) - 1$ is the number of non-exceedances between two consecutive clusters (the size of the inter-cluster).

A geometric distribution has been used as a model of the limiting cluster size distribution π, namely, $\pi(j) = \lim_{n\to\infty} \pi_n(j)$ for $j = 1, 2, \ldots$, where

$$\pi_n(j) = P\{N_{r_n}(u_n) = j \,|\, N_{r_n}(u_n) > 0\} \qquad \text{for} \qquad j = 1, \ldots, r_n,$$

is the cluster size distribution, $r_n = o(n)$, $N_{r_n}(u_n)$ is the number of observations of $\{R_1, \ldots, R_{r_n}\}$ which exceed $u_n = a_n x + b_n$ such that the Leadbetter's mixing condition[1] $D(u_n)$ is satisfied [5, 12]. If the process R_t satisfies the $D''(u_n)$-condition[2] [6], then in our notations

$$\pi(j) = \lim_{n\to\infty} P\{T_2(u_n) - 1 = j\} = (1 - \theta)^{j-1}\theta, \qquad j = 1, 2, \ldots \qquad (15.1)$$

is proposed in [12, p. 126] without rigorous proof.

Definition 1 ([7]). The stationary sequence $\{R_n\}_{n\geq 1}$ is said to have extremal index $\theta \in [0, 1]$ if for each $0 < \tau < \infty$ there is a sequence of real numbers $u_n = u_n(\tau)$ such that

$$\lim_{n\to\infty} n(1 - F(u_n)) = \tau, \qquad \lim_{n\to\infty} P\{M_n \leq u_n\} = e^{-\tau\theta} \qquad \text{hold.} \qquad (15.2)$$

The idea behind the representation (15.1) is that θ may be interpreted as the reciprocal of the limiting mean cluster size $\theta \approx 1/\mathrm{E}T_2(u)$, i.e. the mean number of exceedances over the threshold per cluster [7]. Hence, θ may be used as a probability in the geometric distribution.

The geometric nature of $T_1(u)$ is considered in [1], where $T_1(u)$ is called a "duration between two consecutive violations". It is used to test the independence hypothesis.

In [9] geometric-like asymptotically equivalent distributions of both $T_1(u)$ and $T_2(u)$ with probability corrupted by the extremal index θ are derived. The latter allows us to take into account the dependence in the data. The geometric model of $T_2(u)$ allows us to obtain its asymptotically equal mean. One can consider sums

---

[1] $D(u_n)$ is satisfied if for any $A \in \mathscr{I}_{1,l}(u_n)$ and $B \in \mathscr{I}_{l+s,n}(u_n)$, where $\mathscr{I}_{j,l}(u_n)$ is the set of all intersections of the events of the form $\{R_i \leq u_n\}$ for $j \leq i \leq l$, and for some positive integer sequence $\{s_n\}$ such that $s_n = o(n)$, $|P\{(A \cap B)\} - P\{A\}P\{B\}| \leq \alpha(n, s)$ holds and $\alpha(n, s_n) \to 0$ as $n \to \infty$.

[2] The $D''(u_n)$-condition [3, 6] states that if the stationary sequence $\{R_t\}$ satisfies the $D(u_n)$-condition with $u_n = a_n x + b_n$ and normalizing sequences $a_n > 0$ and $b_n \in R$ such that for all x there exists $\mu \in R$, $\sigma > 0$ and $\xi \in R$, such that

$$n(1 - F(a_n x + b_n)) \to \left(1 + \frac{\xi(x - \mu)}{\sigma}\right)_+^{-1/\xi}, \qquad \text{as} \qquad n \to \infty,$$

holds, where $(x)_+ = \max(x, 0)$, then $\lim_{n\to\infty} \sum_{j=2}^{r_n} P\{R_j \leq u_n < R_{j+1} | R_1 > u_n\} = 0$, where $r_n = o(n)$, $s_n = o(n)$, $\alpha(n, s_n) \to 0$, $(n/r_n)\alpha(n, s_n) \to 0$ and $s_n/r_n \to 0$ as $n \to \infty$.

$S_{T_i(u)} = \sum_{j=1}^{T_i(u)} X_j, i \in \{1, 2\}$, where the sequence $\{X_j\}$ denotes inter-arrival times between events $\{R_t\}$, $S_{T_1(u)}$ and $S_{T_2(u)}$ are interpreted as the return interval and the duration of a cluster, see Fig. 15.1. In [9] it is shown that the limit tail distribution of $S_{T_2(u)}$ that is defined as a sum of random numbers of weakly dependent regularly varying inter-arrival times with tail index $0 < \alpha < 2$ is bounded by the tail of stable distribution.

The paper is organized as follows. In Sects. 15.2 and 15.3 we present the obtained theoretical results and examples of processes satisfying these results. In Sect. 15.4 applications in the teletraffic theory are considered.

15.2 Limit Distributions of $T_1(u)$ and $T_2(u)$

Our results [9] with respect to the distribution of $T_1(u)$ can be considered as an extension of Theorem 1. It is based on the mixing condition $\Delta^*(u_n)$.

Definition 2. For real u and integer $1 \le k \le l$, let $\mathbf{F}_{k,l}(u)$ be the σ-field generated by the events $\{X_i > u\}$, $k \le i \le l$. $\Delta^*(u_n)$ is fulfilled for the mixing coefficients

$$\alpha_{n,q}(u) = \max_{1 \le k \le n-q} \sup |P(B|A) - P(B)|, \tag{15.3}$$

where the supremum is taken over all $A \in \mathbf{F}_{1,k}(u)$ with $\mathbf{P}(A) > 0$ and $B \in \mathbf{F}_{k+q,n}(u)$, if there exist positive integers $r_n = o(n)$, $q_n = o(r_n)$ for which $\alpha_{cr_n,q_n}(u_n) \to 0$ as $n \to \infty$ for all $c > 0$.

Theorem 1 ([4]). *Let the positive integers $\{r_n\}$ and the thresholds $\{u_n\}$, $n \ge 1$, be such that $r_n \to \infty$, $r_n \overline{F}(u_n) \to \tau$ and $\mathbf{P}\{M_{r_n} \le u_n\} \to exp(-\theta\tau)$ as $n \to \infty$ for some $\tau \in (0, \infty)$ and $\theta \in [0, 1]$. If the condition $\Delta^*(u_n)$ is satisfied, then*

$$\mathbf{P}\{\overline{F}(u_n)T_1(u_n) > t\} \to \theta \exp(-\theta t) \tag{15.4}$$

for $t > 0$ as $n \to \infty$, where $\overline{F}(t) = 1 - F(t)$ is the tail function of $\{R_1\}$.

The result (15.4) implies that

$$\overline{F}(u_n)T_1(u_n) =^d T_\theta = \begin{cases} \eta, & \text{with probability} & \theta, \\ 0, & \text{with probability} & 1 - \theta, \end{cases}$$

where η is an exponentially distributed rv. This agrees with the result [5] that the point process of exceedance times has a Poisson process limit. In [13] the result (15.4) was shown for $T_1(u) - 1$ since $T_1(u)$ can provide nonzero values only. In the limit the clusters form single points when thresholds increase [2, Sect. 10.3.1].

In [9] quantiles of the underlying process R_t are taken as thresholds $\{u_n\}$. The following result is derived.

Let us consider the partition of the interval $[1, j]$ for a fixed j, namely,

$$k^*_{n,0} = 1, \qquad k^*_{n,5} = j, \qquad k^*_{n,i} = [jk_{n,i}/n] + 1, \qquad i = \{1, 2\},$$
$$k^*_{n,3} = j - [jk_{n,4}/n], \qquad k^*_{n,4} = j - [jk_{n,3}/n] \tag{15.5}$$

that corresponds to the partition of the interval $[1, n]$

$$\{k_{n,i-1} = o(k_{n,i}), \qquad i \in \{2, 3, 4\}\}, \qquad k_{n,4} = o(n), \qquad n \to \infty, \tag{15.6}$$

where n is the sample size. Here, both $k^*_{n,1}$ and $k^*_{n,2}$ tend to 1, both $k^*_{n,3}$ and $k^*_{n,4}$ tend to j and $\{[k^*_{n,2}, k^*_{n,3}]\}$ is the sequence of extending intervals as $n \to \infty$.

Theorem 2 ([9]). *Let $\{R_n\}_{n \geq 1}$ be a stationary process with the extremal index θ. Let $\{x_{\rho_n}\}$ and $\{x_{\rho^*_n}\}$ be sequences of quantiles of R_1 of the levels $\{1 - \rho_n\}$ and $\{1 - \rho^*_n\}$, respectively, those satisfy the conditions (15.2) if u_n is replaced by x_{ρ_n} or by $x_{\rho^*_n}$, and $q_n = 1 - \rho_n$, $q^*_n = 1 - \rho^*_n$, $\rho^*_n = (1 - q^\theta_n)^{1/\theta}$. Let positive integers $\{k^*_{n,i}\}, i = \overline{0, 5}$, and $\{k_{n,i}\}, i = \overline{1, 4}$, be, respectively, as in (15.5) and (15.6), $p^*_{n,i} = o(\Delta_{n,i}), i \in \{1, 2, \ldots, 5\}$, $\{p^*_{n,3}\}$ is an increasing sequence, $\Delta_{n,i} = k^*_{n,i} - k^*_{n,i-1}$, and $q^*_{n,i} = o(p^*_{n,i})$, such that*

$$\alpha^*_n(x_{\rho_n}) = \max\{\alpha_{k^*_{n,4}, q^*_{n,1}}; \alpha_{k^*_{n,3}, q^*_{n,2}}; \alpha_{\Delta_{n,3}, q^*_{n,3}}; \alpha_{j+1-k^*_{n,2}, q^*_{n,4}};$$
$$\alpha_{j+1-k^*_{n,1}, q^*_{n,5}}; \alpha_{j+1, k^*_{n,4}-k^*_{n,1}}\} = o(1) \tag{15.7}$$

holds as $n \to \infty$, where $\alpha_{n,q} = \alpha_{n,q}(x_{\rho_n})$ is determined by (15.3), then it holds for $j \geq 2$

$$\lim_{n \to \infty} P\{T_1(x_{\rho_n}) = j\}/(\rho_n(1 - \rho_n)^{(j-1)\theta}) = 1, \tag{15.8}$$

$$\lim_{n \to \infty} P\{T_2(x_{\rho^*_n}) = j\}/(q^*_n(1 - q^*_n)^{(j-1)\theta}) = 1, \tag{15.9}$$

and if additionally the sequence $\{R_n\}$ satisfies the $D''(x_{\rho_n})$-condition at $[1, k^*_{n,1} + 2]$ and $[k^*_{n,4} - 1, j + 1]$, then it holds for $j \geq 2$

$$\lim_{n \to \infty} P\{T_1(x_{\rho_n}) = j\}/(\rho_n(1 - \rho_n)^{(j-1)\theta}) \geq \theta^2, \tag{15.10}$$

$$\lim_{n \to \infty} P\{T_2(x_{\rho^*_n}) = j\}/(q^*_n(1 - q^*_n)^{(j-1)\theta}) \geq \theta^2. \tag{15.11}$$

The extremal index θ shows the deviation of the asymptotic distribution from the geometric one. One can rewrite (15.8) and (15.9) in geometric forms as

$$c_n P\{T_1(x_{\rho_n}) = j\} \sim \eta_n(1 - \eta_n)^{j-1}, \qquad d_n P\{T_2(x_{\rho^*_n}) = j\} \sim \chi_n(1 - \chi_n)^{j-1},$$

as $n \to \infty$, using the replacements $(1 - \rho_n)^\theta = 1 - \eta_n, 0 < \eta_n < 1$ and $(1 - q_n^*)^\theta = 1 - \chi_n, 0 < \chi_n < 1$. The following lemma useful in practice is derived in [9].

Lemma 1. *[9] If the conditions of Theorem 2 are satisfied, and $\sup_n E(T_2^{1+\varepsilon}(x_{\rho_n^*}))/\Lambda_n < \infty$ holds for some $\varepsilon > 0$, where $\Lambda_n = q_n^*/(1 - (1 - q_n^*)^\theta)^2$, and the sequence $\{R_n\}$ satisfies the mixing condition (15.7) then it holds*

$$\lim_{n\to\infty} E(T_2(x_{\rho_n^*}))/\Lambda_n = 1.$$

15.3 Examples

We shall consider how Theorem 2 fits to concrete processes.

15.3.1 Autoregressive Maximum (ARMAX) Process

The ARMAX process is determined by the formula $R_t = \max\{\alpha R_{t-1}, (1 - \alpha)Z_t\}$, where $0 \le \alpha < 1$, $\{Z_t\}$ are iid standard Fréchet distributed rvs with the cdf $F(x) = \exp(-(1 - \alpha)/x)$, $x > 0$. The extremal index of the process is given by $\theta = 1 - \alpha$ [2]. The condition (15.7) is fulfilled for $\alpha = 0$. The $D''(x_{\rho_n})$-condition is satisfied. In [9] it is shown that

$$P\{T_1(x_\rho) = j\} = \left(q^{1-\theta} - q\right)^2 q^{\theta j}/(q(1-q)), \qquad j = 2, 3, \ldots, \qquad (15.12)$$

$$P\{T_2(x_\rho) = 2\} = q^\theta (q^{\theta(1-\theta)} - q^\theta), \qquad P\{T_2(x_\rho) = j\} \ge q^\theta (q^{(1-\theta)\theta} - q^\theta)^{j-1}, \tag{15.13}$$

for $j = 3, 4, \ldots$ hold, where $q = 1 - \rho$, and x_ρ is the $(1 - \rho)$th quantile of R_1, i.e. $\mathbf{P}\{R_1 > x_\rho\} = \rho$. The distributions $T_1(x_\rho)$ and $T_2(x_\rho)$ are geometric ones as $\alpha = 0$ (or $\theta = 1$). Moreover, it holds

$$\mathbf{E}(T_1(x_\rho)) = \frac{q^\alpha}{1 - q}, \qquad \mathbf{E}(T_1(x_\rho))^2 = \frac{q^\alpha (1 + q^{1-\alpha})}{(1 - q)(1 - q^{1-\alpha})}.$$

15.3.2 Moving Maxima (MM) Process

The MM process is determined by the formula $R_t = \max_{i=0,\ldots,m}\{\alpha_i \varepsilon_{t-i}\}$, where $\alpha_i \ge 0$ and $\sum_{i=0}^m \alpha_i = 1$. ε_t are iid unit Frechet distributed rvs with the cdf $F(x) = \exp(-1/x)$, $x > 0$. The extremal index of the process is given by $\theta = \max_i\{\alpha_i\}$ [2]. All conditions of Theorem 2 are satisfied. Let $\alpha_0 \ge \alpha_1 \ge \ldots \ge \alpha_m$ hold. Then it holds

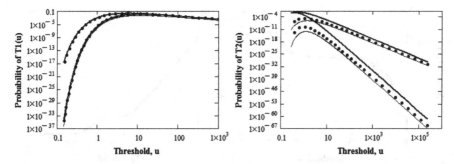

Fig. 15.2 The distribution (15.12) of $T_1(x_\rho)$ of a ARMAX process (*dotted line*), the lower bound (15.14) of an MM process (*thin solid line*) and geometric-like model (15.10) (*solid line*) against the Fréchet quantiles $x_\rho = -1/\ln(1-\rho)$ as threshold u (*left*). Probabilities corresponding to $j = 10$ are located above the ones determined at $j = 20$ for smaller u. The lower bound (15.13) of $T_2(x_\rho)$ of an ARMAX process (*dotted line*), the lower bound (15.14) of a MM process (*thin solid line*) and geometric-like model (15.11) (*solid line*) against the Fréchet quantiles $u = x_\rho = -1/\ln(q^*)$, $q^* = 1 - (1 - (1-\rho)^\theta)^{1/\theta}$ (*right*). Probabilities corresponding to $j = 10$ are located under those ones determined at $j = 20$ for larger u. In both figures $\theta = 0.6$ and $\alpha_1 = 0.3$ were taken

$$P\{T_1(x_\rho) = j\} \geq q^{\theta(j-1)+1-\theta}\left(q^{\alpha_1} - q^\theta\right)(1 - q^\theta)/(1 - q), j = 2, 3, \ldots$$
$$(15.14)$$

$$P\{T_2(x_\rho) = 2\} = q^\theta\left(q^{\alpha_1} - q^\theta\right),$$

$$P\{T_2(x_\rho) = j\} \geq q(q^{\alpha_1} - q^\theta)^{j-1}, \qquad j = 3, 4, \ldots, \qquad (15.15)$$

where $q = 1 - \rho$, x_ρ is the $(1 - \rho)$th quantile of R_1, [9]. The distributions $T_1(x_\rho)$ and $T_2(x_\rho)$ are geometric ones as $\theta = \alpha_0 = 1$. It is shown in [9] that it holds

$$\mathbf{E}(T_1(x_\rho)) \geq \frac{q(q^{\alpha_1-\theta} - 1)}{(1 - q^\theta)(1 - q)}, \qquad \mathbf{E}(T_1(x_\rho))^2 \geq \frac{(1 + q^\theta)q(q^{\alpha_1-\theta} - 1)}{(1 - q^\theta)^2(1 - q)}.$$

The results for both processes are in good agreement with Theorem 2, see Fig. 15.2. Evidently, the probability to get larger $T_2(x_\rho)$ and smaller $T_1(x_\rho)$ is higher for smaller thresholds and vice versa.

15.4   Teletraffic Applications

The main problem in the teletraffic theory concerns the trade-off between available resources and the transmission of information with minimal loss and delay during delivery. We consider the packet traffic in peer-to-peer (P2P) applications like Skype and IPTV where the packet lengths and inter-arrival times between packets are both random. The described theory can be used to evaluate the quality of the packet

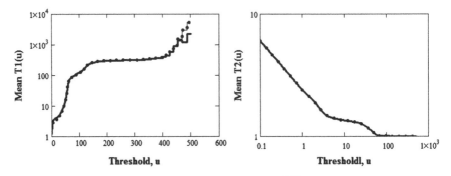

Fig. 15.3 Sample mean of $T_1(u)$ (*solid line*) and the model $\widehat{ET_1(x_\rho)} = \theta^2 \rho/(1 - (1 - \rho)^\theta)^2$ with $\theta = 1$ (*thin solid line*) and with the intervals estimate of θ [4] (*dotted line*) (*left*); sample mean of $T_2(u)$ (*solid line*) and the model $\widehat{ET_2(x_\rho)} = \theta^2 q/(1 - (1 - q)^\theta)^2$ with $\theta = 1$ (*dotted line*) (*right*) against u corresponding to quantiles x_ρ of scaled transmission rates $\{R_n \cdot 10^{-5}\}$ of the SopCast IPTV packet traffic

transmission. We study the loss and delay at the packet layer. The main idea is that the packet loss is caused by exceedances of the rate of a transmission over a threshold u that can be interpreted as a channel capacity [11]. Then the packets can be lost only in the clusters generated by exceedances of the rate above u. In this context, the rates are considered as underlying process $\{R_n\}$, the return intervals as lossless periods and cluster durations as delays between successfully delivered packets, see Fig. 15.1. Since active streams may share the capacity of a channel, one can manage the equivalent capacity (i.e., the part of capacity allocated for each stream) in such a way to minimize the probability of a packet to miss its playout deadline at the receiver. Using geometric distributions of $T_1(u)$ and $T_2(u)$ and Wald's equation one can estimate the means of the lossless time and the delay in the clusters in an on-line regime [8, 11]. The latter can be used to optimize the quality of packet transmissions in P2P overlay networks. Modeling of $\mathbf{E}T_1(u)$ and $\mathbf{E}T_2(u)$ by SopCast IPTV data is shown in Fig. 15.3.

Acknowledgements This work was supported in part by the Russian Foundation for Basic Research, grant 13-08-00744 A.

References

1. Araújo Santos, P., Fraga Alves, M.I.: A new class of independence tests for interval forecasts evaluation. Comput. Stat. Data Anal. **56**(11), 3366–3380 (2012)
2. Beirlant, J., Goegebeur, Y., Teugels, J., Segers, J.: Statistics of Extremes: Theory and Applications. Wiley, Chichester, West Sussex (2004)
3. Chernick, M.R., Hsing, T., McCormick, W.P.: Calculating the extremal index for a class of stationary sequences. Adv. Appl. Prob. **23**, 835–850 (1991)

4. Ferro, C.A.T., Segers, J.: Inference for clusters of extreme values. J. R. Statist. Soc. B. **65**, 545–556 (2003)
5. Hsing, T., Huesler, J., Leadbetter, M.R.: On the exceedance point process for a stationary sequence. Prob. Theory Relat. Fields. **78**, 97–112 (1988)
6. Leadbetter, M.R., Nandagopalan, L.: On exceedance point processes for stationary sequences under mild oscillation restrictions. Lect. Notes Stat. **51**, 69–80 (1989)
7. Leadbetter, M.R., Lingren, G., Rootzén, H.: Extremes and Related Properties of Random Sequence and Processes, chap. 3. Springer, New York (1983)
8. Markovich, N.: Quality assessment of the packet transport of peer-to-peer video traffic in high-speed networks. Perform. Eval. **70**, 28–44 (2013)
9. Markovich, N.: Modeling clusters of extreme values. Extremes **17**(1), 97–125 (2014)
10. Markovich, N.M., Krieger, U.R.: Statistical Analysis and modeling of peer-to-peer multimedia traffic. In: Performance Handbook Next Generation Internet: Performance Evaluation and Applications. Lecture Notes in Computer Science, vol. 5233. Springer, Berlin (2010)
11. Markovich, N.M., Krieger, U.R.: Analyzing measurements from data with underlying dependences and heavy-tailed distributions. In: 2nd ACM/SPEC International Conference on Performance Engineering (ICPE 2011), Karlsruhe, 14–16 March 2011, pp. 425–436 (tutorial)
12. Robinson, M.E., Tawn, J.A.: Extremal analysis of processes sampled at different frequencies. J. R. Stat. Soc. B. **62**(1), 117–135 (2000). doi:10.1111/1467-9868.00223
13. Süveges, M.: Likelihood estimation of the extremal index. Extremes **10**, 41–55 (2007)

Chapter 16
Diagnostic Procedures for Spherically Symmetric Distributions

Simos G. Meintanis

Abstract Goodness-of-fit tests are proposed for the null hypothesis that the distribution of an arbitrary random variable belongs to the general family of spherically symmetric distributions. The test statistics utilize a well-known characterization of spherical symmetry which incorporates the characteristic function of the underlying distribution. An estimated version of this characterization is then employed both in the Kolmogorov–Smirnov sense and the Cramér–von Mises sense and yields corresponding test statistics. Both tests come in convenient forms which are straightforwardy applicable with the computer. Also the consistency of the tests is investigated under general conditions. Since the asymptotic null distribution of the test statistic depends on unknown factors, a resampling procedure is proposed in order to actually carry out the tests.

Keywords Goodness-of-fit • Empirical characteristic function • Resampling procedure

16.1   Introduction

Consider a p-dimensional random vector X with an unknown distribution function F. In the process of performing statistical inference, the distribution of X is often regarded as nuisance and it is then common practice to assume either too much or too little for F. In such cases we are led either to the one extreme of a fully parametric distribution function, or to the other extreme, which is essentially nonparametric, and assumes very little structure for F; for instance, when only the existence of a few moments or, at the most, (reflective) symmetry is postulated. Here we consider a somewhat "intermediate situation," of a semi-parametric null hypothesis. In doing so we focus on the family of spherically

S.G. Meintanis (✉)
Department of Economics, National and Kapodistrian University of Athens,
8 Pesmazoglou Street, 105 59 Athens, Greece

Unit for Business Mathematics and Informatics, North–West University, Potchefstroom,
South Africa (on sabbatical leave from the University of Athens)
e-mail: simosmei@econ.uoa.gr

© Springer Science+Business Media New York 2014   177
M.G. Akritas et al. (eds.), *Topics in Nonparametric Statistics*, Springer Proceedings
in Mathematics & Statistics 74, DOI 10.1007/978-1-4939-0569-0_16

symmetric distributions (SSD) which is known to contain models that constitute building blocks for many of the often employed parametric models, such as the normal, the contaminated normal, the Student t, and the Laplace distribution. Also it is well known that several estimation results in regression are robust with respect to spherical symmetry; see, for instance, [7, 9, 10, 14]. The reader is referred to [4] for a complete treatment of the properties of SSD, and in particular to Sects. 2.7.1–2.7.2 of that book for extension of the well-known t and F statistics from the normal distribution to the class of SSD.

On the basis of i.i.d. copies $X_1, X_2, \ldots, X_n$, of X, we wish to test the null hypothesis,

$\mathcal{H}_0$: the law of X belongs to the family of spherically symmetric laws $\in \mathbb{R}^p$.

(16.1)

Note that the null hypothesis $\mathcal{H}_0$ should be understood in its full generality, meaning that in Eq. (16.1) the specific distribution of X is left unspecified, and we are only interested whether F belongs to the general family of SSD. Therefore our test procedure would also be semi-parametric in nature and, as such, acceptance of $\mathcal{H}_0$ would provide only partial information regarding the structural properties of the underlying theoretical distribution function F.

To outline the proposed procedure, let $\varphi(t)$, $t \in \mathbb{R}^p$, denote the characteristic function (CF) of X, and recall the characterization of the family of SSD, that $\mathcal{H}_0$ holds if and only if, for some univariate function $\phi(\cdot)$ it holds that

$$\varphi(t) = \phi(\|t\|^2), \ \forall t \in \mathbb{R}^p. \tag{16.2}$$

In view of (16.2), it is natural to consider as a test statistic some distance measure involving the discrepancy

$$D_n(t, s) = \varphi_n(t) - \varphi_n(s),$$

computed over points $t, s \in \mathbb{R}^p$, such that $\|t\|^2 = \|s\|^2$, where $\varphi_n(u) = n^{-1} \sum_{j=1}^{n} e^{\iota u' X_j}$, is the empirical CF of X_j, $j = 1, \ldots, n$. A Kolmogorov–Smirnov type distance would follow naturally from the above reasoning as $\sup |D_n(t, s)|$, where the supremum is taken over all possible pairs (t, s) of vectors in $\mathbb{R}^p$ such that $\|t\|^2 = \|s\|^2$. However we shall instead consider a test statistic indexed by a parameter $\mathcal{R}$, which takes into account all possible vectors lying within a sphere of radius $\mathcal{R} \in (0, \infty]$. In particular we suggest the test statistic,

$$\sqrt{n} \sup_{0 \le \rho \le \mathcal{R}} \sup_{(t,s) \in \mathbb{S}_\rho \times \mathbb{S}_\rho} |D_n(t, s)|, \tag{16.3}$$

where $\mathbb{S}_\rho := \{t \in \mathbb{R}^p : \|t\|^2 = \|s\|^2 = \rho^2\}$. Clearly the limiting value $\mathcal{R} \to \infty$ corresponds to the usual (unrestricted) Kolmogorov–Smirnov test. A related approach mimicking a Cramér–von Mises test is, again on the basis of $|D_n(t, s)|$, to consider an integrated distance-squared statistic of the type

$$n \int_0^{\mathcal{R}} \left(\int_{(t,s) \in \mathbb{S}_\rho \times \mathbb{S}_\rho} |D_n(t,s)|^2 d\mu(t,s) \right) W(\rho) d\rho, \tag{16.4}$$

where $\mu(\cdot, \cdot)$ denotes a finite measure on $\mathbb{S}_\rho \times \mathbb{S}_\rho$. Notice that in Eq. (16.4) the option of using the full spectrum of values by letting $\mathcal{R} \to \infty$ necessitates the introduction of a weight function $W(\cdot)$ in order to smooth out the periodic nature of $\varphi_n(\cdot)$, and thereby produce a convergent integral in Eq. (16.4).

Although Eqs. (16.3) and (16.4) allow for arbitrary ways of determining the values $(t,s) \in \mathbb{S}_\rho \times \mathbb{S}_\rho$, for practical computational purposes we shall instead consider test statistics similar to those in (16.3) and (16.4), for which, however, the values $(t,s) := (t(\rho), s(\rho)) \in \mathbb{S}_\rho \times \mathbb{S}_\rho$ have been prespecified. In particular and for fixed integer $J > 0$, we select a grid of values $(t_j(\rho), s_j(\rho)) \in \mathbb{S}_\rho \times \mathbb{S}_\rho$, $j = 1, \ldots, J$, and compute $D_n(\cdot, \cdot)$ at these values. Then the suggested test statistics reduce to

$$\mathrm{KS}_{n,\mathcal{R}} = \sqrt{n} \sup_{0 \le \rho \le \mathcal{R}} \max_{(t_j, s_j)_{j=1}^J \in \mathbb{S}_\rho \times \mathbb{S}_\rho} \left| D_n(t_j, s_j) \right|, \tag{16.5}$$

and

$$\mathrm{CM}_{n,\mathcal{R}} = n \int_0^{\mathcal{R}} \left(\sum_{(t_j, s_j)_{j=1}^J \in \mathbb{S}_\rho \times \mathbb{S}_\rho} |D_n(t_j(\rho), s_j(\rho))|^2 \right) W(\rho) d\rho, \tag{16.6}$$

respectively. Rejection of the null hypothesis $\mathcal{H}_0$ in (16.1) would be for large values of $\mathrm{KS}_{n,\mathcal{R}}$ and $\mathrm{CM}_{n,\mathcal{R}}$.

In the literature there are several tests for spherical symmetry, some of them utilizing the empirical CF. For a review the reader may refer to [13], and the references therein. We mention here the most relevant papers of [3, 8, 16], and the test in Chap. 3 of the monograph of [15]. There exist also some so-called necessary tests, i.e., tests which are based on necessary but not sufficient properties of SSD, and which are very convenient to use; see, for instance, [5, 11].

16.2   Computation of the Test Statistics

Clearly in order to specify the test statistics in Eqs. (16.5) and (16.6) we need to fix the way in which the pairs (t_j, s_j) are selected. Before doing that note that from (16.6) we have by straightforward algebra

$$\mathrm{CM}_{n,\mathcal{R}} = \frac{1}{n} \sum_{j=1}^J \sum_{l,m=1}^n \int_0^{\mathcal{R}} \cos[t_j' X_{lm}] W(\rho) d\rho + \int_0^{\mathcal{R}} \cos[s_j' X_{lm}] W(\rho) d\rho$$
$$- 2 \int_0^{\mathcal{R}} \cos[t_j' X_l - s_j' X_m] W(\rho) d\rho, \tag{16.7}$$

where $X_{lm} = X_l - X_m$, and of course the arguments t_j and s_j depend on ρ, i.e., $t_j = t_j(\rho)$ and $s_j = s_j(\rho)$, $j = 1, \ldots, J$.

A natural way of selecting $t_j = t_j(\rho)$ and $s_j = s_j(\rho)$ is to use the hyperspherical coordinates: Let $t = t(\rho)$ be any point in $\mathbb{R}^p$, which lies in $\mathbb{S}_\rho$, i.e., $\|t\|^2 = \rho^2$. Then this point may be written as $t = \rho(t_1, \ldots, t_p)'$ via the hyperspherical coordinates

$$
\begin{aligned}
t_1 &= \cos \phi_1, \\
t_2 &= \sin \phi_1 \cos \phi_2, \\
t_3 &= \sin \phi_1 \sin \phi_2 \cos \phi_3, \\
&\vdots \\
t_{p-1} &= \sin \phi_1 \cdots \sin \phi_{p-2} \cos \phi_{p-1}, \\
t_p &= \sin \phi_1 \cdots \sin \phi_{p-2} \sin \phi_{p-1},
\end{aligned}
\tag{16.8}
$$

where the angular coordinates satisfy $\{\phi_\ell\}_{\ell=1}^{p-2} \in [0, \pi]$, and $\phi_{p-1} \in [0, 2\pi]$.

With this specification the CM-type test statistic may be computed as follows: Given t as above and any other point $x = (x_1, \ldots, x_p)' \in \mathbb{R}^p$, notice that the arbitrary integral figuring in (16.7) reduces to

$$
\int_0^{\mathcal{R}} \cos[t'(\rho)x]W(\rho)d\rho
\tag{16.9}
$$

$$
= \int_0^{\mathcal{R}} \cos\left[\underbrace{\rho(x_1 \cos \phi_1 + x_2 \sin \phi_1 \cos \phi_2 + \cdots + x_p \sin \phi_1 \cdots \sin \phi_{p-1})}_{u} \right] W(\rho)d\rho
$$

$$
= \int_0^{\mathcal{R}} \cos(\rho u)W(\rho)d\rho := I_W(u).
$$

Typical choices such as $W(\rho) = e^{-a\rho}$ and $W(\rho) = e^{-a\rho^2}$, $a > 0$, lead to the closed-form expressions for this integral as

$$
I_W(u) = \frac{a - e^{-a\mathcal{R}}(a\cos(\mathcal{R}u) - u\sin(\mathcal{R}u))}{a^2 + u^2},
$$

and

$$
I_W(u) = \frac{1}{4}\sqrt{\frac{\pi}{a}}e^{-u^2/4a}\left[\text{Erf}\left(\frac{2a\mathcal{R} - iu}{2\sqrt{a}}\right) + \text{Erf}\left(\frac{2a\mathcal{R} + iu}{2\sqrt{a}}\right)\right],
$$

respectively. The limiting values as $\mathcal{R} \to \infty$ are $I_W(u) = a/(a^2+u^2)$ and $I_W(u) = (1/2)\sqrt{\pi/a}\ e^{-u^2/4a}$.

Our approach, although in the same spirit as the test suggested by Zhu and Neuhaus [16], differs from it in that these authors utilize the property that if the law of X belongs to the SSD class, then $X'e$ is symmetric around zero for every unit vector e. Then the real part of the CF of $X'e$ vanishes identically, and empirical-CF procedures, such as those suggested by Feuerverger and Mureika [6] may be applied. The aforementioned property is already mentioned in [8] and holds true for all distributions in the class of SSD. However it is not equivalent to X being spherically symmetric. In fact, there exist distributions for which $X'e$ is symmetric around zero without X belonging to the family SSD. A typical example is when X follows the uniform distribution on the p-cube $[-a, a]^p$. Consequently it is expected that any test which utilizes this property will have low power against such non-spherically symmetric laws.

16.3   A Consistency Result

The statistics figuring in (16.5) and (16.6) utilize a grid $(t_j, s_j)_{j=1}^J \subset \mathbb{S}_\rho \times \mathbb{S}_\rho$ of points on the p-surface of $\mathbb{S}_\rho$ which varies continuously with ρ. To prove consistency of the tests that reject $\mathcal{H}_0$ for large values of $\mathrm{KS}_{n,\mathcal{R}}$ and $\mathrm{CM}_{n,\mathcal{R}}$ against general alternatives, we impose the following conditions:

- (A1) For some $\rho^* \in (0, \mathcal{R}]$, there is a pair of points (t^*, s^*) such that $\|t^*\| = \|s^*\| = \rho^*$ and $(t^*, s^*) \in \{(t_1, s_1), \dots, (t_J, s_J)\} \subset \mathbb{S}_{\rho^*} \times \mathbb{S}_{\rho^*}$, such that $\varphi(t^*) - \varphi(s^*) \neq 0$.
- (A2) The weight function W figuring in (16.6) satisfies $0 < \int_0^{\mathcal{R}} W(\rho) d\rho < \infty$.

We then have the following result.

Theorem 31. *Suppose that the distribution of X satisfies (A1). We then have*

$$\liminf_{n \to \infty} \frac{\mathrm{KS}_{n,\mathcal{R}}}{\sqrt{n}} > 0 \quad \textit{almost surely.} \tag{16.10}$$

If in addition (A2) holds, we have

$$\liminf_{n \to \infty} \frac{\mathrm{CM}_{n,\mathcal{R}}}{n} > 0 \quad \textit{almost surely.} \tag{16.11}$$

Proof. The almost sure uniform convergence of the empirical CF (see [1, 2, 6, 12]) gives

$$\lim_{n \to \infty} \sup_{\|t\| \leq \mathcal{R}} |\varphi_n(t) - \varphi(t)| = 0 \quad \textit{almost surely}$$

and thus from (A1) we have

$$\liminf_{n\to\infty} \sup_{0\le\rho\le\mathcal{R}} \max_{j=1,\dots,J} \left| D_n(t_j, s_j) \right| \ge \left| \varphi(t^*) - \varphi(s^*) \right| > 0$$

almost surely as $n \to \infty$, which implies (16.10).

To prove (16.11), notice that $|\varphi_n(t)| \le 1$ and thus $|D_n(t,s)|^2 \le 4$. Then the second inequality in (A2) and Lebesgue's dominated convergence theorem gives

$$\lim_{n\to\infty} \frac{\mathrm{CM}_{n,\mathcal{R}}}{n} = \int_0^{\mathcal{R}} \left(\sum_{j=1}^J |\varphi(t_j(\rho)) - \varphi(s_j(\rho))|^2 \right) W(\rho)d\rho := \Delta,$$

almost surely. Now let $m_j = \min_{0\le\rho\le\mathcal{R}} |\varphi(t_j(\rho)) - \varphi(s_j(\rho))|^2$, and $m = \min\{m_j\}_{j=1}^J$. Then

$$\Delta \ge Jm \int_0^{\mathcal{R}} W(\rho)d\rho.$$

By (A1) and the first inequality in (A2) it follows that the last quantity is positive which shows (16.11).

16.4 A Conditional Resampling Test

Both the finite sample and the asymptotic distribution of the test statistics under the null hypothesis of spherical symmetry depend on the unknown distribution of the Euclidean norm of the underlying random vector X [see (16.2)]. To carry out the test in practice, we use the conditional Monte Carlo method proposed by Diks and Tong [3] (see also [16]). The motivation for this resampling procedure is that, provided that $P(X = 0) = 0$, the distribution of X is spherically symmetric if, and only if, $\|X\|$ and $X/\|X\|$ are independent, and $X/\|X\|$ has a uniform distribution over the unit sphere surface $\mathbb{S}_1$.

The resampling scheme, which is conditional on the norms $\|X_1\|, \dots, \|X_n\|$, runs as follows:

(i)   Calculate the test statistic $T := T(X_1, \dots, X_n)$, based on the original observations X_j, $j = 1, \dots, n$.
(ii) Generate vectors u_j^*, $j = 1, \dots, n$, that are uniformly distributed on $\mathbb{S}_1$.
(iii) Compute the new data $X_j^* = u_j^* \|X_j\|$, $j = 1, \dots, n$.
(iv) Compute the test statistic $T^* = T(X_1^*, \dots, X_n^*)$.
(v)   Repeat steps (ii) to (iv) a number of times B, and calculate the corresponding test statistic values $T_1^*, \dots, T_B^*$.
(vi) Reject the null hypothesis if $T > T_{(B-\alpha B)}^*$, where $T_{(1)}^* \le \dots \le T_{(B)}^*$ denote the corresponding order statistics

References

1. Csörgő, S.: Limit behavior of the empirical characteristic function. Ann. Probab. **9**, 130–144 (1981)
2. Csörgő, S.: Multivariate empirical characteristic functions. Z. Wahrscheinlichkeitstheorie und verw. Gebiete **55**, 197–202 (1981)
3. Diks, C., Tong, H.: A test for symmetries of multivariate probability distributions. Biometrika **86**, 605–614 (1999)
4. Fang, K.T., Kotz, S., Ng, K.W.: Symmetric Multivariate and Related Distributions. Chapman and Hall, London (1990)
5. Fang, K.T., Zhu, L.X., Bentler, P.M.: A necessary test of goodness of fit for sphericity. J. Multiv. Anal. **45**, 34–55 (1993)
6. Feuerverger, A., Mureika, R.: The empirical characteristic function and its applications. Ann. Statist. **5**, 88–97 (1977)
7. Fourdrinier, D., Wells, M.T.: Estimation of a loss function for spherically symmetric distributions in the general linear model. Ann. Statist. **23**, 571–592 (1995)
8. Ghosh, S., Ruymgaart, F.H.: Applications of empirical characteristic functions in some multivariate problems. Can. J. Stat. **20**, 429–440 (1992)
9. Jammalamadaka, S.R., Tiwari, R.C., Chib, S.: Bayesian prediction in the linear model with spherically symmetric errors. Econ. Lett. **24**, 39–44 (1987)
10. Li, B., Zha, F., Chiaromonte, F.: Contour regression: a general approach to dimension reduction. Ann. Statist. **33**, 1580–1616 (2005)
11. Liang, J., Fang, K.T., Hickernell, F.J.: Some necessary uniform tests for spherical symmetry. Ann. Inst. Stat. Math. **60**, 679–696 (2008)
12. Marcus, M.B.: Weak convergence of the empirical characteristic function. Ann. Probab. **9**, 194–201 (1981)
13. Meintanis, S.G., Ngatchou–Wandji, J.: Recent tests for symmetry with multivariate and structured data: a review. In: Jiang, J., et al. (eds.) Nonparametric Statistical Methods and Related Topics: A Festschrift in Honour of Professor P.K. Bhattacharya, pp. 35–73. World Scientific, Singapore (2012)
14. Zeng, P., Zhu, Y.: An integral transform method for estimating the central mean and central subspaces. J. Multivar. Anal. **101**, 271–290 (2010)
15. Zhu, L.X.: Asymptotics of goodness–of–fit tests for symmetry. In: NonParametric Monte Carlo Tests and Their Applications. Lecture Notes in Statistics, vol. 20, pp. 27–43. Springer, New York (2005)
16. Zhu, L.X., Neuhaus, G.: Conditional tests for elliptical symmetry. J. Multivar. Anal. **84**, 284–298 (2003)

Chapter 17
Nonparametric Regression Based Image Analysis

P.-A. Cornillon, N. Hengartner, E. Matzner-Løber, and B. Thieurmel

Abstract Multivariate nonparametric smoothers are adversely impacted by the sparseness of data in higher dimension, also known as the curse of dimensionality. Adaptive smoothers, that can exploit the underlying smoothness of the regression function, may partially mitigate this effect. We present an iterative procedure based on traditional kernel smoothers, thin plate spline smoothers or Duchon spline smoother that can be used when the number of covariates is important. However the method is limited to small sample sizes ($n < 2,000$) and we will propose some thoughts to circumvent that problem using, for example, pre-clustering of the data. Applications considered here are image denoising.

Keywords Image sequence denoising • Iterative bias reduction • Kernel smoother • Duchon splines

17.1 Introduction

The recent survey paper [11] presents modern image filtering from multiple perspectives, from machine vision, to machine learning, to signal processing; from graphics to applied mathematics and statistics. It is noteworthy that an entire section of that paper is devoted to an iteratively refined smoother for noise reduction. A similar nonparametric regression estimator was considered independently in [4], who recognized it as a nonparametric iterative bias reduction method. When coupled with a cross-validation stopping rule, the resulting smoother adapts to the underlying smoothness of the regression function. The adaptive property of this iterative bias

P.-A. Cornillon (✉) • E. Matzner-Løber
Université Rennes 2, 35043 Rennes, France
e-mail: pac@uhb.fr; eml@uhb.fr

N. Hengartner
Los Alamos National laboratory, Los Alamos, NM, USA
e-mail: nickh@lanl.gov

B. Thieurmel
Data Knowledge, 75008 Paris, France
e-mail: bt@datak.fr

© Springer Science+Business Media New York 2014   185
M.G. Akritas et al. (eds.), *Topics in Nonparametric Statistics*, Springer Proceedings in Mathematics & Statistics 74, DOI 10.1007/978-1-4939-0569-0_17

corrected smoother helps mitigate the curse of dimensionality, and that smoother has been successfully been applied to fully nonparametric regression with moderately large number of explanatory variables [5].

This paper presents a fully nonparametric regression formulation for denoising images. In Sect. 17.2, we present iterative biased regression (IBR) with kernel smoother and Duchon splines smoother. In Sect. 17.3, we consider image denoising as a regression problem and evaluates IBR in that context. The proposed procedure is compared to BM3D in Sect. 17.4. Section 17.5 gathers concluding remarks.

17.2 Iterative Bias Reduction

For completeness, we recall the definition of the iterative bias corrected smoother. Consider the classical nonparametric regression model

$$Y_i = m(X_i) + \varepsilon_i, \tag{17.1}$$

for the pairs of independent observations $(X_i, Y_i) \in \mathbb{R}^d \times \mathbb{R}$, $i = 1, \ldots, n$. We want to estimate the unknown regression function m assuming that the disturbances $\varepsilon_1, \ldots, \varepsilon_n$ are independent mean zero and finite variance σ^2 random variables. It is helpful to rewrite Eq. (17.1) in vector form by setting $Y = (Y_1, \ldots, Y_n)^t$, $m = (m(X_1), \ldots, m(X_n))^t$, and $\varepsilon = (\varepsilon_1, \ldots, \varepsilon_n)^t$, to get $Y = m + \varepsilon$. Linear smoothers can be written in vector format as

$$\hat{m} = S(X, \lambda)Y,$$

where $S(X, \lambda)$ is an $n \times n$ smoothing matrix depending on the explanatory variables X and a smoothing parameter λ or a vector of smoothing parameters. The latter is typically the bandwidth for kernel smoother or the penalty for splines smoothers. Instead of selecting the optimal value for λ, [4] and [5] propose to start out with a biased smoother that has a large smoothing parameter λ (ensuring that the data are over-smoothed) and then proceed to estimate and correct the bias in an iterative fashion. If one wants to use the same smoothing matrix (denoted it simply by S from now) at each iteration, the smoother $\hat{m}_k$ after $k - 1$ bias correction iterations has a closed form expression:

$$\hat{m}_k = [I - (I - S)^k]Y. \tag{17.2}$$

The latter shows that the qualitative behavior of the sequence of iterative bias corrected smoothers $\hat{m}_k$ is governed by the spectrum of $I - S$ [4]. If the eigenvalues of $I - S$ are in $[0, 1[$, then $\lim_{k \to \infty} \hat{m}_k(X_j) = Y_j$. Thus at the observations, the bias converges to 0 and the variance increases to σ^2.

This convergence of the smoother to the data (as a function of the number of iterations) raises the question of how to select the number of iterations k. For univariate thin-plate spline base smoothers, [3] showed that there exists a k^* such that the iterative bias corrected smoother $\hat{m}_{k^*}$ converges in mean square error to the true regression function at the minimax convergence rate. An extension of that result to the multivariate case is presented in [4]. This optimal number of iterations can be selected from data using classical model selection criterion such as: GCV, AIC, BIC, or gMDL [10]. All these rules are implemented in **ibr** package.

17.2.1   Adaptation to Smoothness

The resulting estimator adapts to the true (unknown) smoothness of the regression function. Since points in higher dimensions are more separated from one another than their lower dimensional projections, smoothers in higher dimensions need average over larger volumes than in lower dimensions to smooth over the same number of points. Thus smoother in higher dimensions have, for similar variances, larger biases.

This effect is at the heart of the curse of dimensionality, which can also be quantified via the minimax Mean Integrated Square Error (MISE): For v-times continuously differentiable regression functions in $\mathbb{R}^d$, the minimax MISE is of order $n^{2v/(2v+d)}$. While this rate degrades as a function of the dimension d, it is improved with increasing smoothness. For example, the minimax MISE rate of convergence of a 40-times differentiable function on $\mathbb{R}^{20}$ is the same as the minimax MISE rate of convergence of a twice differentiable function on $\mathbb{R}$. While in practice, the smoothness of regression function is not known, adaptive smoothers behave (asymptotically) as if the smoothness were known. Thus in higher dimensional smoothing problems, such as those arising in image denoising and image inpainting, adaptive smoothers are desirable as they partially mitigate the curse of dimensionality.

Iterative bias reduction can be applied to any linear smoother. For kernel smothers, the behavior of the sequence of iterative bias corrected kernel smoothers depends critically on the properties of the smoother kernel. Specifically, the smoothing kernel needs to be positive definite [4, 7]. Examples of positive definite kernels include the Gaussian and the triangle densities, and examples of kernel that are not definite positive include the uniform and the Epanechnikov kernels. In this paper we focus only on the Gaussian kernel smoother:

$$ S_{ij} = \frac{K_{ij}}{\sum_{i=1}^{n} K_{ij}}, \quad \text{where } K_{ij} = \exp\left\{ -\sum_{k=1}^{d} (X_{ik} - X_{jk})^2 / (2\lambda_k^2) \right\}, $$

and on the Duchon splines smoother presented in the next section.

17.2.2 Splines Smoothers

The theoretical results given in [4] are given for Thin Plate Splines (TPS) smoother. Suppose the unknown function m from $\mathbb{R}^d \to \mathbb{R}$ belongs to the Sobolev space $\mathcal{H}^{(v)}(\Omega) = \mathcal{H}^{(v)}$, where v is an unknown integer such that $v > d/2$ and Ω is an open subset of $\mathbb{R}^d$, TPS is a solution of the minimization problem

$$\frac{1}{n}\|Y_i - f(X_i)\|^2 + \lambda J_v^d(f),$$

see, for example, [9], where

$$J_v^d(f) = \sum_{\alpha_1 + \cdots + \alpha_d = v} \frac{v!}{\alpha_1! \cdots \alpha_d!} \int \cdots \int \left(\frac{\partial^v f}{\partial x_1^{\alpha_1} \cdots \partial x_d^{\alpha_d}}\right)^2 dx_1 \cdots dx_d.$$

The first part of the functional controls the data fitting while $J_v^d(f)$ controls the smoothness. The trade-off between these two opposite goals is ensured by the choice of the smoothing parameter λ. The main problem of TPS is that the null space of $J_v^d(f)$ consists of polynomials with maximum degree of $(v - 1)$, so its finite dimension is $M = \binom{v+d-1}{v-1}$. As $v > d/2$, the dimension of the null space increases exponentially with d. In his seminal paper, [8] presented a mathematical framework that extends TPS. Noting that the Fourier transform (denoted by $\mathcal{F}(.)$) is isometric, the smoothness penalty $J_v^d(f)$ can be replaced by its squared norm in Fourier space:

$$\int \|D^v f(t)\|^2 dt \quad \text{can be replaced by} \quad \int \|\mathcal{F}(D^v f)(\tau)\|^2 d\tau.$$

In order to solve the problem of exponential growth of the dimension of the null space of $J_v^d(.)$, Duchon introduced a weighting function to define:

$$J_{v,s}^d(f) = \int |\tau|^{2s} \|\mathcal{F}(D^v f)(\tau)\|^2 d\tau.$$

The solution of the new variational problem: $\frac{1}{n}\|Y_i - f(X_i)\|^2 + \lambda J_{v,s}^d(f)$ is now

$$g(x) = \sum_{j=1}^{M_0} \alpha_j \phi_j(x) + \sum_{i=1}^{n} \delta_i \eta_{v,s}^d(\|x - X_i\|),$$

provided that $v + s > d/2$ and $s < d/2$. The $\{\phi_j(x)\}$ are a basis of the subspace spanned by polynomial of degree $v - 1$ and

$$\eta_{v,s}^d(r) \propto \begin{cases} r^{2v+2s-d} \log(r) & d \text{ if } 2v + 2s - d \text{ is even,} \\ r^{2v+2s-d} & d \text{ otherwise} \end{cases}$$

with the same constraint as TPS that is $T\delta = 0$ with the matrix T defined as $T_{ij} = \phi_j(X_i)$. For the special case $s = 0$, Duchon splines reduce to the TPS. But if one wants to have a lower dimension for the null space of $J_{\nu,s}^d$ one has to increase s; for instance, to use a pseudo-cubic splines (with an order $\nu = 2$), one can choose $s = \frac{d-1}{2}$ as suggested by Duchon [8].

17.3 Image Denoising

An image is a matrix of pixels values, in which each pixel measures a grayscale, or a vector of color levels. Pixel are spatially defined by their positional coordinates (i, j) where $(i, j) \in \{1, \ldots, p\}^2$. In order to avoid complex notations we restrict ourselves to squared image. As an example, consider the picture of Barbara below, defined by 512×512 pixels ($p = 512$), each pixel value in $\{0, 1, \ldots, 255\}$, representing 256 gray levels. The left-hand panel of Fig. 17.1 displays the original image, and the right-hand panel shows a noisy version, which we wish to *denoise*. The numerical measure to quantify the error is the **PSNR** (Peak Signal to Noise Ratio):

$$\text{PSNR} = 10 \times \log_{10}\left(\frac{c^2}{\text{MSE}}\right),$$

where c is the maximum possible pixel value of the image and MSE is the mean squared error between the original image and the treated image. The quality of the reconstruction is given in decibel (**dB**) and a well-reconstructed image has a $PSNR$ within [30, 40].

The noisy image given Fig. 17.1 is obtained by adding a Gaussian noise to the original image. Let us denote by Y the vector of all the gray value (for all the pixels).

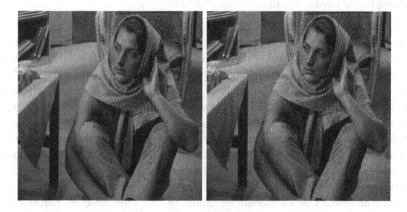

Fig. 17.1 Barbara, original image—noisy image, PSNR=28.14 dB

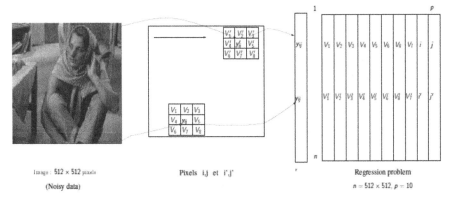

Fig. 17.2 Image analysis as a regression problem

This vector Y is of length $n = p^2$ and its kth coordinate is the gray value of pixel (i, j) where $k = (i - 1) \times p + j$. Let us denote by X the $n \times d$ matrix gathering all the explanatory variables. The kth row of length d (corresponding to pixel (i, j)) is usually made with the gray value of neighboring pixels ; the coordinates i and j can be added as explanatory variables. For instance, when one uses the eight immediate neighbors and the two coordinates, the number of columns of X is $d = 10$ (see Fig. 17.2).

Thus image analysis can be recast as a regression problem, and we can use a nonparametric smoothing approach to fit Y; thereby denoising the image. Taking a nonparametric smoothing approach brings forth two major challenges:

1. the size of the neighboring pixels can be large (one can think to use 24 or 48 neighbors) and thus the dimension d can be large. This problem is known as curse of dimensionality in the statistical literature;
2. the size $n \times n$ of the smoothing matrix S is usually very large. For the Barbara example, we have $n = 512 \times 512 = 262,144$.

Nonparametric smoothing at point Y_k can roughly be thought as doing local averages of data points $Y_{k'}$ measured at covariate locations $X_{k'}$ near a given point X_k. When the number of data points n is fixed and when the dimension d increases, the expected number of points falling into a ball of center Y_k (with a given radius) is decreasing exponentially. Thus to have a constant number of points one has to increase exponentially the size of sample n or to increase the size of neighborhood which leads to biased smoother (and the finding of optimal smoothing parameter λ^* does not alleviate this problem). Thus, the use of classical nonparametric smoother such as kernel regression is far from easy with 24 or 48 neighbors and one has to use smoother that can cope with moderate dimension d such as Duchon Splines or IBR (kernel or Duchon splines).

Concerning the size of smoothing matrix n, we only need to partition the dataset into sub-images. It is obvious that smoothing assumes that the signal varies smoothly with the values of the covariates. Thus, to avoid smoothing over edges which are numerous in an image, we need to partition the image into smaller homogeneous sub-images, leading to tremendous decrease of the sample size n (of sub-image). For instance, when the size of each sub-image is chosen equal to 30×30 the resulting 900 pixels in such sub-image can be used as a dataset for regression and one can apply the iterative regression estimate (Eq. 17.2), with Y being the level of gray of the current pixel and X being the level of gray of its eight (or more) neighbors.

In order to partition the image into homogeneous sub-images, we used the CART algorithm [2] and regression trees. The resulting partition is data dependent and the size and shape of the sub-image vary from one sub-region to another. We used the package **rpart** to define homogeneous regions (so the explanatory variables are only the position i and j and the independent variable is the gray level Y) and within each sub-image, we applied **ibr**. We did modify the **rpart** function in order to control the maximal and minimal size of each sub-region. Specifically, we decide to partition the image of Barbara into sub-regions with at least 100 pixels and at most 700 pixels. Figure 17.3 shows the evolution of the partition with **rpart**; the picture on the rightmost panel has 686 regions.

A direct application of the CART algorithm produces rectangular regions. Such a partition will readily follow horizontal and/or vertical boundaries in the image. Figure 17.4 left shows the result from applying **ibr** within each region, whereas the right side shows the difference between the denoised image and the original image.

While IBR is effective at denoising the image, the denoised image has a PSNR = 33.22dB, one can see the boundaries of the partition. To alleviate this artifact, which is due to the fact that CART uses vertical and horizontal splits, we propose to partition the image several times at various angle of rotation, and for each partition, use **ibr** for denoising. Specifically, consider the following rotation with angle α

$$i' = i \times \cos \alpha + j \times \sin \alpha$$
$$j' = -i \times \sin \alpha + j \times \cos \alpha.$$

Fig. 17.3 Partition of the image of different size using **rpart**. The *right-hand panel* displays the final partition, which is composed of 686 regions

Fig. 17.4 *Left*: Denoised image (i.e. fitted values) using kernel IBR applied to a noisy image. *Right*: the image of the residuals

Fig. 17.5 Partition of the image using rotation with angle $\alpha = 30$ (*left*) and $\alpha = -60$ (*right*)

We now apply **rpart** to the image in that new coordinate system (i', j'). While the partition will still be rectangular (in that coordinate system), we can rotate back the partition into the original coordinate system in which the partitions are now slanted. Figure 17.5 presents two examples of partitions.

We compute a denoised image by averaging, pixel-wise, the smoothers obtained for the various rotated partitions. We found empirically that using a small number of rotations leads to better smoothers than one that uses a large number of rotations. Using three rotations, with maximal size of 300 pixel for each region, we obtain the following result (Fig. 17.6).

The PSNR is 33.72 dB using kernel smoother and 33.60 dB using Duchon spline smoother.

Fig. 17.6 Denoised image (i.e. fitted values) with IBR kernel ($PSNR = 33.72$ dB) and image of the residuals (*right*)

17.4   Comparison with BM3D

BM3D algorithm, developed by [6], is the current state of the art for image denoising. This method mixes block-matching (BM) and 3D filtering. This image strategy is based on an enhanced sparse representation in transform domain. The enhancement of the sparsity is achieved by grouping similar 2D image fragments (or blocks) into 3D data arrays called groups. Then, in order to deal with the 3D groups, a collaborative filtering procedure is developed. The codes and related publications could be found at http://www.cs.tut.fi/~foi/. That algorithm requires knowledge of the standard deviation of the noise. By default, the standard value is 25. Alternatively, when the noise level is unknown, one can use an estimate of the standard deviation in BM3D. We compare our regression based denoting algorithm to BM3D, using the following settings,

- BM3D, with the true standard deviation of the noise. We expect to outperform the classical BM3D for which the standard deviation is unknown and has to be estimated
- BM3D with default standard deviation of the noise
- IBR kernel, three rotations, maximal size $= 700$
- IBR Duchon, three rotations, maximal size $= 700$

on several images with different noise levels. The Matlab code for our comparison is available upon request. This comparison is conducted for five levels of Gaussian noise: $\sigma = 5, 10, 15, 20$, and 30 and replicated ten times.

Figure 17.7 shows that as the noise increases the PSNR of the denoised image decreases. The BM3D algorithm, with the true standard deviation of the noise (in red), is better than IBR (in blue) with a difference of around 2 dB. This case is not very realistic since we never know the true noise level. When the noise level is

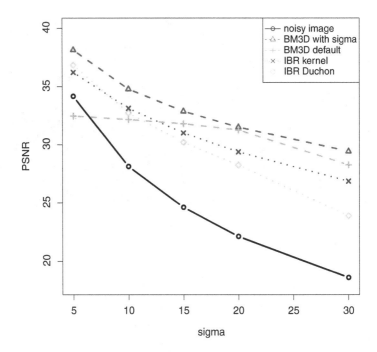

Fig. 17.7 Comparison BM3D versus IBR kernel: true noise level known (*red*) or unknown (*blue*)

unknown, the IBR procedure is better than BM3D (with standard values) whenever the noise level is low (less than 15). IBR with kernel smoother or Duchon splines smoother give similar results.

17.5 Conclusion

General statistical lore suggested that fully nonparametric regression with many covariates (more than 5) should be generally avoided. The recent realization that it is possible to design simple adaptive regression smoothers makes it now practical to smooth data in higher dimensions. While having good statistical properties, the current implementation of the IBR algorithm is limited by number of observations n.

Image denoising presents both problems simultaneously: a large number of covariates and large sample sizes. We present in this paper how to resolve both of these issues in order to apply the IBR algorithm. The initial results are promising, but the simulation study presented in this paper is done only for illustration purpose and somewhat limited. Recall that the noise assumed here is Gaussian and with the same level in the whole image.

In practical situations, the noise is usually connected to the gray level and thus its level is different within the image see, for example, [1]. If the partitioning into sub-images is done in the right way, the level of noise can be thought to be almost the same within sub-images but different from one sub-image to another. Thus, the IBR procedure that makes the assumption of the same level of noise within sub-images can be thought to be realistic and the performance of this method has to be investigated further. The regression formulation also appears useful for image completion (inpainting) and we will investigate this topic in a future work.

References

1. Boulanger, J., Kervrann, C., Bouthemy, P., Elbau, P., Sibarita, J.B., Salamero, J.: Patch-based non-local functional for denoising fluorescence microscopy image sequences. IEEE Trans. Med. Imaging, **29**(2), 442–454 (2009)
2. Breiman, L., Freiman, J., Olshen, R., Stone, C.: Classification and Regression Trees, 4th edn. CRC Press, Boca Raton (1984)
3. Bühlmann, P., Yu, B.: Boosting with the l_2 loss: regression and classification. J. Am. Stat. Assoc. **98**, 324–339 (2003)
4. Cornillon, P.A., Hengartner, N., Matzner-Løber, E.: Recursive bias estimation for multivariate regression. arXiv:1105.3430v2 (2011)
5. Cornillon, P.A., Hengartner, N., Jegou, N., Matzner-Løber, E.: Iterative bias reduction: a comparative study. Stat. Comput. **23**, 77–791 (2012)
6. Dabov, K., Foi, A., Katkovnik, V., Egiazarian, K.: Image denoising by sparse 3d transform-domain collaborative filtering. IEEE Trans. Image Process. **16**(8), 2080–2095 (2007)
7. Di Marzio, M., Taylor, C.: On boosting kernel regression. J. Stat. Plan. Infer. **138**, 2483–2498 (2008)
8. Duchon, J.: Splines minimizing rotation-invariant semi-norms in Sobolev spaces. Lecture Notes in Mathematics. **571**, 85–100 (1977)
9. Gu, C.: Smoothing Spline ANOVA Models. Springer, Berlin (2002)
10. Hansen, M., Yu, B.: Model selection and minimal description length principle. J. Am. Stat. Assoc. **96**, 746–774 (2001)
11. Milanfar, P.: A tour of modern image filtering. IEEE Signal Process. Mag. **30**, 106–128 (2013)

Chapter 18
A Nonparametric Causality Test: Detection of Direct Causal Effects in Multivariate Systems Using Corrected Partial Transfer Entropy

Angeliki Papana, Dimitris Kugiumtzis, and Catherine Kyrtsou

Abstract In a recent work we proposed the corrected transfer entropy (CTE), which reduces the bias in the estimation of transfer entropy (TE), a measure of Granger causality for bivariate time series making use of the conditional mutual information. An extension of TE to account for the presence of other time series is the partial TE (PTE). Here, we propose the correction of PTE, termed Corrected PTE (CPTE), in a similar way to CTE: time shifted surrogates are used in order to quantify and correct the bias, and the estimation of the involved entropies of high-dimensional variables is made with the method of k-nearest neighbors. CPTE is evaluated on coupled stochastic systems with both linear and nonlinear interactions. Finally, we apply CPTE to economic data and investigate whether we can detect the direct causal effects among economic variables.

Keywords Direct causality • Transfer entropy • Multivariate coupled systems

A. Papana (✉)
University of Macedonia, Thessaloniki, Greece
e-mail: angeliki.papana@gmail.com

D. Kugiumtzis
Aristotle University of Thessaloniki, Thessaloniki, Greece
e-mail: dkugiu@gen.auth.gr

C. Kyrtsou
University of Macedonia, Thessaloniki, Greece

BETA, University of Strasbourg, Strasbourg, France

Economix, University of Paris 10, Nanterre, France

ISC-Paris, Ile-de-France, Paris, France
e-mail: ckyrtsou@uom.gr

© Springer Science+Business Media New York 2014   197
M.G. Akritas et al. (eds.), *Topics in Nonparametric Statistics*, Springer Proceedings
in Mathematics & Statistics 74, DOI 10.1007/978-1-4939-0569-0__18

18.1 Introduction

The leading concept of Granger causality has been widely used to study the dynamic relationships between economic time series [4]. In practice, only a subset of the variables of the original multivariate system may be observed and omission of important variables could lead to spurious causalities between the variables. Therefore, the problem of spurious causality is addressed. Moreover, for a better understanding of the causal structure of a multivariate system it is important to study and discriminate between the direct and indirect causal effects.

Transfer entropy (TE) is an information theoretic measure that quantifies the statistical dependence of two variables (or subsystems) evolving in time. Although TE is able to distinguish effectively causal relationships and asymmetry in the interaction of two variables, it does not distinguish between direct and indirect relationships in the presence of other variables. Partial transfer entropy (PTE) is an extension of TE conditioning on the ensemble of the rest of the variables and it can detect the direct causal effects [20]. As reported in [13], using the nearest neighbor estimate, PTE can effectively detect direct coupling even in moderately high dimensions. The corrected transfer entropy (CTE) was proposed as a correction to the TE [12], aiming at reducing the estimation bias of TE. For its estimation, instead of making a formal surrogate data test, the surrogates were used within the estimation procedure of the measure, and the CTE was estimated based on correlation sums.

We introduce here the corrected partial transfer entropy (CPTE) that combines PTE and CTE, which reduces the bias in the estimation of TE, so that TE goes to the zero level when there is no causal effect. Similarly to CTE, the surrogates are used within the estimation procedure of CPTE, instead of performing a significant test for PTE. Further, for the estimation of CPTE, the nearest neighbor estimate is implemented since it has been shown to be robust to the time series length and to its free parameter (number of neighbors) and efficient in high dimensional data (e.g., see [21]).

The paper is organized as follows. In Sect. 18.2, the information causality measures, transfer entropy and partial transfer entropy are introduced and the suggested measure, corrected partial transfer entropy (CPTE) is presented. In Sect. 18.3, CPTE is evaluated on a simulation study using coupled stochastic systems with linear and nonlinear causal effects. As an example of a real application, the direct causal effects among economic variables are investigated in Sect. 18.4. Finally, in Sect. 18.5, the results from the simulation study and the application are discussed, while the usefulness and the limitations of the nonparametric causality test are addressed.

18.2   Methodology

In this section, we introduce the information causality measures transfer entropy (TE) and partial transfer entropy (PTE), and define the corrected partial transfer entropy (CPTE), a measure able to detect direct causal effects in multivariate systems. Transfer entropy (TE) is a nonlinear measure that quantifies the amount of information explained in Y at h time steps ahead from the state of X accounting for the concurrent state of Y [19]. Let x_t, y_t be two time series and $\mathbf{x}_t = (x_t, x_{t-\tau}, \ldots, x_{t-(m-1)\tau})'$ and $\mathbf{y}_t = (y_t, y_{t-\tau}, \ldots, y_{t-(m-1)\tau})'$, the reconstructed vectors of the state space of each system, where τ is the delay time and m is the embedding dimension. TE from X to Y is defined as

$$\begin{aligned} \text{TE}_{X \to Y} &= -H(y_{t+h}|\mathbf{x}_t, \mathbf{y}_t) + H(y_{t+h}|\mathbf{y}_t) \\ &= -H(y_{t+h}, \mathbf{x}_t, \mathbf{y}_t) + H(\mathbf{x}_t, \mathbf{y}_t) + H(y_{t+h}, \mathbf{y}_t) - H(\mathbf{y}_t), \quad (18.1) \end{aligned}$$

where $H(x)$ is the Shannon entropy of the variable X. For a discrete variable X, the Shannon entropy is defined as $H(X) = -\sum p(x_i) \log p(x_i)$, where $p(x_i)$ is the probability mass function of the outcome x_i, typically estimated by the relative frequency of x_i. The partial transfer entropy (PTE) is the extension of TE accounting for the causal effect on the response Y by the other observed variables of a multivariate system besides the driving X, let us denote them Z. PTE is defined as

$$\text{PTE}_{X \to Y|Z} = -H(y_{t+h}|\mathbf{x}_t, \mathbf{y}_t, \mathbf{z}_t) + H(y_{t+h}|\mathbf{y}_t, \mathbf{z}_t). \qquad (18.2)$$

where $\mathbf{z}_t$ is the stacked vector of the reconstructed points for the variables in Z.

The information measure PTE is more general than partial correlation since it is not restricted to linear inter-dependence and relates presence and past (vectors $\mathbf{x}_t, \mathbf{y}_t, \mathbf{z}_t$) with future ($y_{t+h}$). Following the definition of Shannon entropy for discrete variables, one would discretize the data of X, Y, and Z first, but such binning estimate is inappropriate for high dimensional variables ($m > 1$). Instead we consider here the estimate of nearest neighbors. The joint and marginal densities are approximated at each point using the k-nearest neighbors and their distances from the point (for details see [6]). k-nearest neighbor estimate is found to be very robust to time series length, insensitive to its free parameter k and particularly useful for high dimensional data [11, 21].

Asymptotic properties for TE and PTE are mainly known for their binning estimate, which stem from the asymptotic properties of the estimates of entropy and mutual information for discrete variables (e.g., see [5, 10, 17]). Thus parametric significance testing for TE and PTE is possible assuming the binning estimate, but it was found to be less accurate than resampling testing making use of appropriate surrogates [7]. The nearest neighbor estimates of TE and PTE do not have parametric approximate distributions, and we employ resampling techniques in this study.

Theoretically, both PTE and TE should be zero when there is no driving-response effect ($X \rightarrow Y$). However, any entropy estimate gives positive TE and PTE at a level depending on the system, the embedding parameters and the estimation method. We introduce the Corrected Partial Transfer Entropy (CPTE), designed to give zero values in case of no causal effects and positive values otherwise. In order to define $\text{CPTE}_{X \rightarrow Y|Z}$, we compute M surrogate PTE values by randomizing the driving time series X using time shifted surrogates [15]. These M values form the null distribution of PTE for a significance test. We denote by q_0 the PTE value on the original set of time series and $q(1 - \alpha)$ the $(1 - \alpha)$-percentile value from the M surrogate PTE values, where α corresponds to the significance level for an one-sided test. The $\text{CPTE}_{X \rightarrow Y|Z}$ is defined as follows:

$$
\begin{aligned}
\text{CPTE}_{X \rightarrow Y|Z} &= 0, & \text{if } q_0 < q(1 - \alpha) \\
&= q_0 - q(1 - \alpha), & \text{if } q_0 \geq q(1 - \alpha)
\end{aligned}
\tag{18.3}
$$

In essence, we correct for the bias given by $q(1 - \alpha)$ and either obtain a positive value if the null hypothesis of direct causal effect is rejected or obtain a zero value if CPTE is found statistically insignificant.

18.3   Evaluation of CPTE on Simulated Systems

CPTE is evaluated on Monte Carlo simulations on different multivariate stochastic coupled systems with linear and nonlinear causal effects. In this section, we present the simulation systems we used and display the results from the simulation study.

18.3.1   Simulation Setup

CPTE is computed on 100 realizations of the following coupled systems, for all pairs of variables conditioned on the rest of the variables and for all directions.

1. A VAR(1) model with three variables, where X_1 drives X_2 and X_2 drives X_3

$$
\begin{aligned}
x_{1,t} &= \theta_t \\
x_{2,t} &= x_{1,t-1} + \eta_t \\
x_{3,t} &= 0.5x_{3,t-1} + x_{2,t-1} + \epsilon_t,
\end{aligned}
$$

   where θ_t, η_t, ϵ_t are Gaussian white noise with zero mean, diagonal covariance matrix, and standard deviations 1, 0.2, and 0.3, respectively.

2. A VAR(5) model with four variables, where X_1 drives X_3, X_2 drives X_1, X_2 drives X_3, and X_4 drives X_2 [22, Eq. 12]

$$x_{1,t} = 0.8x_{1,t-1} + 0.65x_{2,t-4} + \epsilon_{1,t}$$

$$x_{2,t} = 0.6x_{2,t-1} + 0.6x_{4,t-5} + \epsilon_{2,t}$$

$$x_{3,t} = 0.5x_{3,t-3} - 0.6x_{1,t-1} + 0.4x_{2,t-4} + \epsilon_{3,t}$$

$$x_{4,t} = 1.2x_{4,t-1} - 0.7x_{4,t-2} + \epsilon_{4,t}$$

3. A VAR(4) model of variables, where X_1 drives X_2, X_1 drives X_4, X_2 drives X_4, X_4 drives X_5, X_5 drives X_1, X_5 drives X_2, X_5 drives X_3 [18]

$$x_{1,t} = 0.4x_{1,t-1} - 0.5x_{1,t-2} + 0.4x_{5,t-1} + \epsilon_{1,t}$$

$$x_{2,t} = 0.4x_{2,t-1} - 0.3x_{1,t-4} + 0.4x_{5,t-2} + \epsilon_{2,t}$$

$$x_{3,t} = 0.5x_{3,t-1} - 0.7x_{3,t-2} - 0.3x_{5,t-3} + \epsilon_{3,t}$$

$$x_{4,t} = 0.8x_{4,t-3} + 0.4x_{1,t-2} + 0.3x_{2,t-3} + \epsilon_{4,t}$$

$$x_{5,t} = 0.7x_{5,t-1} - 0.5x_{5,t-2} - 0.4x_{4,t-1} + \epsilon_{5,t}$$

4. A coupled system of three variables with linear and nonlinear causal effects, where X_1 drives X_2, X_2 drives X_3, and X_1 drives X_3 [3, Model 7]

$$x_{1,t} = 3.4x_{1,t-1}(1 - x_{1,t-1})^2 \exp{-x_{1,t-1}^2} + 0.4\epsilon_{1,t}$$

$$x_{2,t} = 3.4x_{2,t-1}(1 - x_{2,t-1})^2 \exp{-x_{2,t-1}^2} + 0.5x_{1,t-1}x_{2,t-1} + 0.4\epsilon_{2,t}$$

$$x_{3,t} = 3.4x_{3,t-1}(1 - x_{3,t-1})^2 \exp{-x_{3,t-1}^2} + 0.3x_{2,t-1} + 0.5x_{1,t-1}^2 + 0.4\epsilon_{3,t}$$

The three first simulation systems are stochastic systems with only linear causal effects, while the fourth one has both linear and nonlinear causal effects. For all simulations systems, the time step h for the estimation of CPTE is set to one (as originally defined for TE in [19]) or m. The embedding dimension m is adapted to the system complexity, the delay time τ is set to one, and we use $\alpha = 0.05$. The number of neighbors k is set to 10 and we note that the choice of k has been found not to be crucial in the implementation of TE or PTE, e.g., see [6, 11, 13]. We consider the time series lengths $n = 512$ and 2,048, in order to examine the performance of the measure for both short and large time series length.

18.3.2   Results from Simulation Study

In order to evaluate the performance of CPTE, we display the percentages of rejection of the null hypothesis of no causal effect from the 100 realizations of the coupled systems.

For the first simulation system, if we set $h = 1$ and $m = 1$, the percentages of statistically significant CPTE at the directions of direct causal effects $X_1 \rightarrow X_2$

Table 18.1 Percentages of statistically significant CPTE for system 1, $h = 1$, $m = 1x$

	$X_1 \to X_2$	$X_2 \to X_1$	$X_2 \to X_3$	$X_3 \to X_2$	$X_1 \to X_3$	$X_3 \to X_1$
$n = 512$	100	6	100	2	6	5
$n = 2,048$	100	4	100	11	5	4

Table 18.2 Percentage of statistically significant CPTE for system 2, $h = 1$, $m = 5$

	$X_1 \to X_2$	$X_2 \to X_1$	$X_1 \to X_3$	$X_3 \to X_1$	$X_1 \to X_4$	$X_4 \to X_1$
$n = 512$	0	100	100	0	0	2
$n = 2,048$	0	100	100	1	0	6
	$X_2 \to X_3$	$X_3 \to X_2$	$X_2 \to X_4$	$X_4 \to X_2$	$X_3 \to X_4$	$X_4 \to X_3$
$n = 512$	22	0	4	100	0	7
$n = 2,048$	62	1	2	100	0	5

and $X_2 \to X_3$ are 100 %, while for the other directions of no causal effects the percentages vary from 2 % to 11 % (see Table 18.1). The choice $h = 1$ and $m = 1$ is favorably suited for this system and only direct causal effects are found significant. For different h or m values, indirect effects are detected by CPTE. For example, if we set $h = 1$ and $m = 2$, the indirect causal effect $X_1 \to X_3$ is detected by CPTE. In this case however, this effect is indeed direct if two time lags are considered. The expression of x_3 after substituting x_2 becomes: $x_{3,t} = 0.5x_{3,t-1} + x_{1,t-2} + \epsilon_t + \eta_{t-1}$. The same holds for $h = 2$ and $m = 1$, and here the direct causal effect $X_1 \to X_2$ cannot be detected as the expression of $x_{2,t}$ for two steps ahead is $x_{2,t} = \theta_{t-1} + \eta_t$.

Concerning the second system, the largest lag in the equations is 5, and therefore by setting $h = 1$ and $m = 5$, CPTE correctly detects the direct causal effects $X_1 \to X_3$, $X_2 \to X_1$, and $X_4 \to X_2$. For the true direct effect $X_2 \to X_3$ being under-valued in the system, the percentages of significant CPTE values increase with n, indicating that larger time series lengths are required to detect this interaction (see Table 18.2). By increasing h, indirect effects become statistically significant, e.g. for $h = 5$, CPTE correctly detects again all the direct interactions, even for small time series lengths, but it also indicates the indirect driving of X_4 to X_1 (with 50 % percentage for $n = 512$, and 100 % for $n = 2,048$) and of X_4 to X_3 (35 % for $n = 512$, 74 % for $n = 2,048$).

The third simulation system is on 5 variables and the largest lag is 4, so we set $m = 4$. For $h = 1$, CPTE correctly detects all the direct causal effects with a confidence increasing with n, e.g. the percentage of detection changes from 34 % for $n = 512$ to 96 % for $n = 2,048$ for the weakest direct causal effect $X_2 \to X_4$. However, for larger n, CPTE also indicates the indirect driving of $X_5 \to X_4$ with percentage 52 % (see Table 18.3). For $h = 4$, the performance of CPTE worsens and it fails to detect some direct causal effects. For example, the percentages of significant CPTE values at the direction $X_1 \to X_4$ are 11 % and 24 % for $n = 512$ and 2,048, respectively. For other couplings, the improvement of the detection from $n = 512$ to $n = 2,048$ is larger: 17 % to 53 % for $X_2 \to X_4$, 18 % to 47 % for $X_5 \to X_2$, and 45 % to 98 % for $X_4 \to X_5$.

Table 18.3 Percentage of statistically significant CPTE for system 3, $h = 1, m = 4$

	$X_1 \rightarrow X_2$	$X_2 \rightarrow X_1$	$X_1 \rightarrow X_3$	$X_3 \rightarrow X_1$	$X_1 \rightarrow X_4$	$X_4 \rightarrow X_1$	$X_1 \rightarrow X_5$
$n = 512$	91	2	6	4	68	3	7
$n = 2,048$	100	2	13	8	100	2	12
	$X_5 \rightarrow X_1$	$X_2 \rightarrow X_3$	$X_3 \rightarrow X_2$	$X_2 \rightarrow X_4$	$X_4 \rightarrow X_2$	$X_2 \rightarrow X_5$	$X_5 \rightarrow X_2$
$n = 512$	100	8	5	34	10	9	100
$n = 2,048$	100	13	8	96	3	7	100
	$X_3 \rightarrow X_4$	$X_4 \rightarrow X_3$	$X_3 \rightarrow X_5$	$X_5 \rightarrow X_3$	$X_4 \rightarrow X_5$	$X_5 \rightarrow X_4$	
$n = 512$	5	5	5	71	100	29	
$n = 2,048$	8	4	6	100	100	52	

Table 18.4 Percentages of statistically significant CPTE values of system 4, for $h = 1, 2, m = 2$, $\tau = 1, k = 10$, and $n = 512, 2,048$, conditioned on the third variables, respectively

	$X_1 \rightarrow X_2$	$X_2 \rightarrow X_1$	$X_2 \rightarrow X_3$	$X_3 \rightarrow X_2$	$X_1 \rightarrow X_3$	$X_3 \rightarrow X_1$
$n = 512$	98	11	88	3	95	9
$n = 2,048$	100	7	100	5	100	10

The last simulation system involves linear interactions ($X_2 \rightarrow X_3$) and nonlinear interactions ($X_1 \rightarrow X_2$ and $X_1 \rightarrow X_3$), all at lag one. For $h = 1$ and $m = 2$, CPTE correctly detects these causal effects for both small and large time series lengths, while the percentage of detection remains low at the absence of coupling, as shown in Table 18.4. Again, if h is larger than 1, false detections are observed. However, increasing n enhances the performance of CPTE, and for $h = 2$ and $n = 4,096$ the percentage of significant CPTE for $X_1 \rightarrow X_2$, $X_2 \rightarrow X_3$, and $X_1 \rightarrow X_3$ are 97 %, 100 %, and 77 %, respectively. Therefore, the effect of the selection of the free parameters h and m on CPTE gets larger for shorter time series.

18.4 Application on Economic Data

As a real application, we investigate the causal effects among economic time series. Specifically, the goal of this section is to investigate the impact of monetary policy into financial uncertainty and the long-term rate by taking the direct effects of this relationship into account. The data are daily measurements from $05/01/2007$ up to $18/5/2012$. They consist of the 3-month Treasury Bill returns as a monetary policy tool, denoted as X_1, the 10-year Treasury Note to represent long-term behavior, denoted as X_2, and the option-implied expected volatility on the $S\&P500$ returns index (VIX), $X3$, in order to take financial uncertainty into consideration.

In similar studies instead of using the 3-month TBill, the changes in monetary policy are mirrored in the evolution of the Fed Funds which is directly controlled by FED. However, as it is pointed out in [1, 8], the 3-month TBill rate can adequately reflect the Fed Funds movements.

An in-depth investigation of the interrelations among the three variables starts by estimating CPTE for all pairs of variables conditioned on the third variable. In the aim to smooth away any linear interdependence from the returns series the CPTE is applied on the VAR filtered variables. As it is shown in [2], information theoretic quantities, such as transfer entropy, perform better when VAR residuals are used. CPTE indicates the nonlinear driving of X_1 on X_2 (CPTE$_{X_1 \to X_2}$ = 0.0024) for $h = 1$, $m = 1$, $\tau = 1$, and $k = 10$. Regarding the "stability" of the results, it is expected to be lost by increasing the embedding dimension m. Clearly, CPTE for larger m values does not indicate any causal effect.

In order to further analyze the directions of those causal effects, PTE values from the VAR filtered returns are also calculated. The statistical significance of PTE is assessed with a surrogate data test. The respective p-values of the two-sided surrogate test are obtained with means of shifted surrogates. If the original PTE value is on the tail of the empirical distribution of the PTE surrogate value, then the "no-causal effects" hypothesis is rejected. It is worth noticing that the two-sided surrogate test for PTE indicates the same causal effects as CPTE, revealing that $X_1 \to X_2$ (p-value = 0.03). The corresponding PTE values for this direction of the causality are much larger compared with the rest of relationships.

18.5 Conclusions

Corrected Partial Transfer Entropy (CPTE) is a nonparametric causality measure able to detect only the direct causal effects among the components (variables) of a multivariate system. CPTE is defined exploiting the concept of surrogate data in order to reduce the bias in Partial Transfer Entropy (PTE), giving zero values in case of no causal effects and otherwise positive values.

CPTE correctly detected the direct causal effects for all tested stochastic simulation systems, but only for the suitable selection of the free parameters. CPTE is sensitive to the selection of the free parameters h and m, especially for short time series. The selection of the step ahead $h = 1$ turns out to be more appropriate than $h = m$ at all cases. The suitable selection of the free parameters seems to be crucial at most cases in order to avoid spurious detections of causal effects. The more complicated a system is, the larger the time series are needed.

In the real application, CPTE indicated the direct driving of the 3-month TBill returns on the 10-year TNote returns, without, however, excluding the presence of indirect dependencies among these interest rate variables and the VIX. Determining the 3-month TBill as the "node" variable, of our 3-dimensional system, highlights the interest in examining its underlying dynamics jointly with the transmission mechanisms of monetary policy. Although the transfer entropy (TE) method has been recently applied in financial data, the partial transfer entropy is a relatively new technique in this field. TE is estimated on the returns of the economic variables (log-returns) and does not rely upon cointegration aspects (e.g., see [9, 14, 16]). On the basis of the well-documented long-term comovement between the 3-month

TBill and the 10-year TNote, the impact of non-stationarity on the performance of the above tests is an important issue meriting further investigation. This point reveals new insights about the informational content of Granger-causality type tests. The results from real data should be handled with care due to their high degree of sensitivity to the specific properties of the under-study variables.

Acknowledgements The research project is implemented within the framework of the Action "Supporting Postdoctoral Researchers" of the Operational Program "Education and Lifelong Learning" (Action's Beneficiary: General Secretariat for Research and Technology), and is co-financed by the European Social Fund (ESF) and the Greek State.

References

1. Garfinkelm, M.R., Thornton, M.R.: The information content of the federal funds rate: Is it unique? J. Money Credit Bank. **27**, 838–847 (1995)
2. Gomez-Herrero, G.: Brain connectivity Analysis with EEG. PhD thesis, Tampere University of Technology (2010)
3. Gourévitch, B., Le Bouquin-Jeannés, R., Faucon, G.: Linear and nonlinear causality between signals: Methods, examples and neurophysiological applications. Biol. Cybern. **95**, 349–369 (2006)
4. Granger, J.: Investigating causal relations by econometric models and cross-spectral methods. Acta Phys. Pol. B **37**, 424–438 (1969)
5. Grassberger, P.: Finite sample corrections to entropy and dimension estimates. Phys. Lett. A **128**(6, 7), 369–373 (1988)
6. Kraskov, A., Stögbauer, H., Grassberger, P.: Estimating mutual information. Phys. Rev. E **69**(6), 066138 (2004)
7. Kugiumtzis, D.: Partial transfer entropy on rank vectors. Eur. Phys. J. Spec. Top. **222**, 401–420 (2013)
8. Kyrtsou, C., Vorlow, C.: Modelling non-linear comovements between time series. J. Macroecon. **31**(1), 200–211 (2009)
9. Marschinski, M., Kantz, H.: Analysing the information flow between financial time series. an improved estimator for transfer entropy. Eur. Phys. J. B **30**, 275–281 (2002)
10. Miller, G.A.: Note on the Bias of Information Estimates. The Free Press, Monticello (1955)
11. Papana, A., Kugiumtzis, D.: Evaluation of mutual information estimators for time series. Int. J. Bifurcat. Chaos **19**(12), 4197–4215 (2009)
12. Papana, A., Kugiumtzis, D., Larsson, P.G.: Reducing the bias of causality measures. Phys. Rev. E **83**(3), 036207 (2011)
13. Papana, A., Kugiumtzis, D., Larsson, P.G.: Detection of direct causal effects and application to epileptic electroencephalogram analysis. Int. J. Bifurcat. Chaos **22**(9), 1250222 (2012)
14. Peter, F.J.: Where is the market? Three econometric approaches to measure contributions to price discovery. PhD thesis, Eberhard-Karls-Universität Tübingen (2011)
15. Quian Quiroga, R., Kraskov, A., Kreuz, T., Grassberger, P.: Performance of different synchronization measures in real data: A case study on electroencephalographic signals. Phys. Rev. E **65**(4), 041903 (2002)
16. Reddy, Y.V., Sebastin, A.: Interaction between forex and stock markets in india: An entropy approach. In: VIKALPA, vol. 33, No. 4 (2008)
17. Roulston, M.S.: Estimating the errors on measured entropy and mutual information. Physica D **125**, 285–294 (1999)

18. Schelter, B., Winterhalder, M., Hellwig, B., Guschlbauer, B., Lucking, C.H., Timmer, J.: Direct or indirect? graphical models for neural oscillators. J. Physiol. **99**, 37–46 (2006)
19. Schreiber, T.: Measuring information transfer. Phys. Rev. Lett. **85**(2), 461–464 (2000)
20. Vakorin, V.A., Krakovska, O.A., McIntosh, A.R.: Confounding effects of indirect connections on causality estimation. J. Neurosci. Methods **184**, 152–160 (2009)
21. Vlachos, I., Kugiumtzis, D.: Non-uniform state space reconstruction and coupling detection. Phys. Rev. E **82**, 016207 (2010)
22. Winterhalder, M., Schelter, B., Hesse, W., Schwab, K., Leistritz, L., Klan, D., Bauer, R., Timmer, J., Witte, H.: Comparison of linear signal processing techniques to infer directed interactions in multivariate neural systems. Signal Process. **85**, 2137–2160 (2005)

Chapter 19
A High Performance Biomarker Detection Method for Exhaled Breath Mass Spectrometry Data

Ariadni Papana Dagiasis, Yuping Wu, Raed A. Dweik, and David van Duin

Abstract Selected-ion flow-tube mass spectrometry, SIFT-MS, technology seems nowadays very promising to be utilized for the discovery and profiling of biomarkers such as volatile compounds, trace gases, and proteins from biological and clinical samples. A high performance biomarker detection method for identifying biomarkers across experimental groups is proposed for the analysis of SIFT-MS mass spectrometry data. Analysis of mass spectrometry data is often complex due to experimental design. Although several methods have been proposed for the identification of biomarkers from mass spectrometry data, there has been only a handful of methods for SIFT-MS data. Our detection method entails a three-step process that facilitates a comprehensive screening of the mass spectrometry data. First, raw mass spectrometry data are pre-processed to capture true biological signal. Second, the pre-processed data are screened via a random-forest-based screening tool. Finally, a visualization tool is complementing the findings from the previous step. In this paper, we present two applications of our method; a control-asthma case study and an H1N1 Flumist time-course case study.

Keywords Biomarker • Random forest • Mass spectrometry • SIFT-MS

A. Papana Dagiasis (✉) • Y. Wu
Department of Mathematics, Cleveland State University, 2121 Euclid Avenue, Cleveland, OH 44115, USA
e-mail: a.papanadagiasis@csuohio.edu; Y.WU88@csuohio.edu

R.A. Dweik
Department of Pulmonary, Allergy and Critical Care Medicine and Pathobiology/Lerner Research Institute, Cleveland Clinic, Cleveland, OH, USA

D. van Duin
Department of Infectious Disease, Cleveland Clinic, Cleveland, OH, USA

© Springer Science+Business Media New York 2014

M.G. Akritas et al. (eds.), *Topics in Nonparametric Statistics*, Springer Proceedings in Mathematics & Statistics 74, DOI 10.1007/978-1-4939-0569-0_19

19.1 Introduction

Biomarker identification is a common task of several researchers nowadays especially in the science of "omics" such as genomics and proteomics. Selected-ion flow-tube mass spectrometry (SIFT-MS) technology seems nowadays very promising to be utilized for the discovery and profiling of biomarkers such as volatile compounds, trace gases, and proteins from biological and clinical samples. Although several methods have been proposed for the identification of biomarkers from mass spectrometry data, there has been no uniform method that extends to complex experimental settings. The overall goal of the following studies is to understand lung physiology and pathology and the pathobiology of lung diseases through the study of exhaled biomarkers. In this paper, we present a high performance method for analyzing SIFT-MS datasets via a nonparametric adaptive technique while focusing on biomarker identification associated with diagnosis and prognosis of asthma from human exhaled breath samples.

Several methods have been developed in the literature for the classification of mass-spectrometry data [1, 5, 7, 20, 21], but these methods mostly focus on the characterization of peptide or protein ions in a sample while utilizing the matrix-assisted laser ionization (MALDI-MS) or surface-enhanced laser ionization (SELDI) technology. However, the detection and quantification of trace gases and volatile compounds via the SIFT-MS technology has received minimal attention. The diagnosis of tuberculosis from above serum samples from wild badgers was studied in [17] with principal components and partial least squares discriminant analysis, while the distribution of several volatile compounds in breath was studied in [18] with the help of SIFT-MS. In this paper, we use SIFT-MS to propose a high-performance biomarker detection method for exhaled breath mass spectrometry data. This method has two key components; first, this method can be used for the analysis of SIFT-MS data as well as other mass spectrometry data and second, it is applicable to fixed-time and time-course experimental data.

19.2 Materials

Mass spectrometry (MS) is an analytical technique widely used for the determination of molecular weights, elemental compositions, and structures of unknown substances. MS is based on the ionization principle where the quantification of the mass-to-charge ratio of charged particles is key. Ionization techniques use ions created in an ion source injected into a tube.[1]

---

[1] SIFT-MS is a chemical ionization technique allowing calculations of analyte concentrations from the known reaction kinetics without the need for internal standard or calibration [14].

19.2.1   SIFT-MS Instrumentation

SIFT-MS is an MS technique used for simultaneous quantification of several trace gases and volatile compounds in biological samples such as exhaled breath, urine, cell cultures, and non-biological samples such as humid air, exhaust gases and rumen gases [14]. SIFT-MS relies on the chemical ionization of sample trace gases by selected positive ions. The selected positive ions, called the "precursor" ions, are H_3O^+, NO^+, and O_2^+. These ions are suitable for breath analysis since they react slowly with N_2, O_2, H_2O, CO_2, or Ar [12]. The neutral analyte of a sample vapor reacts with the precursor ions and forms product ions. The product ions are sorted by their mass-to-charge ratio (m/z) and a detector is used to measure their abundances.[2]

Mass spectrometry data are pairs of abundance measurements and m/z, ordered according to m/z. A mass spectrum is a graph of intensity against m/z. Figure 19.1 shows a sample mass spectrum from an asthma patient using H_3O^+ as the precursor ion. The location and abundance measurements associated with the peaks represent the significant biological information in the spectrum.

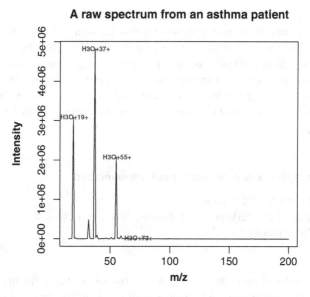

Fig. 19.1 A sample mass spectrum from an asthma patient using H_3O^+ as the precursor ion

---

[2]Absolute concentrations of trace gases and vapors in air are calculated based on the flow tube geometry, the ionic reaction time, flow rates and pressure and ion–molecule reaction rate coefficients [14].

19.2.2 Data Description

Today, SIFT-MS has been used in studies that check for urinary infection [15], cancer [16, 19], renal failure [3, 4], Helicobacter pylori infection [13], and substance abuse. SIFT-MS has also been successfully used in exhaled breath due to its non-expensive and pain-free nature [10] and therefore it can be useful for clinical diagnosis and therapeutic monitoring. In this paper, we detected biomarkers from the SIFT-MS analysis of breath samples for two studies; the H1N1 Flumist study and the Asthma study. The H1N1 2009 monovalent live intranasal vaccine was administered in nine healthy healthcare workers.[3] Each subject was sampled a total of four to six times over four to six sampling days (one sample per day). This was a repeated measures study using healthy controls. Subjects received the H1N1 vaccine on day 1, thus day 0 is baseline. In the second study, we gathered exhaled breath samples from 33 Asthma patients and compared those with 19 healthy individuals. Each subject was sampled one time only.

19.3 Methodology

A high performance biomarker detection method for identifying biomarkers across experimental groups is proposed for the analysis of SIFT-MS data. Our detection method entails a three-step process that facilitates a comprehensive screening of the data. First, raw mass spectrometry data are pre-processed to capture true biological signal. Second, the pre-processed data are screened via a random-forest-based screening tool. Finally, a visualization tool is complementing the findings from the previous step.

The high performance biomarker detection method

1. Pre-processing of MS data
2. Detection of biomarkers via a random-forest approach
3. Visualization tool

Our biomarker detection method has two key components; first, the method can be used for the analysis of SIFT-MS data as well as other MS data and second, it is applicable to fixed-time and time-dependent multi-group experimental data.

---

[3]The dose administered was 0.2 ml. All individuals had previously received trivalent seasonal 2009 vaccine (intramuscular). All subjects underwent nasal pharyngeal swab for influenza (Prodesse PCR) immediately prior to H1N1 vaccination to rule out the presence of sub-clinical influenza prior to vaccination and on day 1 after vaccination to determine viral load.

19.3.1 Data Pre-processing

All samples used in this study were analyzed by SIFT-MS with limit of detection being typically 1 ppb. The resolution of the instrument was set to 1 amu and each scan was composed of two cycles. The counts per second (cps) of each cycle to sort out unusable scans was used. The data were averaged across cycles 1 and 2. We will refer to this set of counts as the "Raw" data.

Pre-processing of the "Raw" MS data, that is selecting features of spectra that correspond to a true biological signal, is required prior to analyzing them for biomarker identification. The standard pre-processing steps are: baseline removal to remove systematic artifacts, spectra normalization to correct systematic variation due to experimental noise, peak detection and quantification, and peak alignment. In this paper, although for baseline correction various methods were tried, the removal of the baseline was not necessary due to minimum magnitude of variability. Each spectrum was normalized using the Syft Technologies normalization procedure as follows. First, the sum of counts at H_3O^+ 19, 37, 55, 73 for each scan was computed. The H_3O^+ mass counts were normalized by dividing the observed H_3O^+ mass counts for each H_3O^+ by the total H_3O^+ sum. Similarly, the NO^+ counts were normalized by dividing the observed counts for each NO^+ by the total NO^+ 30, 48 sum. The O_2^+ counts were normalized by dividing the observed O_2^+ counts by the count of O_2^+ 32. Due to the very small intensity values after normalization, all counts were multiplied by 10^7 for computational convenience. Next, a local maxima criterion for peak detection was used. A data point in a spectrum was a local peak if it was a local maximum within a window with an adaptive window size. Once peaks were detected, they were aligned across spectra to find those peaks that represent the same compound.

19.3.2 Detection of Biomarkers Using Forests

Random forests [2, 6, 8] is an ensemble learning method of computational inference that grows a collection of classification trees choosing the most popular classification. More specifically, bootstrap samples are selected from the pre-processed data and for each sample/spectrum a tree is constructed using a random sample of all the available predictors at each node. The forest classifier is constructed by aggregating every tree classifier from the bootstrap samples. For example, in the case of classification, each classifier predicts/votes the status of each spectrum. Then, the votes from all the trees are aggregated to return the final vote for the spectrum. Random forests can be run for regression or classification [8]. We used the statistical program *R project* [11] for the analysis of our datasets and the package *randomForest* [8].

Our forest-based biomarker detection method utilizes the variable importance (VIMP) values calculated from the random forest analysis of the pre-processed data.

VIMP values measure the change in prediction error on a new test case if a variable were removed from the analysis [8] and therefore large VIMP values indicate the predictiveness of a variable. VIMP values are calculated for every product ion and product ions are sorted based on the magnitude of their corresponding VIMP values. Product ions with the highest VIMP values (above zero) are marked as biomarkers. The proposed biomarker detection method is applicable under two distinct experimental settings; a fixed-time two-group setting and a multi-group time-course setting.

For the **Case-Control** setting, random forest is run in a classification mode and VIMP values are computed. Biomarkers, product ions with the highest VIMP values, are considered to be associated with a disease. For the **Time-course** setting, random forest is run in a regression mode, utilizing time as the predictor, and VIMP values are computed. Biomarkers, product ions with the highest VIMP values, reflect intensities changing over time. This approach is also applicable in the case of comparing several groups and the case of incorporating the effects of additional predictors while trying to track change over time [9]. Overall, our detection method works by first, reducing the dimensionality of the data (peaks detection) and second, by detecting significant biomarkers in the collection of spectra.

19.3.3 Visualization Tool

Finally, a visualization tool, called the VIMP plot, is utilized to complement the findings from the previous step of the biomarker detection method. The VIMP plot is a line plot that shows the VIMP values in ascending (or descending) order for each biomarker detected in the previous step. The proposed graph may be used independently of the experimental setting. In the case of classification, the VIMP plot may be used together with side-by-side boxplots and in the case of regression, the VIMP plot may be accompanied by a time-profile plot for the detected biomarkers.

19.4 Applications

Our high performance detection method was applied to various SIFT-MS exhaled breath data from various samples and diseases such as asthma, liver disorders, pulmonary hypertension, sleep apnea, and influenza infection. Here, we present results from the H1N1 Flumist and Asthma study. Figure 19.2 shows the VIMP values for the NO detected H1N1 Flumist biomarkers: NO+143,169, 77, 99, 113, 118, 85, 117, 55, 107, 19. The time profile for NO+117+ is given in Fig. 19.3. Similar results were obtained from our analysis for the H_3O and O_2 precursors. For the Asthma study, the VIMP values are given in Fig. 19.4 for the H_3O biomarkers. The side-by-side boxplots for the top five detected biomarkers are displayed in Fig. 19.5.

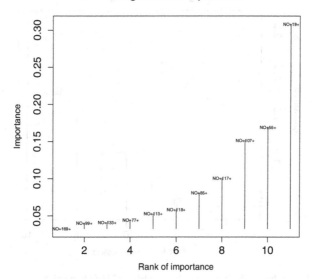

Fig. 19.2 VIMP plot for the H1N1 Flumist study. Detected biomarkers: NO+143,169, 77, 99, 113, 118, 85, 117, 55, 107, 19+

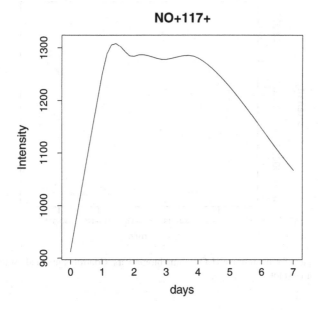

Fig. 19.3 Time profile for the detected biomarker NO+117+ for the H1N1 Flumist study

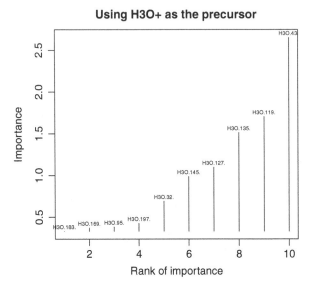

Fig. 19.4 VIMP plot for the Asthma study for the detected H_3O biomarkers

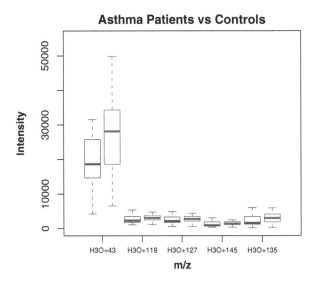

Fig. 19.5 Boxplots for the top five H_3O asthma biomarkers. Control is first and asthma is second for every pair of boxplots

19.5   Discussion

Our biomarker detection method is a high performance detection method due to its multiple-step nature, as well as, its non-parametric and data-adaptive nature. Initially, the dimensionality of the data is reduced and information about potential biomarkers of interest is gathered. Then, a nonparametric classifier, setting-adaptive, is used to detect significant biomarkers in the collection of spectra. The random-forest-based classifier has several good properties; it promotes good accuracy, it is robust to outliers and noise, it is faster than other ensemble methods, it provides internal estimates of error, strength, correlation, and variable importance, it is simple and it works efficiently for large datasets.

References

1. Barrett, J.H., Cairns, D.A.: Application of the random forest classification method to peaks detected from mass spectrometric profiles of cancer patients and controls. Stat. Appl. Genet. Mol. Biol. **7** (2008)
2. Breiman, L.: Random forests. Mach. Learn. **45**, 5–32 (2001)
3. Davies, S., Španel, P., Smith, D.: Quantitative analysis of ammonia on the breath of patients in end-stage renal failure. Kidney Int. **52**, 223–228 (1997)
4. Davies, S., Španel, P., Smith, D.: A new 'online' method to measure increased exhaled isoprene in end-stage renal failure. Nephrol. Dial. Transplant. **16**, 836–839 (2001)
5. Hand, D.J.: Breast cancer diagnosis from proteomics mass spectrometry data: a comparative evaluation. Stat. Appl. Genet. Mol. Biol. **7** (2008)
6. Hastie, T., Tibshirani, R., Friedman, J.: The Elements of Statistical Learning. Springer, New York (2001)
7. Heidema, A.G., Nagelkerke, N.: Developoing a discrimination rule between breast cancer patients and controls using proteomics mass spectrometric data: A three-step approach. Stat. Appl. Genet. Mol. Biol. **7**, (2008)
8. Liaw, A., Wiener, M.: Classification and regression by randomForest. R News **2**(3), 18–22 (2002)
9. Papana, A., Ishwaran, H.: Gene hunting with forests for multigroup time course data. Stat. Probab. Lett. **79**, 1146–1154 (2009)
10. Paschke, K.M., Mashir, A., Dweik, R.A.: Clinical applications of breath testing. Medicine Reports Ltd **2**, 56 (2010)
11. R Development Core Team: R: A language and environment for statistical computing. R Foundation for Statistical Computing, Vienna (2011). ISBN 3-900051-07-0
12. Smith, D., Pysanenko, A., Španel, P.: The quantification of carbon dioxide in humid air and exhaled breath by selected ion flow mass spectrometry. Rapid Commun. Mass Spectrom. **23**, 1419–1425 (2009)
13. Smith, D., Španel, P.: The novel selected ion flow tube approach trace gas analysis of air and breath. Rapid Commun. Mass Spectrom. **10**, 1183–1198 (1996)
14. Smith, D., Španel, P.: Selected ion flow tube mass spectrometry (SIFT-MS) for on-line trace gas analysis. Mass Spectrom. Rev. **24**, 661–700 (2005)
15. Smith, D., Španel, P., Holland, T.A., Al Singari, W., Elder, J.B.: Selected ion flow tube mass spectrometry of urine headspace. Rapid Commun. Mass Spectrom. **13**, 724–729 (1999)

16. Smith, D., Wang, T., Sule-Suso, J., Španel, P.: Quantification of acetaldehyde released by lung cancer cells *in vitro* using selected ion flow tube mass spectrometry. Rapid Commun. Mass Spectrom. **17**, 845–850 (2003)

17. Spooner, A.D., Bessant, C., Turner, C., Knobloch, H., Chambers, M.: Evaluation of a combination of SIFT-MS and multivariate data analysis for the diagnosis of *mycobacterium bovis* in wild badgers. Analyst **134**, 1922–1927 (2009)

18. Turner, C., Španel, P., Smith, D.: A longitudinal study of ammonia, acetone and propanol in the exhaled breath of 30 subjects using selected ion flow tube mass spectrometry, SIFT-MS. Physiol. Meas. **27**, 321–337 (2006)

19. Španel, P., Smith, D.: Sift studies of the reactions of H_3O^+, NO^+, and O_2^+ with some chloroalkanes and chloroalkanes. Int. J. Mass Spectrom. **184**, 175–181 (1999)

20. Valkenborg, D., Van Sanden, S., Lin, D., Kasim, A., Zhu, Q., Haldermans, P., Jansen, I., Shkedy, Z., Burzykowski, T.: A cross-validation study to select a classification procedure for clinical diagnosis based on proteomic mass spectrometry. Stat. Appl. Genet. Mol. Biol. **7** (2008)

21. Wu, B., Fishman, T., McMurray, W., Mor, G., Stone, K., Ward, D., Williams, K., Zhao, H.: Comparison of statistical methods for classification of ovarian cancer using mass spectrometry data. Bioinformatics **19**(13), 1636–1643 (2003)

Chapter 20
Local Polynomials for Variable Selection

Francesco Giordano and Maria Lucia Parrella

Abstract Nonparametric estimators are particularly affected by the *curse of dimensionality*. An interesting method has been proposed recently, the RODEO, which uses the nonparametric local linear estimator for high dimensional regression, avoiding the curse of dimensionality when the model is sparse. This method can be used for variable selection as well, but it is blind to linear dependencies. For this reason, it is suggested to use the RODEO on the residuals of a LASSO. In this paper we propose an alternative solution, based on the adaptation of the well-known asymptotic results for the local linear estimator. The proposal can be used to complete the RODEO, avoiding the necessity of filtering the data through the LASSO. Some theoretical properties and the results of a simulation study are shown.

Keywords Local polynomials • Variable selection • Nonparametric regression models

20.1 Introduction

Nonparametric analysis is often accused to be too much cumbersome from a computational point of view and too much complicated from an analytical perspective, compared with the gains obtained in terms of more general assumptions on the analysed framework. For this reason, in recent years many efforts have been done to propose nonparametric procedures which are simple to analyse and automatic to implement. The RODEO method of [1] is an example of this tendency. It has been proposed to perform nonparametric high dimensional regression for sparse models, avoiding the curse of dimensionality problem which generally affects nonparametric estimators. The theoretical results shown in their paper are innovative and promising, but some drawbacks of the computational procedure restrict its applicability. In particular, the RODEO is based on a hard-threshold variable selection procedure which isolates the nonlinear covariates from the others (linear and irrelevant). In this paper, we call *nonlinear covariates* those variables

F. Giordano (✉) • M.L. Parrella
Department of Economics and Statistics, University of Salerno, Salerno, Italy
e-mail: giordano@unisa.it; mparrella@unisa.it

© Springer Science+Business Media New York 2014

M.G. Akritas et al. (eds.), *Topics in Nonparametric Statistics*, Springer Proceedings in Mathematics & Statistics 74, DOI 10.1007/978-1-4939-0569-0_20

having a nonlinear effect on the dependent variable of the regression model, and *linear covariates* those variables having a linear effect (the remaining ones are irrelevant). The advantage of the RODEO is that the rate of convergence of the smoothing estimator depends on the number of nonlinear covariates, which is much smaller than the number of total covariates when the model is sparse. This reduces the problem of the curse of dimensionality considerably. On the other side, the drawback of the RODEO is that it fails to identify the linear dependencies, because the statistic used to test the relevance of the covariates does not distinguish the irrelevant covariates from the linear ones (it is equal to zero in both cases). Thus the method is ineffective for variable selection. To overcome this, [1] suggest to use the RODEO method on the residuals of a LASSO, but the consequences of this deserve to be further explored.

The aim of this work is to propose an alternative solution for the identification of the relevant linear covariates. It is based on well-known asymptotic results on the multivariate local polynomial estimators, derived by [3] and [2], but these results are adapted to the particular analysed framework in order to have a better performance of the variable selection procedure. Our proposal can be used as a final step of the RODEO procedure, to allow the RODEO to be applied to the original data with no necessity of using the LASSO.

Let $(X_1, Y_1), \ldots, (X_n, Y_n)$ be a set of $\mathbb{R}^{d+1}$-valued random vectors, where the Y_i are the dependent variables and the X_i are the $\mathbb{R}^d$-valued covariates of the following nonparametric regression model

$$Y_i = m(X_i) + \varepsilon_i, \qquad \varepsilon_i \sim N(0, \sigma_\varepsilon^2). \qquad (20.1)$$

Here $m(X_i) = E(Y|X_i) : \mathbb{R}^d \to \mathbb{R}$ is the multivariate conditional mean function. The errors ε_i are *i.i.d*, and they are supposed to be independent of X_i. We use the notation $X_i = (X_i(1), \ldots, X_i(d))$ to refer to the covariates and $x = (x_1, \ldots, x_d)$ to denote the target point at which we want to estimate m. We indicate with $f_X(x)$ the density function of the covariates, having support $\text{supp}(f_X) \subseteq \mathbb{R}^d$. In general, throughout this paper we use the same notation as in [1].

20.2   The Multivariate Local Linear Estimator

The Local Linear Estimator (LLE) is a nonparametric tool whose properties have been studied deeply (for example, see [2]). It corresponds to perform a locally weighted least squares fit of a linear function, being equal to

$$\arg\min_{\beta_0, \beta_1} \sum_{i=1}^{n} \left\{ Y_i - \beta_0(x) - \beta_1^T(x)(X_i - x) \right\}^2 K_H(X_i - x), \qquad (20.2)$$

where the function $K_H(u) = |H|^{-1} K(H^{-1}u)$ gives the local weights and $K(u)$ is the Kernel function, a d-variate probability density function. The $d \times d$ matrix

H represents the smoothing parameter, called the *bandwidth matrix*. We assume it is a diagonal, positive definite matrix. It controls the variance of the Kernel function and regulates the amount of local data used on each dimension, and so the local smoothness of the regression function. Denote with $\boldsymbol{\beta}(x) = (\beta_0(x),$ $\beta_1(x), \ldots, \beta_d(x))^T$ the vector of coefficients to estimate. Using the matrix notation, the solution of the minimization problem in the (20.2) can be written as $\hat{\boldsymbol{\beta}}(x; H) = (\mathbf{X}^T \mathbf{W} \mathbf{X})^{-1} \mathbf{X}^T \mathbf{W} \mathbf{Y}$, where $\hat{\boldsymbol{\beta}}(x; H)$ is the estimator of the vector $\boldsymbol{\beta}(x)$ and

$$\mathbf{X} = \begin{pmatrix} 1 & (X_1 - x)^T \\ \vdots & \vdots \\ 1 & (X_n - x)^T \end{pmatrix}, \quad \mathbf{W} = \begin{pmatrix} K_H(X_1 - x) & \ldots & 0 \\ \vdots & \ddots & \vdots \\ 0 & \ldots & K_H(X_n - x) \end{pmatrix}.$$

It is extremely difficult and inefficient to use local linear estimators in high dimension, due to the crucial effects of the bandwidth matrix. The optimal bandwidth matrix H^{opt} must take account of the trade-off between bias and variance of the LLE, and can be defined by minimizing the asymptotic mean square error

$$H^{\mathrm{opt}} = \arg \min_H \left[\mathrm{Abias}^2 \{ \hat{\beta}_j(x; H) \} + \mathrm{Avar} \{ \hat{\beta}_j(x; H) \} \right], \tag{20.3}$$

where *Abias* and *Avar* denote the asymptotic bias and variance. The appeal of the RODEO procedure is that it exploits the properties of the bandwidth matrix for local linear estimators, under the assumption that model (20.1) is a sparse model. The RODEO procedure includes a practical way to select the bandwidth matrix (see [1] for the details). Denote with $H^* = \mathrm{diag}(h_1^*, \ldots, h_d^*)$ such estimated bandwidth matrix. Its peculiarity is that the bandwidths associated with the linear and irrelevant covariates are automatically set to a high value, say h^U. As a result, the amount of usable data significantly increases. Denote with $C \subset \{1, 2, \ldots, d\}$ the set including the indexes of the nonlinear covariates. It means that $h_j^* = h^U, \forall j \in \overline{C}$. On the other side, the bandwidths associated to the nonlinear covariates are shrinked towards zero, so that $h_j^* < h^U, \forall j \in C$. Thus, the RODEO method identifies the nonlinear covariates (indexed by C), but it does not distinguish the relevant linear covariates from the irrelevant ones (all included in $\overline{C}$). The test proposed in the following section tries to solve this problem.

20.3 Using LLE's Asymptotics for Variable Selection

Let $\mathbb{D}_m(x) = \left(\mathbb{D}_m^{(1)}(x), \ldots, \mathbb{D}_m^{(d)}(x) \right)^T$ denote the gradient of $m(x)$. Note from the (20.2) that $\hat{\boldsymbol{\beta}}(x; H)$ gives an estimation of the function $m(x)$ and its gradient. In particular,

$$\hat{\beta}(x; H) = \begin{pmatrix} \hat{\beta}_0(x; H) \\ \hat{\beta}_1(x; H) \\ \vdots \\ \hat{\beta}_d(x; H) \end{pmatrix} \equiv \begin{pmatrix} \hat{m}(x; H) \\ \hat{\mathbb{D}}_m^{(1)}(x; H) \\ \vdots \\ \hat{\mathbb{D}}_m^{(d)}(x; H) \end{pmatrix}. \tag{20.4}$$

The asymptotic behaviour of such statistics has been studied by [2], who proved the joint asymptotic normality of the vector $\hat{\beta}(x; H)$ and derived the explicit asymptotic expansions for the conditional bias and the conditional covariance matrix, given observations of predictor variables. Our proposal is to use the asymptotic normal distribution of the estimator $\hat{\mathbb{D}}_m(x; H)$ in order to test the relevance of the locally linear covariates. The idea is based on the fact that the partial derivatives $\mathbb{D}_m^{(j)}(x)$ are zero for irrelevant covariates. For the moment, we consider a sequence of marginal tests instead of a unique conjoint test, but we are working on a variant of the method based on multiple testing or based on the hard-threshold technique. So, we consider the following hypotheses

$$H_0 : \mathbb{D}_m^{(j)}(x) = 0 \qquad H_1 : \mathbb{D}_m^{(j)}(x) \neq 0 \qquad \text{for } j \in \overline{C}. \tag{20.5}$$

Rejecting the hypothesis H_0, for a given j, means to state the relevance of the covariate $X(j)$ for the estimation of $m(x)$. What makes this testing procedure new is the fact that the bandwidth matrix is selected in order to allow the total power of the test to be improved. We explain why in the following lines.

It is easy to show (see the proof of Lemma 1) that the bias of the statistic $\hat{\mathbb{D}}_m^{(j)}(x; H)$ is independent from h_j if $X(j)$ is a linear covariate, that is when $\partial^2 m(x)/\partial x_j^2 = 0$. Moreover, it is independent from h_j under the alternative hypothesis H_1. So, the statistic test is always independent from h_j, under both the hypotheses, whereas its asymptotic variance is equal to

$$\text{Avar}\{\hat{\mathbb{D}}_m^{(j)}(x; H)\} = \frac{\sigma_\varepsilon^2 \rho_2}{n f_X(x)|H|h_j^2}, \tag{20.6}$$

where $\rho_2 = \int u_1^2 K^2(u) du$. So, in order to minimize the mean square error of the test statistic under the null hypothesis, we only have to minimize the variance in (20.6) by choosing a bandwidth h_j extremely large. This implies some benefits on the power of the test, since the distance among the two distributions (under the null and alternative hypotheses) increases when the variance decreases (*ceteris paribus*). However, in order to have such benefits, we must be sure that the test is performed in the class of the locally linear covariates and that the bandwidth is estimated large for all the linear and irrelevant covariates. This is guaranteed using the matrix H^* output by the RODEO method and considering the set of covariates included in $\overline{C}$.

Finally, we use the suggestion of [1] to estimate the variance functional in (20.6), so we reject the hypothesis H_0 for the covariate $X(j)$ if

$$\left|\hat{\mathbb{D}}_m^{(j)}(x; H^*)\right| > z_{1-\alpha/2}\sqrt{\hat{\sigma}_\varepsilon^2 \mathbf{e}_{j+1}^T \mathbf{B}\mathbf{B}^T \mathbf{e}_{j+1}},$$

where $\hat{\sigma}_\varepsilon^2$ is some consistent estimator of σ_ε^2 (for example, the one suggested in [1]), $\mathbf{e}_{j+1}$ is the unit vector with a one in position $j + 1$, $\mathbf{B} = (\mathbf{X}^T \mathbf{W}\mathbf{X})^{-1}\mathbf{X}^T \mathbf{W}$ and $z_{1-\alpha/2}$ is the $1 - \alpha/2$ percentile of the normal distribution, with α the size of the test.

20.4 Theoretical Results

In this section we show the consistency of our proposal. In particular, Proposition 1, states the rate of convergence of our proposed statistic-test. First, we state the following assumptions, which are also used in [1].

(A1) The bandwidth H is a diagonal and positive definite matrix.
(A2) The multivariate Kernel function K is a product kernel, based on a univariate kernel function with compact support, which is nonnegative, symmetric, and bounded; this implies that all the moments of the Kernel exist and that the odd-ordered moments of K and K^2 are zero.
(A3) The second derivatives of $m(x)$ are $\left|m_{jj}(x)\right| > 0$, for each $j = 1, \ldots, k$.
(A4) All derivatives of $m(\cdot)$ are bounded up to and including order 4.
(A5) $(h^U, \ldots, h^U) \in \mathbb{B} \subset \mathbb{R}^d$, that is $\mathbb{B}$ is a compact subset in $\mathbb{R}^d$.
(A6) The density function $f(x)$ of $(X_1, \ldots, X_d)$ is uniform on the unit cube.

Lemma 1. *Using the estimated bandwidth matrix H^*, under the assumptions (A1)–(A6), the bias and the variance of the estimator $\hat{\mathbb{D}}_m^{(j)}(x; H)$ defined in (20.4) are*

$$\text{bias}\{\hat{\mathbb{D}}_m^{(j)}(x; H^*)|X_1, \ldots, X_n\} = O_p\left(\tau_C^2\right),$$

$$\text{var}\{\hat{\mathbb{D}}_m^{(j)}(x; H^*)|X_1, \ldots, X_n\} = \frac{\sigma_\varepsilon^2 \rho_2(K)}{n|H^*|(h^U)^2}(1 + o_p(1))$$

where $j \in \overline{C}$ and $\tau_C = \max\{h_i^*, i \in C\}$.

Proof (sketch). We assume that there are k nonlinear covariates in C, $r - k$ linear covariates in A and $d - r$ irrelevant variables in the complementary set $U = \overline{A \cup C}$. To be simple, we rearrange the covariates as follows: nonlinear covariates for $j = 1, \ldots, k$, linear covariates for $j = k + 1, \ldots, r$, and irrelevant variables for $j = r + 1, \ldots, d$. Moreover, we suppose that the set of linear covariates A can be furtherly partitioned into two disjoint subsets: the covariates from $k + 1$ to $k + s$ belong to the subset A_c, which includes those linear covariates which are multiplied to other nonlinear covariates, introducing nonlinear mixed effects in model (20.1); the covariates from $k + s + 1$ to $k + r$ belong to the subset A_u, which includes those linear covariates which have a linear additive relation in model (20.1) or which are multiplied to other linear covariates. Therefore, $A = A_c \cup A_u$ and $C \cup A \cup U = \{1, \ldots, d\}$. Under such conditions, we can partition the gradient of $m(x)$ as follows:

$$\mathbb{D}_m(x) = \left(\mathbb{D}_m^C(x), \mathbb{D}_m^{A_c}(x), \mathbb{D}_m^{A_u}(x), \mathbf{0} \right)^T,$$

where $\mathbf{0}$ denotes a vector with $d - r$ zeroes, and

$$\mathbb{D}_m^C(x) = \left(\frac{\partial m(x)}{\partial x_j} \right)^T_{j \in C} \quad \mathbb{D}_m^{A_c}(x) = \left(\frac{\partial m(x)}{\partial x_j} \right)^T_{j \in A_c} \quad \mathbb{D}_m^{A_u}(x) = \left(\frac{\partial m(x)}{\partial x_j} \right)^T_{j \in A_u}.$$

Moreover, it is also useful to take account of the partial derivatives of $m(x)$ of total order 3. To this end, we define the following matrix

$$\mathbb{G}_m(x) = \begin{pmatrix} \frac{\partial^3 m(x)}{\partial x_1^3} & \frac{\partial^3 m(x)}{\partial x_1 \partial x_2^2} & \cdots & \frac{\partial^3 m(x)}{\partial x_1 \partial x_d^2} \\ \frac{\partial^3 m(x)}{\partial x_2 \partial x_1^2} & \frac{\partial^3 m(x)}{\partial x_2^3} & \cdots & \frac{\partial^3 m(x)}{\partial x_2 \partial x_d^2} \\ \vdots & \vdots & \ddots & \vdots \\ \frac{\partial^3 m(x)}{\partial x_d \partial x_1^2} & \frac{\partial^3 m(x)}{\partial x_d \partial x_2^2} & \cdots & \frac{\partial^3 m(x)}{\partial x_d^3} \end{pmatrix} = \begin{pmatrix} \mathbb{G}_m^C(x) & \mathbf{0}\,\mathbf{0}\,\mathbf{0} \\ \mathbb{G}_m^{A_c C}(x) & \mathbf{0}\,\mathbf{0}\,\mathbf{0} \\ \mathbf{0} & \mathbf{0}\,\mathbf{0}\,\mathbf{0} \\ \mathbf{0} & \mathbf{0}\,\mathbf{0}\,\mathbf{0} \end{pmatrix}, \quad (20.7)$$

where $\mathbf{0}$ denotes a vector/matrix with all zeroes. Note that the matrix $\mathbb{G}_m(x)$ is not symmetric. Note also that, for additive models, matrix $\mathbb{G}_m^{A_c C}(x)$ is null while matrix $\mathbb{G}_m^C(x)$ is diagonal. In the last case, the analysis is remarkably simplified.

Let us partition in the same way the bandwidth matrix H. Following the classic approach used in [2], we can derive the asymptotic bias and the asymptotic variance of the LLE in our specific setup, using the uniformity assumption for f_X and remembering that the estimated bandwidth matrix H^* assigns, with probability tending to one, the value h^U to the locally linear covariates $X(j)$, with $j \in \overline{C}$. The bias is

$$E\left\{ \left(\begin{pmatrix} \hat{\mathbb{D}}_m^C(x; H) \\ \hat{\mathbb{D}}_m^{A_c}(x; H) \\ \hat{\mathbb{D}}_m^{A_u}(x; H) \\ \hat{\mathbb{D}}_m^U(x; H) \end{pmatrix} - \begin{pmatrix} \mathbb{D}_m^C(x) \\ \mathbb{D}_m^{A_c}(x) \\ \mathbb{D}_m^{A_u}(x) \\ \mathbf{0} \end{pmatrix} \right) \middle| X_1, \ldots, X_n \right\} = B_m(x, H) + o_p\left(\tau_C^2 \right), \quad (20.8)$$

where, denoted with $\mathbf{1}$ a vector of ones and defined $\mu_r = \int u_1^r K(u) du$, it is

$$B_m(x, H) = \frac{1}{2} \mu_2 \begin{pmatrix} \left[\mathbb{G}_m^C(x) H_C^2 + \left(\frac{\mu_4}{3\mu_2^2} - 1 \right) \mathrm{diag}\{\mathbb{G}_m^C(x) H_C^2\} \right] \mathbf{1} \\ \mathbb{G}_m^{A_c C}(x) H_C^2 \mathbf{1} \\ \mathbf{0} \\ \mathbf{0} \end{pmatrix}.$$

On the other side, using Lemma 7.4 in [1] and the assumptions of this lemma, the conditional covariance matrix is

$$
\text{Cov}\left\{\left.\begin{pmatrix}\hat{\mathbb{D}}_m^C(x;H)\\ \hat{\mathbb{D}}_m^{A_c}(x;H)\\ \hat{\mathbb{D}}_m^{A_u}(x;H)\\ \hat{\mathbb{D}}_m^U(x;H)\end{pmatrix}\right| X_1,\dots,X_n\right\} = \frac{\sigma_\varepsilon^2 \rho_2}{n|H|}\begin{pmatrix}H_C^{-2} & 0 & 0 & 0\\ 0 & H_{A_c}^{-2} & 0 & 0\\ 0 & 0 & H_{A_u}^{-2} & 0\\ 0 & 0 & 0 & H_U^{-2}\end{pmatrix}(1+o_p(1)),
$$

where $\rho_2 = \int u_1^2 K^2(u)\,du$. The result of the lemma follows. $\qquad\square$

Proposition 1. *Under the assumptions (A1)–(A6), for all $\epsilon > 0$ and $j \in \overline{C}$, it is*

$$
\left(\hat{\mathbb{D}}_m^{(j)}(x;H^*) - \mathbb{D}_m^{(j)}(x)\right)^2 = O_p\left(n^{-4/(4+k)+\epsilon}\right). \tag{20.9}
$$

Proof. Using Theorem 5.1 in [1] and Lemma 1, the conditional squared bias is

$$
\text{bias}^2\left(\hat{\mathbb{D}}_m^{(j)}(x;H^*)|X_1,\dots,X_n\right) = O_p\left(n^{-4/(4+k)+\epsilon}\right) \quad \forall j \in \overline{C}.
$$

Using, again, Lemma 1, the conditional variance is

$$
\text{var}\{\hat{\mathbb{D}}_m^{(j)}(x;H^*)|X_1,\dots,X_n\} = \frac{\sigma_\varepsilon^2 \rho_2(K)}{n(h^U)^{d+2-k}\prod_{i=1}^k h_i^*}(1 + o_p(1)) \quad \forall j \in \overline{C}.
$$

By Theorem 5.1 in [1], it is $\text{var}\{\hat{\mathbb{D}}_m^{(j)}(x;H^*)|X_1,\dots,X_n\} = O_p\left(n^{-4/(4+k)+\epsilon}\right)$, $\forall j \in \overline{C}$. Finally, using the same arguments as in the proof of Corollary 5.2 in [1], the result follows. $\qquad\square$

Remark 1. the rate of convergence shown in Proposition 1 is faster than $n^{-2/(4+k)}$, which is the optimal rate for the first derivative estimator (see [2]). This is a consequence of the linearity conditions.

Remark 2. suppose a transform on the covariate values is such that the nonlinear dependencies are transform into linear ones. Then $k = 0$ and Proposition 1 implies that the rate of convergence is $n^{-1+\varepsilon}$, which is very close to the parametric case.

20.5 Some Results from a Simulation Study

In this section we briefly investigate the empirical performance of our proposal. We generate datasets from three different models, reported in the following table.

Model	$m(x)$	r	k
1	$5x_8^2 x_9^2$	2	2
2	$2x_{10} + 5x_8^2 x_9^2$	3	2
3	$2x_{10}x_2 + 5x_8^2 x_9^2$	4	2

Model 1 has been used by [1]. The other models are variants of the first one, with the addition of some linear mixed effects. We simulate 200 Monte Carlo replications for each model, considering different configurations of settings. The number of relevant covariates varies from $r = 2$ to $r = 4$, as shown in the table, while for all the models, the number of nonlinear covariates is $k = 2$. The remaining $d - r$ covariates are irrelevant, so they are generated independently from Y. Note that the linear, the nonlinear, and the irrelevant covariates are not sequentially sorted, but they are presented randomly in the models. Finally, all the covariates are uniformly distributed, $f_X \sim U(0, 1)$, and the errors are normally distributed, $f_\varepsilon \sim N(0, 0.5^2)$, as in [1]. In Fig. 20.1, for models 1–3, we show the percentages of rejection of the null hypotheses H_0 in (20.5), for $j \in \overline{C}$, $x = (1/2, \ldots, 1/2)$ and $d = 10$. For the nonlinear covariates in C (i.e., covariates 8 and 9), the percentages plotted show the result of the RODEO nonlinear hard-threshold test. To show the consistency of our

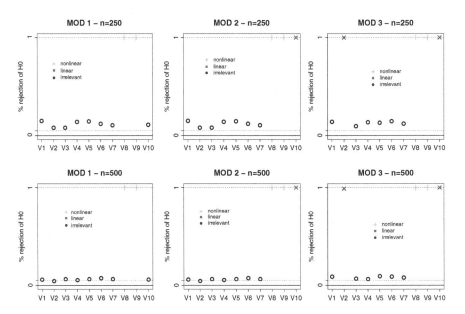

Fig. 20.1 For each covariate of models 1–3, we show the percentages of rejection for the null hypothesis H_0 of test (20.5), when $x = (1/2, \ldots, 1/2)$ and $d = 10$. To show the consistency of the test, we report $n = 250$ on the *top* and $n = 500$ on the *bottom*. The nonlinear covariates are represented with the symbol "+," the linear covariates with the symbol "×," and the irrelevant covariates with the symbol "o." The size of each single test, $\alpha = 0.05$, is shown by a *dashed line*

test, we consider two different sample sizes, $n = (250, 500)$. We report $n = 250$ on the top and $n = 500$ on the bottom of the figure. For each model, the nonlinear covariates are represented with the symbol "+," the linear covariates with the symbol "×," and the irrelevant covariates with the symbol "o." Note that each percentage plotted may represent the size of the test (for the irrelevant covariates) or the power of the test (for the relevant covariates). As desired, the realized power tends to one while the size tends to $\alpha = 0.05$, as $n \to \infty$. Both these limits are shown in the plots by a dashed line.

References

1. Lafferty, J., Wasserman, L.: RODEO: sparse, greedy nonparametric regression. Ann. Stat. **36**, 28–63 (2008)
2. Lu, Z.-Q.: Multivariate locally weighted polynomial fitting and partial derivative estimation. J. Multivar. Anal. **59**, 187–205 (1996)
3. Ruppert, D., Wand, P.: Multivariate locally weighted least squares regression. Ann. Stat. **22**, 1346–1370 (1994)

Chapter 21
On the CLT on Low Dimensional Stratified Spaces

Leif Ellingson, Harrie Hendriks, Vic Patrangenaru, and Paul San Valentin

Abstract Noncategorical observations, when regarded as points on a stratified space, lead to a nonparametric data analysis extending data analysis on manifolds. In particular, given a probability measure on a sample space with a manifold stratification, one may define the associated Fréchet function, Fréchet total variance, and Fréchet mean set. The sample counterparts of these parameters have a more nuanced asymptotic behaviors than in nonparametric data analysis on manifolds. This allows for the most inclusive data analysis known to date. Unlike the case of manifolds, Fréchet sample means on stratified spaces may stick to a lower dimensional stratum, a new dimension reduction phenomenon. The downside of stickiness is that it yields a less meaningful interpretation of the analysis. To compensate for this, an extrinsic data analysis, that is more sensitive to input data is suggested. In this paper one explores analysis of data on low dimensional stratified spaces, via simulations. An example of extrinsic analysis on phylogenetic tree data is also given.

Keywords Central limit theorem • Stratified space • Frechet means • Intrinsic means • Inrinsic means

21.1 Data Analysis on Manifolds

The question of studying random elements (nowadays called objects) was first raised by Fréchet [7]. As an example, Fréchet suggested analyzing the shape of a contour of a closed curve and the shape of an egg selected at random from a

L. Ellingson (✉)
Texas Tech University, Lubbock, TX, USA
e-mail: leif.ellingson@ttu.edu

H. Hendriks
Radboud University Nijmegen, Nijmegen, The Netherlands
e-mail: H.Hendriks@math.ru.nl

V. Patrangenaru • P.S. Valentin
Florida State University, Tallahassee, FL, USA
e-mail: vic@stat.fsu.edu; gsanvale@math.fsu.edu

© Springer Science+Business Media New York 2014
M.G. Akritas et al. (eds.), *Topics in Nonparametric Statistics*, Springer Proceedings in Mathematics & Statistics 74, DOI 10.1007/978-1-4939-0569-0__21

wire egg basket. Fréchet's approach to Analysis of Object Data (AoOD) consists of identifying an object with a point in a complete metric space (M, d). Next, given a random object X on M, he defined what we call today the *Fréchet function* on M, given by $F_d(p) = E(d^2(X, p))$. A minimizer of F_d is called a *Fréchet mean* and the minimum value of F_d is the *Fréchet total variance*. Ziezold [23] showed that the *Fréchet sample mean set* is a consistent estimator of Fréchet population mean set. Going beyond consistency, Hendriks and Landsman [10] and, independently, Patrangenaru [19] assumed the metric space from which the data is sampled is a submanifold of the numerical space, or from an abstract *manifold*, thus allowing for a description of the asymptotics of the extrinsic sample mean on a manifold embedded in a numerical space, thus providing key examples of limit behavior of Fréchet sample means. In addition, Hendriks and Landsman [10] and, independently, Patrangenaru [19] used different consistent estimators of the *population extrinsic covariance matrix* to studentize the extrinsic sample mean vector and estimate the extrinsic population mean. The large sample behavior of the Fréchet sample mean on a manifold in general was given in Bhattacharya and Patrangenaru [4]. A manifold modeled allows for one to describe on a Hilbert space $\mathbb{H}$ is locally diffeomorphic to $\mathbb{H}$, which, due to terms of a representation consistency, allows to describe the asymptotics of the Fréchet sample means in terms of a representation in the *tangent space* at the Fréchet mean, if the latter exists. A 2D manifold, immersed in $\mathbb{R}^3$ is shown in Fig. 21.1.

Two types of distances were considered on the sampling manifold M : a *chord distance* $d =_j d$, where $j : M \to L$ is an *embedding* of M into a vector space over the reals L, or an *arc distance* $d = d_g$, where g is a Riemannian tensor on M. The nonparametric methods for studying data of this sort has led to an *extrinsic* and an *intrinsic* data analysis.

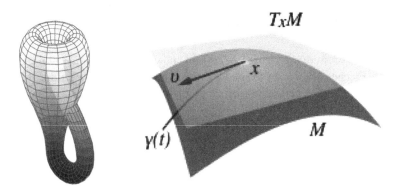

Fig. 21.1 *Left*: A 2D manifold—Klein bottle. *Right*: a tangent space at a point of a 2D manifold M

21.2   Data Analysis on One Dimensional Stratified Spaces

While analysis of Fréchet means on manifolds is now an established area, sample spaces in modern data analysis, including Kendall shape spaces in dimension 3 or higher, are not all manifolds. However, they do have a *manifold stratification*; they are *stratified spaces* (see Verona [21]).

Definition 1. A filtration by closed subsets $F_i, i = 0, 1, \ldots$ of a metric space M, such that the difference between successive members F_i and F_{i-1} of the filtration is either empty or a smooth manifold of dimension i, is called a stratification. The connected components of the difference $F_i \setminus F_{i-1}$ are the *strata* of dimension i.

The *regular part* of M is the highest dimensional stratum. At each regular point, the stratified space has a tangent space. The dimension of the stratified space is m if $M = F_m \neq F_{m-1}$, otherwise dim $M = \infty$. All other points are *singular*. The analysis of data on stratified spaces is still in very early stage. Examples of stratified sample spaces, which are not themselves manifolds include similarity shape spaces (see Kendall et al. [15]), affine shape spaces (see Groisser and Tagare [9]), and projective shape spaces (see Mardia and Patrangenaru [17]). Spaces of positive semidefinite matrices, which arise as data points in Diffusion Tensor Imaging (for example, see Schwartzman et al. [20]), and tree spaces (see Billera et al. [5]; Wang and Marron [22]) are additional examples of stratified sample spaces.

21.2.1   Phylogenetic Trees

The data that biologists use usually come from one homogenous sequence, which in the biologist's language concerns the relationship between gene trees and genes that are made from one tree. The gene sequence might be about 200 base pairs long. One of the problems that has occurred in the last 40 years is that biologists believe that the way evolution works is that there would only be one *species tree*. Different genes have different histories, so you get different gene trees. Putting them together is a statistical problem that helps the study of the evolutionary process. A phylogenetic tree with p leaves is an equivalence class based on a certain equivalence, of a DNA-based connected directed graph of species with no loops, having an unobserved *root* (common ancestor) and p observed *leaves* (current observed species of a certain family of living creatures). A tree with p leaves is a simply connected graph with a distinguished vertex, labeled o, called the root, and p vertices of degree 1, called leaves, that are labeled from 1 to p. In addition, we assume that all interior edges have positive lengths. An edge of a p-tree is called interior if it is not connected to a leaf. Now consider a tree T, with interior edges $e_1, \ldots, e_r$ of lengths $l_1, \ldots, l_r$, respectively. If T is binary, then $r = p - 2$, otherwise $r < n - 2$. The vector $(l_1, \ldots, l_r)^T$ specifies a point in the positive open *orthant* $(0, \infty)^r$. That is to say that a binary p-tree has the maximal possible number of interior edges and thus determines the largest possible dimensional orthant; in this case, the orthant is

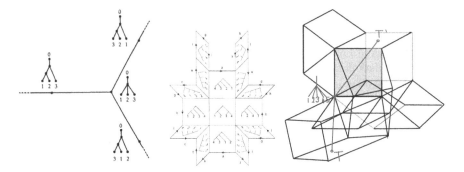

Fig. 21.2 Tree spaces T_3, T_4, T_5

$p - 2$-dimensional. The orthant corresponding to each non-binary tree appears as a boundary face of the orthants corresponding to at least three binary trees. In particular, the origin of each orthant corresponds to the (unique) tree with no interior edges, which is known as the *star tree*. The space T_p is constructed by taking one $p - 2$-dimensional orthant for each of the $(2p - 3)!!$ possible binary trees and gluing them together along their common faces. Note that tree spaces are *not* manifolds. Singularities (points where the space does not have a tangent space) are present in the tree space structure. For further detail on phylogenetic trees and the construction of the tree space, see Billera et al. [5]. Phylogenetic trees with p leaves are points on a metric space T_p that has $p - 2$ dimensional stratification. In particular, the space of trees with three leaves is $T_3 = S_3$, a 3-spider, which is the union of three line segments with a common end (see Fig. 21.2, left). For a probability measure on S_p, if none of the the "legs" of the p-spider has a dominant expected mean distance to the center of the spider, then the Fréchet mean is the star tree. This result will be stated more formally in the following section and extended to more general spaces. T_4 is a two dimensional stratified space obtained from $15 = (2 \times 4 - 3)!!$ 2D quadrants glued according to tree identification rules (see Billera et al. [5]). Interior points of these quadrants are combinatorial binary trees with four leaves, the coordinates of an interior point being given by the two interior edges of a binary tree in one of these combinatorial binary trees. Points on the boundaries of the quadrants correspond to combinatorial trees with four leaves, which are obtained from a combinatorial binary tree by shrinking one of the interior edges to zero length. Therefore a representation of T_4 as a surface with singularities can be obtained from the polyhedral surface given in Fig. 21.3 by identifying the edges labeled with the same letter. While this is a 3D pictorial representation only, in fact, as mentioned in Sect. 21.3, given that $2^4 - 4 - 2 = 10$, T_4 is embedded in $\mathbb{R}^{10}$ having the *star tree* at the origin. In this representation, the intersection of a small sphere in $\mathbb{R}^{10}$ centered at the origin with T_4 is the so-called *Petersen graph*. An edge of Petersen graph is the transverse intersection of one quadrant with a sphere, thus there are 15 edges, and a vertex is the point where one of the coordinate axes pierces the sphere, therefore there are 10 vertices (see Fig. 21.4).

Fig. 21.3 A 2D stratified space—T_4, space of trees with four leaves

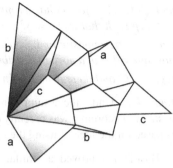

Fig. 21.4 Petersen graph

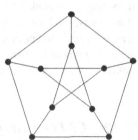

The remainder of this paper will focus on data analysis for low dimensional tree spaces and simple graphs.

21.2.2 CLTs on Trees

Assume $X_i, i = 1, \ldots, n$ are i.i.d. random objects on a spider S_p, having legs $L_a, a = 1, \ldots p$, and center C. Further, assume the intrinsic mean $\mu_{X_1, I}$ exists and the intrinsic variance is finite. Any probability measure Q on S_p decomposes uniquely as a weighted sum of probability measures Q_k on the legs L_k and an atom Q_0 at C (Hotz et al. [13]). More precisely, there are nonnegative real numbers $\{w_k\}_{k=0}^p$ summing to 1 such that, for any Borel set $A \subseteq S_p$, the measure Q takes the value

$$Q(A) = w_0 Q_0(A \cap C) + \sum_{k=1}^{p} w_k Q_k(A \cap L_k). \tag{21.1}$$

We will consider the nontrivial case when the moments $v_a = E(Q_a), a = 1, \ldots, p$ are all positive.

Theorem 1 (Hotz et al. [14]). *Assume $w_0 = 0$. (i) If there exists $a \in \overline{1, p}$ such that $w_a v_a > \sum_{b \neq a} w_b v_b$, then $\mu_{X_1, I} \in L_a$ and, for n large enough $\bar{X}_{n, I} \in L_a$*

and $\sqrt{n}(\bar{X}_{n,I} - \mu_{X_1,I})$ has asymptotically a normal distribution. (ii) If there exists $a \in \overline{1, p}$ such that $w_a v_a = \sum_{b \neq a} w_b v_b$, then, after folding the legs $L_b, b \neq a$, into one half line opposite to L_a, $\sqrt{n}(\bar{X}_{n,I})$ has asymptotically the distribution of the absolute value of a normal distribution. (iii) If $\forall a \in \overline{1, p}, w_a v_a < \sum_{b \neq a} w_b v_b$, then $\mu_{X_1,I} = C$ and there is n_0 s.t. $\forall n \geq n_0$, then $\bar{X}_{n,I} = 0$ a.s..

Remark 1. Under the assumptions of Theorem 1 (iii), we say that the sample mean is *sticky*. Theorem 1 was recently extended to C. L. T. on *open books* [13]. Its proof is based on the idea of using the so-called *Tits metrics* (see Gromov [8], p.11).

Basrak [2] proved a similar result for distributions on metric binary trees. According to Basrak [2]: "Limit theorems on (binary) trees will need minor adjustments, since on a general tree, the barycenter can split the tree into more than three subtrees." Nevertheless, asymptotically, the inductive mean will have one of the three types of behavior described in Theorem 3 in Basrak [2], meaning that the stickiness phenomenon is still present for distributions on trees.

21.2.3 CLTs on Graphs Including Cycles

The above comment from Basrak [2] does not extend to arbitrary finite, connected graphs, since graphs usually have cycles (see Fig. 21.5). The key property of metric acyclic graphs (trees) is that they are CAT(0) and that a random object on such a space with finite intrinsic variance has an intrinsic mean. There are random objects on a circle having an intrinsic mean set with at least two points. In fact the intrinsic mean set may contain an arbitrary number of points, or be even the entire circle. However, Basrak's condition for trees can be extended to the case of graphs in *some* general cases even if the graph G has cycles with positive mass on any arc of a cycle. In our approach, we have in mind graphs with the structure of path metric spaces in the sense of Gromov [8, p. 11].

Consider the case that the Fréchet mean is unique, to be denoted by μ_I. In this case, by consistency of the intrinsic sample mean set [23], any measurable intrinsic sample mean $\hat{v}_{I,n}$ selected from the sample mean set converges a.s. to μ_I, and since around μ_I the graph is homeomorphic to $\mathbb{R}$, the intrinsic sample mean behaves like in Theorem 1(i).

Assume μ_I is a vertex, and there is a small neighborhood of μ_I that looks like a spider S_p with center $C = \mu_I$ and with p legs, $p \geq 2$, such that each leg belongs to

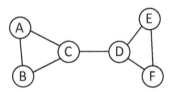

Fig. 21.5 Bicyclic graph

a different connected component of $G \backslash \mu_I$. Then the sample Fréchet mean behaves like in Theorem 1. More generally, given a unique Fréchet mean, μ_I, there is a small neighborhood of μ_I that looks like a spider S_p with center $C = \mu_I$ and legs L_a oriented such that C is the starting point of the leg.

For each $x \in G$ and leg L_a, $d(x, \xi)$ is either increasing or decreasing for $\xi \in L_a$. We define $\varepsilon_a(x) = +1$ if increasing and $\varepsilon_a(x) = -1$ if decreasing. Then μ_I is sticky if for all legs L_a

$$E(d(X, \mu_I)\varepsilon_a(X)) > 0. \tag{21.2}$$

This is easily identified as a condition that implies that μ_I is a local minimum of the Fréchet variance. The quantity $M(a) = E(d(X, \mu_I)\varepsilon_a(X))$ is called the *net moment in the direction of* L_a.

As an aside, one can consider for any $\xi \in G$ its star neigborhood S_ξ and the "derivative" in the direction of leg L_a of S_ξ, given by

$$D_{L_a} F_d = E(2d(X, \xi)\varepsilon_a(X)). \tag{21.3}$$

Remark 2. It would be useful to consider the case when the probability distribution of X is concentrated in a dense countable subset of the graph G. (For example, in the circle S^1 at rational angles from some base point. At every rational angle the Fréchet function cannot be minimal, since the antipodal point carries mass.)

Remark 3. One may note that in the case of trees, and in particular in case of data on a spider, the condition of stickiness in this section is equivalent with the condition in Theorem 1. Details are given in Hendriks and Patrangenaru [11]. Consider the situation where μ_I is a vertex, and C_a is a connected component of $G \backslash \{\mu_I\}$ that has a unique edge joined to μ_I. Then $G_a = C_a \cup \mu_I$ is a subgraph of G. For each $x \in G_a$ there is the intrinsic distance from x to μ_I, and with respect to the probability Q_a conditional to be in C_a there is an expected intrinsic distance, called the moment (of Q_a or G_a with respect to μ_I), which we label m_a. The condition for stickiness is equivalent in this case to $m_a < \sum_{b \neq a} m_a$, which in the case of a star tree (spider), neighborhood of the Fréchet mean, is the right notion to study the empirical limit behavior, equivalent to the condition in Theorem 1 (iii).

Remark 4. Certain complications arise in describing stickiness phenomena, in case when the graph includes a cycle of positive probability mass. Consider, for example, the case of a simple graph such as the circle, realizable as a one vertex, one edge graph. See Hotz and Huckemann [12]. In this case, if μ_I is a Fréchet mean, then the antipodal point, $-\mu_I$, must have probability 0. Even, if the density in a neighborhood of $-\mu_I$ is continuous, then the density at $-\mu_I$ cannot exceed $(2\pi)^{-1}$. If the density is below $(2\pi)^{-1}$, Hotz and Huckemann [12] prove a central limit theorem to normal distribution.

In the next section we consider concrete simulations that on data on simple graphs, justifying the results in this section.

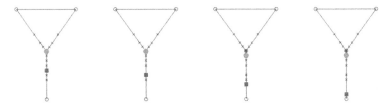

Fig. 21.6 Graph exhibiting a family of distributions having non-sticky means (*light dots*)

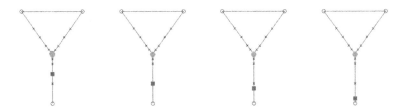

Fig. 21.7 Graph exhibiting a family of distributions having sticky means (*light dots*)

Fig. 21.8 Graph exhibiting a family of distributions having both sticky and non-sticky means (*light dots*)

21.2.4 Simulations of Sticky and Non-sticky Sample Means on Graphs

Figure 21.6 shows a graph exhibiting non-sticky intrinsic sample mean (green dot) as we pull an observation (red dot) away. On the other hand, Fig. 21.7 shows how the intrinsic sample mean sticks to its original location even after pulling an observation away from it.

A similar behavior of distributions having either sticky and regular means on bicyclic graphs is displayed in Fig. 21.8.

21.3 Computational Examples on 2D Stratified Spaces

For computational algorithms for intrinsic sample means on T_p see Owen and Provan [18]. Here we consider some computational examples to illustrate the behavior of extrinsic means for simulated data on T_4 (for the asymptotic distribution

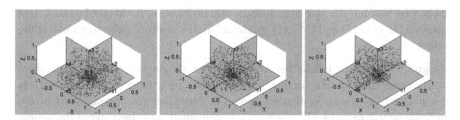

Fig. 21.9 Extrinsic mean set (*red*) and intrinsic mean (*green*) on various simulations in T_4

of the intrinsic sample means on this space, see Barden et al. [1]). For both examples, our embedding j is $j : T_4 \rightarrow \mathbb{R}^{10}$. In general T_p can be embedded in $\mathbb{R}^k, k = 2^p - p - 2$.

21.3.1   Simulations on T_4

Figure 21.9 shows data simulated on a portion of T_4, which is gray shaded in Fig. 21.2. This portion of T_4 consists of five orthants arranged radially around the origin. This portion was labeled Q_5 in Barden et al. [1]. Please note that while the region is displayed in three dimensions, this is for display purposes only. The intrinsic means, displayed in green, are computed according to the distance discussed in Billera et al. [5]. The extrinsic mean sets, displayed in red, are calculated by embedding this region into $\mathbb{R}^{10}$, as described above. For the purposes of these simulations, we assume that the data is distributed only on this region, not on the remaining portions of T_4.

In the left image, the simulated data are distributed identically over all five quadrants with equal probability of being in each quadrant, resulting in the intrinsic sample mean being the star tree and the extrinsic sample mean set consisting of five trees radially symmetric about the origin. To show the behavior of the intrinsic and extrinsic sample means, we perturbed the initial distribution by reducing the maximum possible distance in the $x1$ direction, adjusting the probability of an observation in the $(x1, x2)$ and $(x5, x1)$ quadrants accordingly.

In the middle image, the distribution has been perturbed slightly in this manner. As a result, the extrinsic mean set now consists of only two trees, but the intrinsic mean is still the star tree because there is still a substantial concentration of mass on the $(x1, x2)$ and $(x5, x1)$ quadrants. We continue to perturb the distribution in the above manner until the intrinsic mean was no longer the star tree. The results of this are shown in the image on the right. In this case, not only is the intrinsic mean no longer the star tree, but the extrinsic mean is also now unique. Please note that we had to greatly perturb the original distribution in this manner to obtain an intrinsic mean that was not the star tree. Indeed, this distribution no longer has sufficient mass on the $(x1, x2)$ and $(x5, x1)$ quadrants for the intrinsic mean to remain at the origin.

Remark 5. Note that unlike the extrinsic mean set on Q_5 which is immediate to compute, the formula for the intrinsic mean of a random object on Q_5 is complicated, as shown in Barden et al. [1]. Moreover the computations of the intrinsic sample mean on Q_5 is increasingly computationally intensive as the sample size grows larger. The advantage of the intrinsic analysis on Q_5 is the uniqueness of the intrinsic mean, leading to a nice asymptotic behavior of the intrinsic sample mean [1]. The problem of the large sample behavior of the extrinsic sample mean set on Q_5 is still open and is nontrivial.

Remark 6. There are distributions on Q_5 having sticky intrinsic means (see Fig. 21.9). For such a distribution the sample intrinsic mean coincides with the population intrinsic means, which is not very informative. In such cases it is preferred to use an extrinsic approach, and construct confidence regions for the extrinsic mean (Fig. 21.10).

21.3.2 Extrinsic Sample Mean for RNA Data

In this example we compute the extrinsic sample mean of ten simulated phylogenetic trees obtained using the Mesquite software [16] and the concatenated *Parkinsea* RNA data taken from UT-Austin CRW website [6], such that there will only be four leaves in the phylogenetic tree. In Fig. 21.11 we show the location of the one *Parkinsea* point in the tree space using its link of the origin. Figure 21.12 shows the resulting extrinsic sample mean tree and its relative location in the tree space.

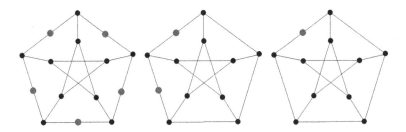

Fig. 21.10 Petersen graph representation of the extrinsic mean set (*light*) associated with the T_4 simulated data in Fig. 21.9

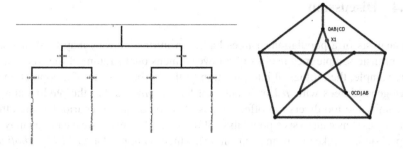

Fig. 21.11 *Left*: A simulated phylogenetic tree of *Parkinsea*. *Right*: Location of the tree in the tree space using the link of the origin

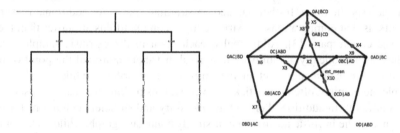

Fig. 21.12 *Left*: The extrinsic sample mean tree. *Right*: Location of the extrinsic mean tree relative to the sample trees

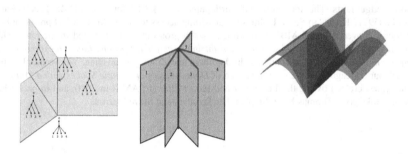

Fig. 21.13 Embeddings of open books: part of T_4, hard cover book and paperback book

21.3.3 Extrinsic CLTs on Open Books

The extrinsic mean set of a distribution on a carton open book (see Fig. 21.13) is on the spine if and only the distribution is entirely concentrated on the spine; such embeddings do not present the stickiness phenomenon. On the other hand, paperback open books present a stickiness phenomenon for a large family of distributions. In general the CLT for an embedding of stratified spaces is now known (see [3]).

21.4 Discussion

An analysis of a population of trees based on intrinsic means may be difficult if the intrinsic sample mean sticks to a lower dimensional stratum of the tree space. For example, the biological interpretation of the intrinsic mean of a population of phylogenetic trees with p leaves being at the *star tree* is that the phylogenies in the family are too diverse to offer a plausible evolutionary scenario. On the other hand, an extrinsic analysis a population of trees based on extrinsic mean sets may be helpful since, unlike the intrinsic mean tree, which is unique due to the *hyperbolicity* of (T_p, ρ_0), if we canonically embedded T_p in $\mathbb{R}^k$ as a $p - 2$ dimensional stratified space, the *extrinsic mean set* reflects all average evolutionary trees from a given family of phylogenetic trees, revealing more mean evolutionary scenarios. Extrinsic data analysis should be further pursued on stratified spaces for additional reasons. The first is that computations of extrinsic means on manifolds are faster than their intrinsic counterparts [3]. Additionally, each point in the extrinsic sample mean set has, asymptotically, a multivariate normal distribution around the point of the extrinsic population mean set in the corresponding orthant, while the intrinsic sample mean might often time stick to a vertex, thus making the data analysis on a graph difficult. In addition, there are no necessary and sufficient conditions for the existence of the intrinsic mean on a non-simply connected graph, while on the other hand, there are such conditions for the existence of extrinsic means on graphs and tree spaces.

Acknowledgements The first, third, and fourth authors thank NSF for awards DMS-1106935 and DMS-0805977. Thanks to Susan Holmes for providing access to her SAMSI-AoOD presentations and useful conversations at MBI. This work is part a group effort that started in the late 1990s with the study of *asymptotics of extrinsic and intrinsic sample means and function estimation on arbitrary manifolds* by the second and third authors and Zinoviy Landsman. We would also like to thank our colleagues on the working group of *Data Analysis on Sample Spaces with a Manifold Stratification* created by the third coauthor and Ezra Miller at SAMSI in 2010, and in particular, special thanks go to Thomas Hotz, Stephan Huckemann, and Steve Marron.

References

1. Barden, D., Le, H., Owen, M.: Central limit theorems for Fréchet means in the space of phylogenetic trees. Electron. J. Probab. **18**(25), 1–25 (2013)
2. Basrak, B.: Limit theorems for the inductive mean on metric trees. J. Appl. Probab. **47**, 1136–1149 (2010)
3. Bhattacharya, R.N., Ellingson, L., Liu, X., Patrangenaru, V., Crane, M.: Extrinsic analysis on manifolds is computationally faster than intrinsic analysis, with applications to quality control by machine vision. Appl. Stoch. Models. Bus. Ind. **28**, 222–235 (2012)
4. Bhattacharya, R.N., Patrangenaru, V.: Large sample theory of intrinsic and extrinsic sample means on manifolds- Part II. Ann. Stat. **33**(3), 1211–1245 (2005)
5. Billera, L.J., Holmes, S.P., Vogtmann, K.: Geometry of the space of phylogenetic trees. Adv. Appl. Math. **27**(4), 733–767 (2001)

6. Cannone, J.J., Subramanian, S., Schnare, M.N., Collett, J.R., D'Souza, L.M., Du, Y., Feng, B., Lin, N., Madabusi, L.V., Müller, K.M., Pande, N., Shang, Z., Yu, N., Gutell, R.R.: The comparative RNA web (CRW) site: an online database of comparative sequence and structure information for ribosomal, intron, and other RNAs. BMC Bioinforma. **3**, 2 (2002) [Correction: BMC Bioinforma. **3**, 15]
7. Fréchet, M.: Les élements aléatoires de nature quelconque dans un espace distancié. Ann. Inst. H. Poincaré **10**, 215–310 (1948)
8. Gromov, M.: Metric Structures for Riemannian and Non-Riemannian Spaces, With appendices by Katz, M., Pansu, P., Semmes, S. Translated from the French by Sean Michael Bates. Progress in Mathematics, vol. 152. Birkhäuser, Boston (1999)
9. Groisser, D., Tagare, H.D.: On the topology and geometry of spaces of affine shapes. J. Math. Imag. Vis. **34**(2), 222–233 (2009)
10. Hendriks, H., Landsman, Z.: Asymptotic behaviour of sample mean location for manifolds. Stat. Probab. Lett. **26**, 169–178 (1996)
11. Hendriks, H., Patrangenaru, V.: Stickiness of Fréchet Means on Finite Graphs. In preparation (2013)
12. Hotz, T., Huckemann, S.: Intrinsic Means on the Circle: Uniqueness, Locus and Asymptotics, arXiv:1108.2141 (2011)
13. Hotz, T., Huckemann, S., Le, H., Marron, J. S., Mattingly, J.C., Miller, E., Nolen, J., Owen, M., Patrangenaru, V., Skwerer, S.: Sticky central limit theorems on open books. Acccepted at Annals of Applied Probability, arxiv:1202.4267 (2013)
14. Hotz, T., Marron, S., Moriarty, J., Patrangenaru, V., Skwerer, S., Huckemann, S., Bharat, K.: Asymptotic behaviour of intrinsic means on stratified spaces. Preprint, SAMSI-AOD Program, 2010–2011 (2010) (currently available to group members only)
15. Kendall, D.G., Barden, D., Carne, T.K., Le, H.: Shape and Shape Theory. Wiley, New York (1999)
16. Maddison, W.P., Maddison, D.R.: Mesquite: a modular system for evolutionary analysis. Version 2.75. http://mesquiteproject.org (2011)
17. Mardia, K.V., Patrangenaru, V.: Directions and projective shapes. Ann. Stat. **33**(4), 1666–1699 (2005)
18. Owen, M., Provan, S.: A fast algorithm for computing geodesic distances in tree spaces. Comput. Biol. Bioinforma. **8**, 2–13 (2011)
19. Patrangenaru, V.: Asymptotic statistics on manifolds and their applications. Ph.D. Thesis, Indiana University (1998)
20. Schwartzman, A., Mascarenhas, W., Taylor, J.: Inference for eigenvalues and eigenvectors of Gaussian symmetric matrices. Ann. Stat. **36**(6), 2886–2919 (2008)
21. Verona, A.: Stratified mappings—structure and triangulability. Lecture Notes in Mathematics, vol. 1102. Springer, Berlin (1984)
22. Wang, H., Marron, J.S.: Object oriented data analysis: Sets of trees. Ann. Statist. **35**(5), 1849–1873 (2007)
23. Ziezold, H.: On expected figures and a strong law of large numbers for random elements in quasi-metric spaces. Transactions of the Seventh Prague Conference on Information Theory, Statistical Decision Functions, Random Processes and of the Eighth European Meeting of Statisticians (Tech. Univ. Prague, Prague, 1974), vol. A, pp. 591–602. Reidel, Dordrecht (1977)

Chapter 22
Kernel Density Outlier Detector

M. Pavlidou and G. Zioutas

Abstract Based on the widely known kernel density estimator of the probability function, a new algorithm is proposed in order to detect outliers and to provide a robust estimation of location and scatter. With the help of the Gaussian Transform, a robust weighted kernel estimation of the density probability function is calculated, referring to the whole of the data, including the outliers. In the next step, the data points having the smallest values according to the robust pdf are removed as the least probable to belong to the clean data. The program based on this algorithm is more accurate even on greatly correlated outliers with the data, and even with outliers with small Euclidean distance from the data. In case the data have many variables, we can use Principal Component algorithm by (Introduction to Multivariate Statistical Analysis in Chemometrics. CRC press, 2008) [1] with the same efficiency in the detection of outliers.

Keywords Robust kernel density estimation • Outlier detector

22.1 Introduction

In statistics, an outlier is an observation that is numerically distant from the rest of the data. Grubbs defined an outlier as: An outlying observation, or outlier, is one that appears to deviate markedly from other members of the sample in which it occurs. Hawkins [2] suggested that:

An outlier is an observation which deviates so much from the other observations as to arouse suspicions that it was generated by a different mechanism. Outliers can occur by chance in any distribution, but they are often indicative either of measurement error or that the population has a heavy-tailed distribution.

Outlier detection refers to the problem of finding patterns in data that do not conform to expected behaviour. Depending on the application domain, these non-conforming patterns can have various names, e.g., outliers, anomalies, exceptions, discordant observations, novelties or noise.

M. Pavlidou (✉) • G. Zioutas
Aristotle University of Thessaloniki, Thessaloniki, Greece
e-mail: mepa@eng.auth.gr; zioutag@eng.auth.gr

© Springer Science+Business Media New York 2014   241
M.G. Akritas et al. (eds.), *Topics in Nonparametric Statistics*, Springer Proceedings
in Mathematics & Statistics 74, DOI 10.1007/978-1-4939-0569-0_22

There have been introduced many outlier detecting methods, such as the Minimum Covariance Determinant, MCD, by Rousseeuw and Van Driessen [4], the Stahel Donoho estimate, SDE, proposed by Maronna et al. [3]. Most of them use the Euclidean distance and the covariance matrix of the data, or the Mahalanobis distance, as the MCD, or use projections as in SDE, in order to mark data as outliers.

Distance based techniques compare distances, Euclidean at the beginning and in the Robust version of Rousseeuw and Van Driessen Mahalanobis distances, in order to categorize clean data and outliers. The Minimum Covariance Determinant scatter estimator (MCD) is a highly Robust estimator for the dispersion matrix of a multivariate elliptically symmetric distribution. It is very efficient when the clean data follow the Gaussian Distribution. As for a profile it usually is a vector of numbers that together capture important aspects of the user's behaviour. Profiles are learned from the data, thereby eliminating the need for defining "normal" behaviour. By comparing profiles we can find points with unusual behaviour. The model based processes are somewhat similar with the exception that they focus on learning abnormalities in the sense of semi-supervised learning, and not focusing on normal patterns. These both methods tend to require a fairly large training data set. So far, the underlying idea of all our outlier detection procedures is that the majority of the data, which are supposed to be the clean data, follow a normal probability density function with unknown mean and standard deviation values. In an outlier infected data set, here is a method though, to estimate the probability density function nonparametrically. This is called Kernel Density Estimator, KDD [6]. The problem is that the pdf estimated will not be accurate or cannot be used as a criteria, because it is based both on the clean and the outlier data. The above methods in most of the cases assume gaussian distributions for the majority of the data, i.e. the clean data, thus trying to detect outliers using scatter matrices or fitting measures to gaussian distributions or other known distributions such as Cauchy or Laplace which are widely used in Telecommunications. However, not every data is produced according to the Gaussian distribution or not every healthy data distribution is known. Our proposition is that for more robust and accurate measurements, we should use the actual density of the data. One way to do this is through Kernel Density Probability Estimates. After the distribution of the whole of the data is estimated, trying to emphasize healthy data and penalize possible noisy data, the data points with the larger probabilities estimated with the help of our modified weighted version of the Kernel Density Probability function are chosen as more possible to belong to the healthy data set and those with reduced probabilities according to the estimated model are considered to be the outliers.

22.2 Weighted Kernel Density Probability Estimate

In statistics, kernel density estimation, KDD, is a nonparametric way to estimate the probability density function of a random variable. Kernel density estimation is a fundamental data smoothing problem where inferences about the population are made, based on a finite data sample.

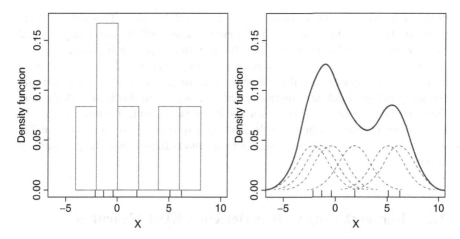

Fig. 22.1 Example of the histogram of a dataset compared to the Kernel Density Estimate using Gaussian Kernels, KDE. The KDE is more smooth than the histogram. Local maxima correspond to bigger probabilities of the data points

A range of kernel functions are commonly used: uniform, triangular, biweight, triweight, Epanechnikov, normal, and others. Due to its convenient mathematical properties, the normal kernel is often used $K(x) = \phi(x)$, as in this paper, where ϕ is the standard normal density function.

The probability estimated at the point x with the help of a Kernel Density Estimator is the superposition of the contribution of the (multivariate) Gaussians of all the other points of the data set, centred at each of these points (Fig. 22.1).

The use of the Kernel Density estimate for representing feature distributions as well as for the detection of outliers may also be motivated by the fact that the individual component densities may model some underlying of hidden classes. Since every single data point contributes to the estimation of the sample distribution, we can exclude certain data points, by comparing the a-posteriori possibility of the estimated kernel density function. The data points that are assigned with the smallest probability values are considered to be the least stable ones and are filtered as possible outliers.

As we mentioned before, the estimated density at each intersection is essentially the average of the densities of all intersections that overlap that point. This means that when calculating the probability at any intersection, the contribution of every other data point density, or every data point gaussian in our case, is the same. This way, only the data points that have the density pattern of the majority of the data contribute more to the estimation of the pdf and have therefore larger probability. But what if a large enough density of outliers are gathered in a high density or even around a healthy data point? The consequence will be for the pdf to have a peak at that point and these outliers to show greater probability. A new proposition would be use a weighted version of the Kernel Density estimate using the the Euclidean distances between each data point and all the rest data points. This way, outlying points far from the intersection, even in great density, will be underweighted and therefore have a smaller impact on the pdf and a smaller probability value.

More explicitly, we have combined two data properties in order to form a robust probability density estimate, the difference in density, based on the Kernel Density estimate, and the Euclidean distance used as a weight to the Kernel Probability Estimate. This combination, as we will see in the experimental results, proves to be more efficient in the detection of outliers, which will simply have a small probability value in this kernel pdf estimate, whether data and outliers follow the gaussian pdf or not.

22.3 Kernel Density Outlier Detection KDD Algorithm

As we mentioned before, we believe that it is more accurate to use the Probability Estimate of the dataset in order to detect outliers than to suppose a priori that data and outliers follow the Gaussian distribution and to use metrics relying on this assumption. However, when data are corrupted with outliers or noise, the Probability estimate based on our dataset does not reflect the true Probability Estimate of the datapoints. In our effort to make the Density Probability Estimate more robust, we used the Kernel Probability Estimate with a Bandwidth that is selected with the help of the Mahalanobis Distance, as well as weights for this Kernel Probability Estimate, which are based on the Euclidean Distance between the data. The equation of the weighted Kernel Density Probability Estimate, using Gaussian kernels, has the following form:

$$p(x) = \frac{1}{nh^D} \sum_i K \frac{x - xi}{h} \qquad (22.1)$$

where wij are the weights of each gaussian kernel, D is the Bandwidth of the gaussian kernel and xi are the data points whose probability estimate is being calculated. The weights satisfy the constraint that

$$\sum_i wi = 1.$$

The weight wij will be of the form

$$wi = (a - \sum_i \text{Eucl}(xi, xj))/a \qquad (22.2)$$

where a is the largest Euclidean distance between our data points and it is used for normalization, including the outliers, and $\Sigma \text{Eucl}(xi, xj)$ is the sum Euclidean distance between each data point xj of the jth Gaussian and the point xi.

The algorithm for the KDD Kernel Density Outlier Detection Method, in order
to remove c % of the data as possible outliers, is presented:

1. Compute sum of Euclidean distances (a) of each point from the others and next,
   according to Eq. (1.6), the weights wi.
2. With the help of Eq. (1.5), compute p(x) several times, around 12 is usually
   enough, each time with different Bandwidth D, with D ranging around the
   Standard deviation of the dataset. Choose the bandwidth of D, for which the
   largest probability values p(x) values of the data correspond to the minimum
   sum of Mahalanobis distances.
3. Choose the 0.6*N biggest values of p(x), with the chosen bandwidth.
4. Compute Robust Mean and Covariance matrix based on these most probable data
   points.
5. Execute C step of MCD Algorithm, chi-square recruiting.

We should note that if either data features are highly correlated or high in number,
it is desirable first to perform Principal Component Analysis algorithm (PCA) of
Varmuza and Filzmoser [1]. The efficiency and the results present the same and
even greater accuracy, as we will see later on.

22.4 Experimental Results

In this simulation, as in the paper of Maronna and Zamar [7] we generate random
data of the same distribution and contamination. The methods which we compare
with our KDD proposal are the best [5] methods of the above paper like: Principal
Component Outlier Detector (PCOut), Minimum Volume Ellipsoid (MVE) and fast
Minimum Covariance Determinant (MCD). The sampling situations were p-variate
normal ϵ contaminated distributions, with p taking the values 5 and 10, and $n=10p$
(Table 22.1). We generated correlated data as follows. Let $m = [n \epsilon$ (where [.]
denotes the integer part) we generated y_i as p-variate normals $N_p(0, I)$ for $i =
1, \ldots, n-m$, and as $N_p(y_o, d^2 I)$ for some y_0 and $i > n - m$; we chose $\delta = .1$. The
choice of a normal distribution with a small dispersion, rather than exact point-mass
contamination, is due to the fact that exactly repeated points may cause problems
with the subsampling algorithms (Table 22.2).
Put $x_i = Ry_i$, where R is the matrix with

$$R_{jj} = 1 \quad \text{and} \quad R_{jk} = \rho \quad \text{for } i \neq j \quad (12)$$

Then for $\epsilon = 0$, X has covariance matrix R_2, and the multiple correlation ρmult
between any coordinate of X and all of the others is easily calculated as a function
of ρ. We chose ρmult $= 0.999$. This is a very collinear situation.
In this section also, we show results as in the paper of Filzmoser Maronna and
Werner (2008) for location outliers with $k = 2$ to $k=10$ and for scatter outliers with δ^2
ranging from $\delta^2=0.1$ to $\delta^2=5$. We present our results table with (1) The percentage of

Table 22.1 KDD comparative performance evaluation for Gaussian distributed data with several parameter values

Method	k	$\delta^2 = 0.1$		$\delta^2 = 0.5$		$\delta^2 = 1$		$\delta^2 = 2$		$\delta^2 = 5$	
		%FN	%FP	%FN	%FP	%FN	%FP	%FN	%FP	%FN	%FP
LTED	0	100.00	6.61	100.00	4.3	–	2.96	64.2	1.55	7.69	1.35
PCOut	0	100.00	7.15	99.96	6.81	–	5.30	61.16	4.00	8.84	3.49
MCD	0	100.00	5.51	100.00	3.74	–	2.60	63.94	1.64	7.65	1.40
KDD	0	100.00	9.17	99	7.3	–	3.4	39	1.5	1	2.41
LTED	2	100.00	4.71	98	3	90.00	2.23	49.00	1.45	6.4	1.37
PCOut	2	100.00	7.21	99.44	6.29	82.05	4.19	45.61	3.10	8.90	
MCD	2	100.00	4.69	99.94	3.20	95.58	2.21	50.42	1.54	5.96	1.40
KDD	2	0	5.46	0	5.39	0	2.81	0	1.64	0	1.7
LTED	5	65.8	1.55	12	1.4	6.9	1.45	3.3	1.41	0.00	1.35
PCOut	5	67.27	1.60	15.29	1.59	7.25	1.65	3.77	1.74	5.92	2.11
MCD	5	100.00	4.00	32.16	1.43	16.46	1.41	7.20	1.40	1.52	1.40
KDD	5	0.00	1.7	0.00	1.65	0.00	1.6	0.00	1.67	0.00	1.61
LTED	10	0.00	1.35	0.00	1.34	0.00	1.35	0.00	1.34	0.00	1.38
PCOut	10	0.00	1.49	0.00	1.69	0.00	1.79	0.00	1.87	0.03	1.93
MCD	10	0.00	1.40	0.00	1.40	0.00	1.40	0.00	1.39	0.01	1.40
KDD	10	0.00	1.45	0.00	2.11	0.00	1.9	0.00	1.75	0.00	1.3

We can see the false negative and false positive results
%FN and %FP $\rho = 10, n = 1,000, n = 100, \epsilon = 10\%, \rho = 0.5, 500$ simulations

Table 22.2 KDD comparative performance evaluation for Gaussian distributed data and larger percentage of outliers

Method	k	$\delta^2=0.1$		$\delta^2=0.5$		$\delta^2=1$		$\delta^2=2$		$\delta^2=5$	
		%FN	%FP	%FN	%FP	%FN	%FP	%FN	%FP	%FN	%FP
LTED	0	90.00	5.62	90.00	3.8	–	2.93	44.2	1.51	6.24	1.46
PCOut	0	100.00	10.15	100.00	10.01	–	7.70	76.14	5.60	10.89	5.24
MCD	0	100.00	7.8	100.00	4.78	–	3.90	70.02	2.8	9.2	2.30
KDD	0	100	8.4	99.3	19.5	–	3.8	51.6	4.3	6.6	2.9
LTED	2	0.00	4.45	0.00	2.8	0.00	2.28	0.00	1.34	0.00	1.23
PCOut	2	100.00	8.5	100.00	7.91	85.6	6.04	51.01	4.15	9.16	4.18
MCD	2	100.00	5.82	100.00	4.15	94.04	3.31	55.45	2.10	6.91	2.24
KDD	2	0	2.14	0	2.43	0	2.14	0	3.14	0	2.29
LTED	5	0.00	1.58	0.00	1.32	0.00	1.33	0.00	1.40	0.00	1.36
PCOut	5	69.62	1.85	17.28	1.94	8.36	1.86	4.1	1.88	7.9	1.97
MCD	5	100.00	4.81	39.12	1.88	22.15	1.81	11.50	1.40	5.6	1.54
KDD	5	0.00	2.71	0.00	3.57	0.00	3.14	0.00	3.14	0.00	3.57
LTED	10	0.00	1.33	0.00	1.33	0.00	1.32	0.00	1.33	0.00	1.36
PCOut	10	0.00	1.44	0.00	1.56	0.00	1.65	0.00	1.77	0.00	1.85
MCD	10	0.00	1.39	0.00	1.40	0.00	1.39	0.00	1.39	0.01	1.40
KDD	10	0.00	2.43	0.00	3.29	0.00	3.57	0.00	3.14	0.00	3.57

%FN and %FP, $\rho = 10, n = 1,000, n = 300, \epsilon = 30\%, \rho = 0.5, 500$ simulations

Table 22.3 Performance Evaluation of KDDw on larger dimensions

Method Dimension			
	p	%FN	%FP
LTED	50	0	0
PCOut	50	49.5	5.92
KDD	50	0	2.3
LTED	100	0	0
PCOut	100	31.8	2.31
KDD	100	0	2.2
LTED	200	0	0
PCOut	200	18.3	3.98
KDD	200	0	2.3
LTED	500	0	0
PCOut	500	12.9	3.10
KDD	500	0	1.5
LTED	1,000	2.1	1.36
PCOut	1,000	6.06	3.39
KDD	1,000	0	1.8
LTED	2,000	0.26	1.50
PCOut	2,000	0.38	2.54
KDD	2,000	0	2

Outliers generated with a slightly larger covariance matrix ($\delta^2 = 1.2$)

Table 22.4 Application of MCD, MVE, PCOut and KDD on UCI breast cancer data

Method	Correlation	Cosine similarity
UCI BC Data	3.9×10^{-4}	4.0×10^{-4}
KDD	1.5×10^{-4}	4.7×10^{-4}
MCD	3×10^{-5}	3.2×10^{-5}
MVE	3.7×10^{-5}	6×10^{-7}
PCout	6×10^{-5}	5×10^{-5}

Evaluation Metrics on remaining clean data

false negatives(FN)-outliers that were not identified, or masked outliers, and (2) the percentage of false positives (FP)- non-outliers that were classified as outliers, or swamped non-outliers. Table 22.3 shows results for percentage of outliers ϵ=0.1 as a compromise between low and high levels of contamination, the results are comparable. Note that the case k=0, $\delta = 1$ corresponds to no outliers in the data, which reduces to a measurement of only the false positives. Examination of Table 22.4 reveals that KDD performs well at identifying outliers (low False negatives) and it has satisfactory percentage of False positive than most of the methods. Also, we observe that KDD does well both for location outliers, i.e. $k = 5$, and for scatter outliers $\delta^2 = 2$ and $\delta^2 = 5$.

As we increase the contamination up to ϵ=0.30 KDD does exceptionally well for all type of outliers, as we can see in Table 22.2. Its performance is getting better as we increase contamination, and this is a big advantage among all the other robust estimators.

In Table 22.3 we can also evaluate KDD's performance in different feature numbers, as the variance has increased to 1.2 and multiple correlation coefficient to 0.7. In this case, KDD has been performed on the principal components derived from PCA.

In the next step, we applied our method to real data, Breast Cancer Data from UCI Repository. The dataset gives information on breast cancer patients, who at the time being were reported stable in terms of their health condition. At the same time we used the three widely known outlier detectors fast MCD, MVE and PCOut again. The performance of each detector is evaluated with the help of two metrics, the average of the data points pairwise correlation as well as the cosine similarity metric.

As seen in the results of Table 22.5, the average pairwise correlation of the clean data using the KDD detector is quite higher than the other methods and is close to the initial subset's correlation. This may be a hint that KDD kept as clean data the data points that had a bigger similarity between them, as we can see in the cosine similarity measure, which is even bigger than the initial dataset.

Similar are the comparative results on Ionosphere data from UCI Data Repository. This radar data was collected by a phased array of 16 high-frequency antennas system and measurements took place in Goose Bay, Labrador. Once more, we applied our four techniques for detecting outliers and we present the results regarding the two metrics, the average pairwise correlation and the average cosine similarity measure. Again, in table we notice that our method gives those subset of data as clean, that have the biggest average cosine similarity as well as the bigger correlation after the removal of the outliers. The average mahalanobis distance has no big difference (Table 22.5).

In order to introduce KDD's performance on our last data set, Parkinson's Telemonitoring Data of UCI Repository, we applied all our previous four outlier detection methods and we chose randomly two features to display. In the first scatterplot, the two features of the original dataset are presented. As we can see, the majority of the points lie on the left lower values. However, there seems to be a positive correlation on greater values, although some data points in the middle seem to exceed the supposed ellipsoid boundary (Fig. 22.2).

Finally, here are the results of the metrics on UCI Parkinson's Telemonitoring Data in Table 1.9. The cosine similarity measure and the average pairwise

Table 22.5 KDD, MCD, PCOut and MVE similarity measures on UCI ionosphere datasets

Method	Mahalanobis	Correlation	Cosine
KDD	7.1158	0.6660	0.7472
fastMCD	7.6393	0.5527	0.6606
pcout	7.0820	0.6307	0.7136
MVE	7.0901	0.6200	0.7222

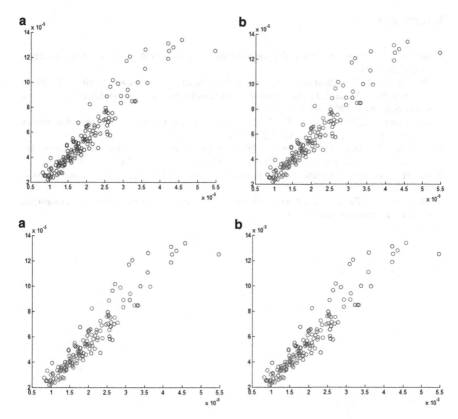

Fig. 22.2 (**a**) Two-dimensional scatterplot of original dataset. (**b**) Data of KDD comparing to original. (**c**) Data of MCD comparing to original. (**d**) Data of PCOut comparing to original

Table 22.6 KDD, MCD, PCOut and MVE similarity measures on UCI Parkinson Telemonitoring dataset

Method	Correlation	Cosine
UCI Data	0.1057	0.0877
KDD	0.1098	0.0910
fastMCD	0.1065	0.0883
pcout	0.1089	0.0902

correlation of the KDD data have higher values than the other measures as well as than the original data. This could imply that the KDD produced data subset is supposed to be more clean, more correlated and supposedly belonging to the same pattern (Table 22.6).

References

1. Varmuza, K., Filzmoser, P.: Introduction to Multivariate Statistical Analysis in Chemometrics. CRC press (2008)
2. Hawkins, D.M.: Identification of Outliers. Springer-Science and Business Media, B.V. (1980)
3. Maronna, R.A., Yohai, V.J.: The behavior of the Stahel-Don oho robust multivariate estimator. J. Am. Stat. Assoc. **90**(429), 330–341 (1995)
4. Rousseeuw, P.J., Van Driessen, K.: A fast algorithm for the minimum covariance determinant estimator. Technometr. **41**(3), 212–223 (1999)
5. Filzmoser, P., Hron, K., Reimann, C.: Principal component analysis for compositional data with outliers, Vienna University of Technology, Forschungsbericht, 2007
6. Kernel Density Estimation: Wikipedia.org, http://wikipedia.org/wiki/Kernel_density_estimation
7. Maronna, R.A., Zamar, R.H.: Robust estimates of location and dispersion for high-dimensional datasets. Technometr. **44**(4) (2002)

Chapter 23
Model Selection for Classification with a Large Number of Classes

Justin Davis and Marianna Pensky

Abstract In the present paper, we study the problem of model selection for classification of high-dimensional vectors into a large number of classes. The objective is to construct a model selection procedure and study its asymptotic properties when both, the number of features and the number of classes, are large. Although the problem has been investigated by many authors, we research a more difficult version of a less explored random effect model where, moreover, features are sparse and have only moderate strength. The paper formulates necessary and sufficient conditions for separability of features into the informative and noninformative sets. In particular, the surprising conclusion of the paper is that separation of features becomes easier as the number of classes grows.

Keywords High dimensional data • Low sample size • Multivariate analysis • Classification

23.1 Introduction

It is a well-established result that addition of "noninformative" dimensions in data, by which we mean any dimensions which do not improve the accuracy of a generic classifier, eventually makes accurate classification impossible even with just two classes. Additional noninformative dimensions are especially problematic when the total number of dimensions exceeds the number of samples in a data set; this is one aspect of the high-dimension, low sample size (HDLSS) problem. The difficulty becomes even more acute when the number of dimensions exceeds the number of observations as it often happens in many practical applications. Indeed, Fan and Fan [6] argue that when the dimension of vectors to be classified tends to infinity

J. Davis (✉)
Tulane University, New Orleans, LA, USA
e-mail: jdavis37@tulane.edu

M. Pensky
University of Central Florida, Orlando, FL, USA
e-mail: marianna.pensky@ucf.edu

© Springer Science+Business Media New York 2014   251
M.G. Akritas et al. (eds.), *Topics in Nonparametric Statistics*, Springer Proceedings in Mathematics & Statistics 74, DOI 10.1007/978-1-4939-0569-0__23

and the sample size remains finite, almost all linear discriminant rules can perform as badly as the random guessing. The objective of the present paper is to study dimension reduction which is designed specifically for classification.

A number of authors studied model selection in the HDLSS problem, see, e.g., [1–3, 5, 6, 8] and [9], among others. The difference between the present paper and studies and methodologies cited above is that, first, we consider model selection which is specifically designed for classification; second, we do not assume that the difference between class means is asymptotically large, or even, large at all; third, we consider the novel case when the number of classes is large—in particular, the number of classes grows approximately as a logarithm of the number of features. This setup is motivated by classification of communication signals recorded from South American knife fishes considered in Davis and Pensky [4] as well as classification of genetics data where the number of features is enormous and the number of classes (e.g., various biological conditions) can be large while the sample is relatively limited.

The rest of the paper is organized as follows. In Sect. 23.2, we briefly review the results of Davis and Pensky [4] and introduce the general framework of the paper. Section 23.3 presents main results of the paper on model selection and separability of features into informative and non-informative sets. Finally, Sect. 23.4 concludes the paper with discussion.

23.2 General Framework

Consider the problem of classification of p-dimensional vectors into L classes $\omega_1, \cdots, \omega_L$ based on n_i training samples from class i, $i = 1, \cdots, L$, where $\sum_{i=1}^{L} n_i = n$. For convenience, we arrange samples as row vectors

$$\mathbf{D}^\top = [\mathbf{d}^1, \mathbf{d}^2, \cdots, \mathbf{d}^n],$$

yielding the $(n \times p)$ matrix $\mathbf{D}$; i.e. the first n_1 rows of $\mathbf{D}$ are samples from the first class, the next n_2 rows are samples from the second, etc. We denote the columns of $\mathbf{D}$ by $\mathbf{d}_i \in \Re^n$, $i = 1, \cdots, p$; hence, the first column of $\mathbf{D}$ contains the first component of each of the n samples, the second contains the second component of all samples, etc.

The objective is to select a sparse subset of these p vectors which enable classification of vectors $\mathbf{d}^1, \mathbf{d}^2, \cdots, \mathbf{d}^n$ into classes $\omega_1, \cdots, \omega_L$. For this purpose, we introduce a binary vector $\mathbf{x} \in \Re^p$ with $x_i = 1$ if vector component $\mathbf{d}_i$ is "informative" and should be retained in subsequent discriminatory analysis, or $x_i = 0$ if $\mathbf{d}_i$ should be discarded. The goal of the analysis, then, is to draw conclusions about vector $\mathbf{x}$ on the basis of matrix $\mathbf{D}$. For this purpose, we introduce the following notations. Let $p_1 = \sum_{i=1}^{p} x_i$ and $p_0 = p - p_1$ represent the number of informative and noninformative dimensions, respectively. Let $\mathbf{e} \in \Re^n$ be a column vector with all components being equal to one and $\mathbf{g}_l \in \Re^n$, $l = 1, \cdots, L$, be column vectors with the j-th components $(g_l)_j = 1$ if it corresponds to class l, i.e. if

$$n_1 + \cdots, n_{l-1} + 1 \le j \le n_1 + \cdots, n_{l-1} + n_l,$$

and $(g_l)_j = 0$ otherwise; i.e. $(g_l)_j$ is nonzero iff sample j is from class l and g may be thought of as a matrix of indicator functions. We also define $\mathbf{G} \in \mathfrak{R}^{n \times L}$ with columns $\mathbf{g}_l, l = 1, \cdots, L$.

The idea is to search for components $\mathbf{d}_i$ which are constant within classes but vary in between classes. This was accomplished by the CONstant FEature Selection Strategy (CONFESS) proposed by Davis and Pensky [4]. In particular, CONFESS assumes that $\mathbf{d}_i$, a noisy measurement of the "true" i-th component $\boldsymbol{\mu}_i$, i.e.

$$\mathbf{d}_i = \boldsymbol{\mu}_i + \boldsymbol{\varepsilon}_i$$

where $\boldsymbol{\varepsilon}_i$ are multivariate normal $\boldsymbol{\varepsilon}_i \sim N(\mathbf{0}, \sigma_i^2 \mathbf{I}_n)$ and $\boldsymbol{\mu}_i$ can be partitioned into a sum of three components. The first is constant across all components and contributes nothing to discrimination among classes; this component is from S_C, described below. Another is constant within classes but varies between classes and, were one able to extract this component exactly, would allow for direct classification; these are contained in S_G and one of its subspaces, S_1, in which all vectors have componentwise sum 0. Finally, the third varies within classes but does not provide any useful information for classification; these are contained in S_0.

For this purpose, CONFESS introduces the following subspaces: $S_C = \text{Span}(\mathbf{e})$, the one-dimensional subspace of scalar multiples of $\mathbf{e}$, $S_G = \text{Span}(\mathbf{g}_1, \cdots, \mathbf{g}_L)$, $S_1 = S_G \setminus S_C$, and $S_0 = \mathfrak{R}^n \setminus S_G$, so that $\mathfrak{R}^n = S_C \oplus S_0 \oplus S_1$. Denote by $\mathbf{P}_0$ and $\mathbf{P}_1$ matrices of orthogonal projections onto S_0 and S_1. Then

$$\boldsymbol{\mu}_i = n^{-1/2} m_i \mathbf{e} + \mathbf{u}_i + \mathbf{v}_i, \quad i = 1, \cdots, p, \qquad (23.1)$$

where it is assumed that

$$m_i \sim N(0, \sigma_i^2 \tau^2),$$
$$(\mathbf{u}_i | x_i = 0) \sim \delta(\mathbf{0}),$$
$$(\mathbf{u}_i | x_i = 1) \sim N(\mathbf{0}, \sigma_i^2 \boldsymbol{\Sigma}_u) \qquad (23.2)$$
$$\mathbf{v}_i \sim N(\mathbf{0}, \sigma_i^2 \boldsymbol{\Sigma}_v).$$

Here matrices $\boldsymbol{\Sigma}_u = \mathbf{P}_1 \boldsymbol{\Sigma} \mathbf{P}_1$ and $\boldsymbol{\Sigma}_v = \mathbf{P}_0 \boldsymbol{\Sigma} \mathbf{P}_0$ ensure that $\mathbf{u}_i \in S_1$ and $\mathbf{v}_i \in S_0$ and it is assumed that vectors $\mathbf{u}_i$ and $\mathbf{v}_i$ are independent. Here, $\mathbf{u}_i$ represents the groupwise contribution to variance and $\mathbf{v}_i$ the within-group variation.

Remark 1. Conditions (23.1) and (23.2) require that each vector $\boldsymbol{\mu}_i$ can be partitioned into the sum of three independent components, $n^{-1/2} m_i \mathbf{e}$, $\mathbf{u}_i$ and $\mathbf{v}_i$, and feature i is considered to be non-informative if the second component $\mathbf{u}_i$ of vector $\boldsymbol{\mu}_i$ is identically equal to zero. In addition, all components are assumed to be normally distributed. If assumption of independence is violated, in general, results of the paper hold. However, due to dependence between features, vectors $\mathbf{v}_i$ carry some information about $\mathbf{u}_i$, so that the inference should be performed conditionally on vectors $\mathbf{v}_i$.

If normality assumptions do not hold, then Y_i's do not obey chi-squared distribution and, possibly, would have heavier tails. The latter would lead to different separability conditions in Theorem 1.

23.3   Model Selection and Separability

Assume we have a sequence of data sets $\{\mathbf{D}\}$ generated according to model (23.1) and (23.2). In order to study model selection and its precision, assume that $\boldsymbol{\Sigma} = \rho^2 \mathbf{I}$ and that σ_i, $i = 1, \cdots, p$, and ρ are known, i.e., all parameters but x_i are known. We shall also assume that features are sparse, i.e. $p_1 = vp$ and $p_0 = (1 - v)p$ where $p \to \infty$ and v is possibly small. We shall also assume that the number of classes is growing, i.e. $L \to \infty$ as $p \to \infty$.

For each vector $\mathbf{d}_i$, we define $\mathbf{y}_i = \mathbf{H}_1 \mathbf{d}_i$ where $\mathbf{H}_1^\top \mathbf{H}_1 = \mathbf{P}_1$ is the unique Cholesky decomposition of $\mathbf{P}_1$, the unique orthogonal projection into S_1, and denote

$$Y_i = \frac{\|\mathbf{y}_i\|^2}{\sigma_i^2 (1 + \rho^2)}.$$

Then, it can be shown that

$$(Y_i | x_i = 0) \sim \frac{\chi_{L-1}^2}{1 + \rho^2}, \quad (Y_i | x_i = 1) \sim \chi_{L-1}^2. \tag{23.3}$$

It follows from (23.3), that for any $z > 0$, one has

$$P(Y_i \leq z | x_i = 1) \leq P(Y_i \leq z | x_i = 0).$$

Define random variables

$$U = \max\{Y_i | x_i = 0\}_{i=1}^{p_0}, \quad V = \min\{Y_i | x_i = 1\}_{i=1}^{p_1}.$$

If there exists $\lambda > 0$ such that $P(U < \lambda < V) = 1$, then, almost surely, one could select all informative dimensions by retaining only those with $Y_i > \lambda$. However, this is a very stringent condition and we call $\{\mathbf{D}\}$ *separable* if there exists a sequence $\{\lambda\}$ such that

$$\lim_{p \to \infty} P(U \leq \lambda) = 1 \quad \text{and} \quad \lim_{p \to \infty} P(V \leq \lambda) = 0.$$

If $\{\mathbf{D}\}$ is separable, the probability of (retaining) discarding an (non)informative dimension goes to 0 as $p \to \infty$. The following lemma provides necessary and sufficient conditions for separability.

Lemma 1. *Let $\{\mathbf{D}\}$ be a sequence of data sets generated according to model (23.1) and (23.2) with $\boldsymbol{\Sigma} = \rho^2\mathbf{I}$, and all parameters but x_i are known. Then, $\{\mathbf{D}\}$ is separable iff*

$$\lim_{p\to\infty} p_0(1 - F_{L-1}((1 + \rho^2)\lambda)) = 0 \quad \text{and} \quad \lim_{p\to\infty} p_1 F_{L-1}(\lambda) = 0, \quad (23.4)$$

where

$$F_{L-1}(x) = \frac{\gamma\left(\frac{L-1}{2}, \frac{x}{2}\right)}{\Gamma\left(\frac{L-1}{2}\right)}.$$

Here $\gamma(k, z)$ is the incomplete gamma function (see formula 8.350 of [7]).

Using asymptotic expansion 8.327.3 of [7] of the logarithm of the Gamma function and asymptotic expansions of incomplete gamma functions for various relationships between L and x provided in Paris [10], we can rewrite conditions (23.4) as follows.

Lemma 2. *Denote $\nu = p_1/p$, $\tau = (L - 1)/\lambda$ and $\tau_\rho = (1 + \rho^2)/\tau = (1 + \rho^2)\lambda/(L - 1)$. Let*

$$F_1(\tau, L, p, \nu) = (L - 1)(\tau^{-1} + \ln\tau - 1) + \ln(L - 1) + 2\ln(1 - \tau^{-1}) - 2\ln p_1,$$

$$F_2(\tau, \rho, L, p, \nu) = (L - 1)(\tau_\rho - \ln\tau_\rho - 1) + \ln(L - 1) + 2\ln(\tau_\rho - 1) - 2\ln p_0.$$

Then, $\{\mathbf{D}\}$ is separable iff

$$\lim_{p\to\infty} F_1(\tau, L, p, \nu) = \infty \quad \text{and} \quad \lim_{p\to\infty} F_2(\tau, \rho, L, p, \nu) = \infty. \quad (23.5)$$

Note that τ should satisfy inequality

$$1 < \tau < 1 + \rho^2. \quad (23.6)$$

Moreover, since $F_1(\tau, L, p, \nu)$ is increasing, $F_2(\tau, \rho, L, p, \nu)$ decreasing in τ, it suffices to find $\tau = \hat{\tau}$ such that

$$F_1(\tau, L, p, \nu) = F_2(\tau, \rho, L, p, \nu) \quad (23.7)$$

and show that, say, $\lim_{p\to\infty} F_1(\hat{\tau}, L, p, \nu) = \infty$. Equation (23.7) can be simplified to

$$(L - 1)(\ln(1 + \rho^2) - \rho^2\tau^{-1}) + 2\ln\left(\frac{\tau - 1}{(1 + \rho^2) - \tau}\right) + 2\ln\left(\frac{1 - \nu}{\nu}\right) = 0.$$

Consider collections of models with $L = L(p)$ and $v = v(p)$ such that $\lim_{p \to \infty} L(p) = \infty$ and $\lim_{p \to \infty} v(p) \geq 0$ but ρ is a fixed constant. This is a more reasonable setup than the one with $\rho(p) = \log(p)$ that is usually considered in the detection and model selection literature. There are two possible cases here: $\lim_{p \to \infty} v(p) = v_0 \geq 0$ (case 1) and $\lim_{p \to \infty} v(p) = 0$ (case 2).

Case 1. Since $L \to \infty$, one obtains

$$\hat{\tau} = \hat{\tau}_1 \rho^2 / \ln(1 + \rho^2) + o(1) \quad \text{as} \quad p \to \infty, \tag{23.8}$$

and, as $p \to \infty$, conditions (23.5) appear as

$$F_1(\tau, L, p, v) = (L-1)(\hat{\tau}^{-1} + \ln \hat{\tau} - 1) + \ln(L - 1) + 2 \ln(1 - \hat{\tau}^{-1}) - 2 \ln(p) \to \infty. \tag{23.9}$$

Case 2. Since $\ln(1 - v) \approx 0$ for $v \to 0$, one derives

$$\hat{\tau} = \hat{\tau}_2 = \frac{\rho^2}{\ln(1 + \rho^2)} \frac{1}{1 - \frac{2 \ln v}{(L-1) \ln(1+\rho^2)}} \tag{23.10}$$

and separability conditions (23.5) hold iff (23.9) is valid. Moreover, by direct calculations one can check that condition (23.6) is satisfied only if $-2 \ln v / (L-1) < \rho^2 - \ln(1 + \rho^2)$, so that $\hat{\tau}_1 = k \hat{\tau}_2$ for some constant $0 < k < \infty$.

Summarizing both cases, we obtain the following statement.

Theorem 1. *Let* {**D**} *be a sequence of data sets generated according to model (23.1) and (23.2) with* $\boldsymbol{\Sigma} = \rho^2 \mathbf{I}$, *and all parameters but* x_i *are known. Then,* {**D**} *is separable if*

$$(L - 1)(\hat{\tau}^{-1} + \ln \hat{\tau} - 1) \geq 2 \ln p \quad \text{and} \quad (L - 1)(\rho^2 - \ln(1 + \rho^2)) \geq -2 \ln v,$$

and {**D**} *is not separable if*

$$\lim_{p \to \infty} 2 \ln p / (L-1) > \hat{\tau}^{-1} + \ln \hat{\tau} - 1 \quad \text{or} \quad \lim_{p \to \infty} -2 \ln v / (L-1) > \rho^2 - \ln(1+\rho^2).$$

Here $\hat{\tau}$ *is given by expression (23.8) if* $\lim_{p \to \infty} v(p) \geq 0$ *and by expression (23.10) if* $\lim_{p \to \infty} v(p) = 0$.

23.4   Discussion

In the present paper, we studied the problem of model selection for classification of high-dimensional vectors into a large number of classes. The objective is to construct a model selection procedure and study its asymptotic properties when both the number of features and the number of classes are large. Although the problem has been investigated by many authors, we research a more difficult version of a less explored random effect model where, moreover, features are sparse and have only moderate strength. The paper formulates necessary and sufficient conditions for separability of features into the informative and noninformative sets. In particular, the surprising conclusion of the paper is that separation of features becomes easier as the number of classes grows.

The most basic assumption of the models, that the a priori segregation of samples into classes with a certain structure (e.g. (23.2)) can be and has been done correctly, is perhaps the most problematic assumption. It may be that some covariates are "partially informative." For example, in [4], it was noted that a covariate differed from zero for only a single class and therefore served to differentiate that class from all others but was otherwise constant among classes. While that vector was recognized as informative in that application, it is unclear whether partially informative vectors will be retained in general; i.e. if one but only one class differs significantly from the mean, this useful variation may be falsely attributed to random error. It is unlikely that a model with indicators for partial informativity for each class is tractable; CONFESS naively modified would contain pL indicators $\{x_{il}\}_{i=1}^{p}{}_{l=1}^{L}$, with $x_{il} = 1$ iff class l differs significantly from all other classes as measured on vector i.

We note as well that the model does not apply directly to discrete data and that the assumption of a linear model may not represent all forms of between-class variation; e.g. in some settings, classes may differ by multiplicative factors.

Acknowledgements Marianna Pensky was partially supported by National Science Foundation (NSF), grant DMS-1106564.

References

1. Arias-Castro, E.W., Candes, E.J., Plan, Y.: Global testing under sparse alternatives: ANOVA, multiple comparisons and the higher criticism. Ann. Stat. **39**, 2533–2556 (2011)
2. Benjamini, Y., Hochberg, Y.: Controlling the false discovery rate: a practical and powerful approach to multiple testing. J. R. Stat. Soc. Ser. B **57**, 289–300 (1995)
3. Cai, T.T., Jin, J., Low, M.G.: Estimation and confidence sets for sparse normal mixtures. Ann. Stat. **35**, 2421–2449 (2007)
4. Davis, J., Pensky, M., Crampton, W.: Bayesian feature selection for classification with possibly large number of classes. J. Stat. Plan. Inference **141**, 3256–3266 (2011)
5. Donoho, D., Jin, J.: Higher criticism for detecting sparse heterogeneous mixtures Ann. Stat. **32**, 962–994 (2004)
6. Fan, J., Fan, Y.: High-dimensional classification using features annealed independence rules. Ann. Stat. **6**, 2605–2637 (2008)
7. Gradshteyn, I.S., Ryzhik, I.M.: Table of Integrals, Series, and Products, 7th edn. Academic Press, Amsterdam (2007)
8. Hall, P., Jin, J.: Innovated higher criticism for detecting sparse signals in correlated noise. Ann. Stat. **38**, 1686–1732 (2010)
9. Ingster, Y.I., Tsybakov, A.B., Verzelen, N.: Detection boundary in sparse regression. Electron. J. Stat. **4**, 1476–1526 (2010)
10. Paris, R.B.: A uniform asymptotic expansion for the incomplete gamma function. J. Comput. Appl. Math. **148**, 323–339 (2002)

Chapter 24
Semiparametric Bayesian Small Area Estimation Based on Dirichlet Process Priors

Silvia Polettini

Abstract Small area estimation concerns the problem of releasing estimates for domains that are not planned by design in statistical surveys. For such domains the observed sample size may often be too small to allow for accurate estimation of aggregates of interest. To borrow strength from related domains, the vast majority of small area models relies on mixed effects regression models. Whereas inference on the fixed effects is shown to be robust to deviations from normality, estimation of the random effects is crucial for predicting small area quantities. The potential impact of distributional assumptions on the random effects is shown to be important; missing covariates can lead to multimodal distributions for the random effects; the latter may also be skewed. Any parametric assumption, applying to nonobservable quantities, is difficult to check. This contribution examines a Bayesian semiparametric version of the Fay–Herriot model in which the default normality assumption for the random effects is replaced by a nonparametric specification, based on the Dirichlet process. Viability of the approach and the effect of introducing a flexible specification of the random effects are investigated through an application to simulated data.

Keywords Dirichlet process • Fay-Herriot • Gibbs sampling • Hierarchical Bayes • Mixed model • Nonparametric random effects • Small area estimation

24.1 Introduction: The Fay–Herriot Model

Small area estimation deals with the problem of releasing estimates for domains that are not planned by design in statistical surveys. Indeed sample surveys are generally designed to provide estimates of means of variables of interest for pre-specified, large domains. The growing request for estimates of aggregates (like poverty rates, mean income, etc.) defined at increasingly detailed domains (e.g. states, provinces, labour districts, demographic subgroups, etc.) has justified a growing interest for techniques that allow for accurate estimation of aggregates at unplanned domains,

S. Polettini (✉)
Department of Methods and Models for Economics, Territory and Finance,
Sapienza University of Rome, Via del Castro Laurenziano 9, 00161 Rome, Italy
e-mail: silvia.polettini@uniroma1.it

© Springer Science+Business Media New York 2014   259
M.G. Akritas et al. (eds.), *Topics in Nonparametric Statistics*, Springer Proceedings
in Mathematics & Statistics 74, DOI 10.1007/978-1-4939-0569-0_24

called small areas. For such small areas the observed sample size may often be too small to allow for accurate estimation of the aggregates of interest. Modern methods for small area estimation heavily rely on mixed effects models that increase the effective domain sample size by borrowing strength from related areas. The book by Rao [16] contains a thorough analysis of the model-based approach to small area estimation and SAE methods in general.

In this contribution we focus on area-level models. Area-level models rely on aggregated, area-specific quantities and therefore are often the only estimable model under microdata confidentiality protocols. Indeed, due to disclosure limitation procedures, data aggregated to the area-level are currently more readily available to users than are unit-level data, for both the variable of interest and for the auxiliary information. Another advantage of area-level modelling is that it allows one to account for the sampling design by introducing the direct survey estimates and their corresponding (design-based) variance estimates.

In the literature, the first formulation of an area-level model is the Fay–Herriot model [8]. Let m be the number of sampled small areas. In a small area estimation problem the design-unbiased, direct estimators of the target small-area parameters in the sampled areas $\hat{\theta}_i$, $i = 1, \ldots, m$ may present unacceptably high variances due to small sample size in some or all of the small areas. The Fay–Herriot model prescribes a sampling model for the direct survey estimates $\hat{\theta}_i$, supplemented by a linking model for the small area parameters of interest. Under the sampling model, design unbiased, direct survey estimators $\hat{\theta}_i$ of the small area parameters θ_i, $i = 1, \ldots, m$, are assumed to be available, whose sampling error is ϵ_i. Usually the ϵ_i's are assumed to be independent normal random variables, $\epsilon_i \sim N(0, \psi_i)$, so that

$$\hat{\theta}_i | \theta_i, \psi_i \sim N(\theta_i, \psi_i), \quad i = 1, \ldots, m. \tag{24.1}$$

To achieve the desired borrowing strength across areas, a linking model for θ_i is introduced, namely $\theta_i = x_i'\beta + v_i$, where $x_i = (x_{i1}, \ldots, x_{ip})'$ is a vector of auxiliary variables, β is a vector of regression coefficients, and finally the v_i's are area-specific random effects accounting for heterogeneity and lack of fit. Normality of the random effects is usually assumed: $v_i \sim N(0, \sigma_v^2)$, so that

$$\theta_i | \beta, \sigma_v^2 \sim N(x_i'\beta, \sigma_v^2), \quad i = 1, \ldots, m. \tag{24.2}$$

Under the Fay–Herriot model, the sampling variances are assumed to be known. In practice, smoothed estimators of such variances, usually by means of generalized variance function approach [5], are used, and then these are treated as known.

Combining the previous equations, one obtains a mixed effects linear regression model with normal random components, $\hat{\theta}_i = x_i'\beta + \epsilon_i + v_i$. Since areas of interest may not be all sampled in practice, it is assumed that the combined area-level model above also holds for the non-sampled areas. This amounts to assuming no selection-bias for areas.

In a frequentist setting, the Empirical Best Linear Predictor (EBLUP), obtained by replacing the unknown variance components in the BLUP by suitable estimators, is readily available for the Fay–Herriot model, based, e.g., on REML (see, e.g., [4]).

For area-level models, the distributional assumptions on sampling errors ϵ_i are usually justified by the properties of the direct estimators $\hat{\theta}_i$. By contrast, the normality assumption for the random effects v_i has no justification other than computational convenience and is difficult to detect in practice, since it involves unobservable quantities. The problem affects both frequentist and Bayesian analysis, although availability of MCMC techniques makes computational convenience less relevant in the latter framework.

24.2 Proposed Approach

The assumption of normality may fail to represent the distribution of the random effects for several reasons: missing covariates may lead to multimodal distributions; the distribution may be skewed. The effect on model estimates of distributional assumptions on the random effects is shown to be important [7, 10, 11, 15]. For instance, the presence of outliers may affect the precision of estimates of fixed effects and induce bias in estimation of the random effects.

Accurate prediction of the random effects is crucial for predicting small area quantities; although pointwise prediction is robust to deviations from normality, the precision of such predictions is decreased; also, estimation of nonlinear functionals may suffer from misrepresentation of the law of the random effects. For the reasons mentioned above, it would be important to rely on a model that has a flexible specification of the random effects, so as to achieve a greater adaptability and robustness against model misspecifications. In [7] the authors develop two robustified versions the Fay–Herriot model [8] by describing the random effects by either an exponential power (EP) or a skewed EP distribution and investigate robustness of such Fay–Herriot-type models under deviations from normality. Their aim is to understand whether estimates of linear and especially nonlinear functionals such as ranks are sensitive to deviations from normality of the random effects. Although the models proposed in [7] are based on distributions that generalize, and contain, the normal, yet these parametric models may fail to adequately describe the distribution of the random effects, and again the problem of checking the adequacy of these models arises. Datta and Lahiri [3] propose a robust hierarchical Bayesian model generalizing the Fay–Herriot model and constructed for the purpose of accommodating outliers. Heavy tailed distributions, namely a scale mixture of normal distributions, are used for the random effects. The resulting family includes as a special case the EP model proposed by [7]. Our approach has strong relations with the model of [3].

In this paper a different extension of the Fay–Herriot model is considered, based on Dirichlet process priors (DPP). Here the assumption of normality producing the linking model (24.2) is replaced by

$$v_i | \sigma_v^2, M \sim G(\cdot), \text{independently,} \quad G \sim DP(M, N(0, \sigma_v^2)) \quad i = 1, \ldots, m$$
(24.3)

where $DP(M, \phi)$ stands for the Dirichlet process (DP) [1, 9] with precision parameter M and base measure ϕ. In the context of a generalization of the Fay–Herriot model, it is natural to assume ϕ to be a normal distribution.

The representation above not only relaxes the normality assumption but also provides an enlarged model for describing the random effects, thus accommodating for outliers as well as multimodalities and other departures from normality.

The complete model specification reads as follows:

$$\hat{\theta}_i = \theta_i + e_i \quad e_i \sim N(0, \psi_i) \quad \text{independently,} \quad i = 1, \ldots, m \quad (24.4)$$

$$\theta_i = x_i'\beta + v_i, \quad v_i \sim G(\cdot), \quad \text{independently,} \quad i = 1, \ldots, m \quad (24.5)$$

$$G \sim DP(M, N(0, \sigma_v^2)) \quad (24.6)$$

$$\sigma_v^2 \sim IG(a_1, b_1) \quad (24.7)$$

$$\beta \sim N(0, d\mathrm{I}) \quad \text{where I is the identity matrix} \quad (24.8)$$

$$M \sim Gamma(a_2, b_2) \quad (24.9)$$

The hyperparameters in (24.7)–(24.9) are fixed; for comparability with the EBLUP, the hyperprior on the β vector is assumed to be normal with large variance.

The paper [2] provided a Polya-urn scheme representation of the joint distribution of realizations from a $DP(M, \phi)$ process as the product of successive conditional distributions of type

$$v_i | v_1, \ldots, v_{i-1}, M \sim \frac{M}{M+i-1}\phi(v_i) + \frac{1}{M+i-1}\sum_{k=1}^{i-1}\delta(v_k = v_i),$$

with $\delta(\cdot)$ denoting the Dirac delta function. The above representation induces clusters in the random effects due to the existence of a positive probability that a newly generated cluster coincides with a previous one. As a consequence, small areas are partitioned into clusters sharing the same random effect.

It is worth noting that, given n observations from a Dirichlet process with parameters M and ϕ, the marginal distribution of a partition $n_1, n_2, \ldots, n_k$ such that $\sum_1^k n_j = m, n_j > 0$ for all $j = 1, \ldots, k$ is

$$\pi(n_1, \ldots n_k) = \frac{\Gamma(M)}{\Gamma(M+m)}M^k\prod_{j=1}^{k}\Gamma(n_j);$$

thus the introduction of the Dirichlet process amounts to model random partitions of m objects. From the previous equations we see that M, the precision parameter of the DP, affects the number of clusters and therefore it is expected to influence the prediction of small area quantities.

Under the proposed model and following [13, 14] the likelihood function is

$$L(\beta|\hat{\boldsymbol{\theta}}) = \sum_{c=1}^{m} \sum_{C:|C|=c} \frac{\Gamma(M)}{\Gamma(M+m)} M^c \prod_{j=1}^{c} \Gamma(n_j) \int p(\hat{\theta}_{(j)}|\beta, v_j) dG_0(v_j)$$

where C is a partition of cells $\{1, .., K\}$ into c groups (or clusters), n_j is the number of observations in the j-th cluster, $1 \leq n_j \leq m$, $\hat{\theta}_{(j)}$ is the vector of the direct estimates belonging to cluster j and finally

$$p(\hat{\theta}_{(j)}|\beta, v_j) = \prod_{k \in \text{cluster } j} \frac{1}{\sqrt{2\pi\psi_k}} \exp\left\{-\frac{1}{2\psi_k}(\hat{\theta}_k - x_k'\beta - v_j)^2\right\}.$$

As evident from the previous equation, all areas belonging to a given cluster are assigned the same random effect; furthermore, the number of clusters in each partition is unknown.

A matrix representation of clusters is convenient. To each partition C of m objects into c clusters, let us associate an allocation matrix A of size $m \times c$ whose entries $a_{k,j}$ are 1 when the random effect v_k belongs to cluster j and zero otherwise. The random effects can be defined as $v = A\eta$ (and $v_k = a_k'\eta$) by setting $\eta_j = v_k$ when $v_k \in$ cluster j. Under this reparametrization the likelihood becomes

$$L(\beta|\hat{\theta}, A) = \frac{\Gamma(M)}{\Gamma(M+m)} \sum_{c=1}^{m} M^c$$

$$\times \sum_{A \in \mathscr{A}_c} \prod_{j=1}^{c} \Gamma(n_j) \int \prod_{k=1}^{m} p(\hat{\theta}_k|A, \eta, \beta)\phi(\eta_1, \ldots \eta_c) d\eta.$$

where $\mathscr{A}_c$ is the set of all allocation matrices A having c nonempty clusters and $\eta = (\eta_1, \ldots \eta_c)$, $\eta_j \sim N(0, \sigma_v^2)$, independently, and

$$\prod_{k=1}^{m} p(\hat{\theta}_k|A, \eta, \beta)\phi(\eta) =$$

$$\prod_{i=1}^{m} \frac{1}{\sqrt{2\pi\psi_i}} \exp\left\{-\frac{(\hat{\theta}_i - x_i'\beta - a_i'\eta)^2}{2\psi_i}\right\} \left(\frac{1}{2\pi\sigma_v^2}\right)^{c/2} \exp\{-\eta'\eta/2\sigma_v^2\}$$

As already discussed, uncertainty on M, affecting the number of clusters, is included in the model. This amounts to a form of model averaging. For a noninformative specification of the prior for M, we follow [6], where it is suggested to use a Gamma prior with parameters chosen to match the prior information about the number of clusters. Assuming absence of prior information, we specify a Gamma with parameters chosen so that the induced prior for the total number of clusters is closest to the discrete uniform distribution on $\{1, \ldots, m\}$ in terms of the Kullback–Leibler divergence.

Table 24.1 Measures of average absolute (ARB) and squared (ASRB) relative bias under the simulated settings

	ARB	ASRB
Case 1: Standard normal		
Proposed model	0.48	1.36
Standard HB model	0.50	1.64
Case 2: Mixture of normals		
Proposed model	6.87	3993.65
Standard HB model	9.62	8262.80

The posterior distribution of the small area quantities θ_i is analytically intractable, so Markov Chain Monte Carlo techniques are used to perform inference. Specifically, a Gibbs sampler is constructed, repeatedly sampling one set of parameters at a time, namely $\beta|$rest, $v|$rest, $M|$rest, $\sigma_v^2|$rest. Given model specification and the previous equations, sampling the fixed effects parameters given the cluster configuration proceeds as in standard normal hierarchical models; updating $\sigma_v^2|$rest is also standard, whereas the algorithm proposed in [18] is used for generating the random effects. Finally, $M|$rest is updated using a Metropolis step.

The semiparametric linear mixed models is reported in [12] to reduce the variability of the regression parameters estimates, producing uniformly shorter HPD intervals than the standard normal random effects models. It is of interest here to understand the performance of the method in predicting small area quantities under the extended Fay–Herriot model above, primarily the domain means θ_i in (24.5).

24.3 Application

The performance of the method is tested by means of an application to simulated data. Simulation enables us to benchmark the fitted values to the true underlying values. Working in a Bayesian framework, it is natural to compare the proposed model with the standard normal hierarchical Bayesian (HB) model. Such comparison is performed under two different settings for the random effects:

1. normal case: $v_i \sim N(0, 1)$, independently, $i = 1, \ldots, m$
2. mixture of two normals $v \sim 0.9N(0, 1) + 0.1N(0, 25)$, independently, $i = 1, \ldots, m$, as in [17].

A synthetic dataset of $m = 100$ areas was generated from the model $\hat{\theta}_i = \beta_0 + \beta_1 x_{1i} + \beta_2 x_{2i} + \beta_3 x_{3i} + \epsilon_i + v_i$ with $\beta = (1, 2, -3, 4)$; covariates have been generated from independent normal distributions and $\epsilon_i \sim N(0, \psi_i)$ independently, $i = 1, \ldots, m$, with ψ_i ranging from 0.055 to 6.998, with average 1.54 and median 2.18.

In the application, the prior on β was taken to be $N(0, 10^6)$ in both the parametric and the semiparametric model, for comparability with the EBLUP. As regards the proposed model, the prior for M was chosen as suggested in [6] to return a "flat" prior on the number of clusters, as detailed in the previous section, while the prior for the variance of the base measure of the Dirichlet Process was chosen to be IG(0.01, 0.01).

The HB predictor is robust to departures from normality of the random effects and, as expected, the predictions of the small area means based on posterior means do not differ remarkably under either models in both settings (results not shown here). This confirms that the performance of the new model in terms of point predictions is sensible.

Denoting by $\tilde{\theta}_i$ the small area predictions under a given model, the following quantities are considered as measures of prediction error: average relative bias,

$$\text{ARB} = \frac{1}{m} \sum_{i=1}^{m} \frac{|\tilde{\theta}_i - \theta_i|}{|\theta_i|}$$

and average squared relative bias,

$$\text{ARSB} = \frac{1}{m} \sum_{i=1}^{m} \left(\frac{\tilde{\theta}_i - \theta_i}{\theta_i} \right)^2.$$

The above quantities were computed to compare the standard HB Fay–Herriot model and the proposed semiparametric one. As shown in Table 24.1, the summaries indicate a slight advantage in using the proposed method in all cases, even in the normal setting where the standard HB model is optimal.

Besides measures of prediction error, the main interest is whether the semiparametric model may produce more accurate inferences in terms of shorter credible intervals. The equal tail 95 % credible intervals were compared to those obtained under the standard normal hierarchical model. Whereas for the standard normal setting the two models give intervals of comparable size, with average size 2.1 under the proposed model and 2.7 under the standard HB model, under the mixture of normals setting the nonparametric model achieves a slightly better performance, the mean interval size being 2.7 under the proposed model and 3.8 under the standard HB model. The reason for the observed reduction in variability even under normality of the random effects may be ascribed to areas in the same cluster sharing the same random effect; the DP prior effectively achieves flexibility without increasing the dimensionality of the problem to a large extent.

Figures 24.1 and 24.2 show the posterior predictive distributions for a selection of areas under the two assumptions for the model generating the random effects. In most cases the posterior distributions are more concentrated than the under standard HB model and the true value belongs to the 95 % credibility interval.

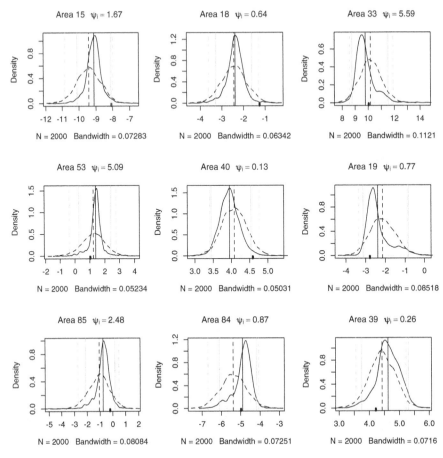

Fig. 24.1 MCMC approximation of the posterior predictive distributions of the area means under the standard normal setting. In each panel the kernel density estimates of the posterior predictive distribution of the small area mean are presented for the standard HB model (*dashed line*) and the proposed semiparametric model (*solid line*). The true area mean (*black dot*), along with its HB (*black dashed line*) and semiparametric (*black solid line*) estimates are depicted. The endpoints of the 95 % credibility intervals are shown in *grey* (HB: *grey dashed lines*, proposed model: *grey solid lines*). For each of the selected areas, ψ_i, the sampling variance of the direct estimator $\hat{\theta}_i$ is reported

24.4 Final Remarks

This contribution investigates a semiparametric version of the Fay–Herriot model based on the DP prior. The model formulation allows to relax parametric assumptions on the random effects using a parsimonious nonparametric representation of such model component, relying on a Dirichlet process prior. The aim is to capture area-specific variability not accounted for by the covariates using a flexible

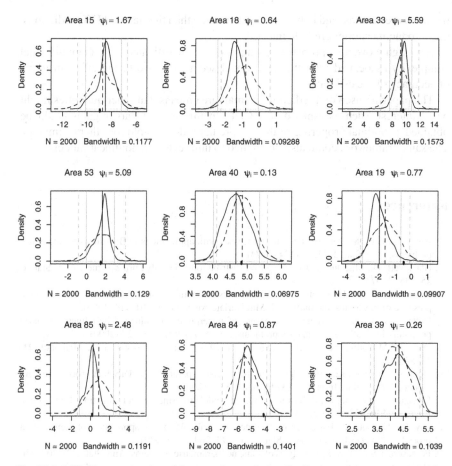

Fig. 24.2 MCMC approximation of the posterior predictive distributions of the area means under the mixture of normals setting. In each panel the kernel density estimates of the posterior predictive distribution of the small area mean are presented for the standard HB model (*dashed line*) and the proposed semiparametric model (*solid line*). The true area mean (*black dot*), along with its HB (*black dashed line*) and semiparametric (*black solid line*) estimates are depicted. The endpoints of the 95 % credibility intervals are shown in *grey* (HB: *grey dashed lines*, proposed model: *grey solid lines*). For each of the selected areas, ψ_i, the sampling variance of the direct estimator $\hat{\theta}_i$ is reported

representation, without excessively increasing the number of model parameters. The characteristics of the DP imply that the random effects are given a mixture structure, with number of components that depends on the DP precision parameter, M. Uncertainty on M is also accounted for in the proposed model.

The experiments performed seem to indicate that the proposed model effectively achieves flexibility in modelling the random effects, without increasing the dimensionality of the problem. As a consequence, the posterior predictive distributions of the small area means tend to be more concentrated than the standard HB predictions, thus producing shorter credible intervals.

Although coverage properties remain to be investigated, the computational complexity associated with the new model seem to be worth the effort in terms of accuracy of the estimates.

The paper focused on prediction of small area means, but other quantities might be of interest, such as ranks or the CDF. In this respect, larger gains could be obtained as optimal properties of the HB predictor do not hold for other estimators. The same framework could also be applied to unit level and nonlinear models.

References

1. Antoniak, C.E.: Mixtures of Dirichlet processes with applications to Bayesian nonparametric problems. Ann. Stat. **2**, 1152–1174 (1974)
2. Blackwell, D., MacQueen, J.B.: Ferguson distributions via Pólya urn schemes. Ann. Stat. **1**, 353–355 (1973)
3. Datta, G.S., Lahiri, P.: Robust hierarchical Bayes estimation of small area characteristics in the presence of covariates and outliers. J. Multivariate Anal. **54**(2), 310–328 (1995).
4. Datta, G.S., Lahiri, P.: A unified measure of uncertainty of estimated best linear unbiased predictors in small area estimation problems. Stat. Sin. **10**(2), 613–627 (2000)
5. Dick, P.: Modelling net undercoverage in the 1991 Canadian Census. Surv. Methodol. **21**, 45–54 (1995)
6. Dorazio, R.M.: On selecting a prior for the precision parameter of Dirichlet process mixture models. J. Stat. Plan. Inference **139**(9), 3384–3390 (2009)
7. Fabrizi, E., Trivisano, C.: Robust linear mixed models for Small Area Estimation. J. Stat. Plan. Inference **140**, 433–443 (2010)
8. Fay, R., Herriot, R.: Estimates of income for small places: an application of James-Stein procedures to census data. J. Am. Stat. Assoc. **74**, 269–277 (1979)
9. Ferguson, T.S.: A Bayesian analysis of some nonparametric problems. Ann. Stat. **1**(2), 209–230 (1973)
10. Ghidey, W., Lesaffre, E., Eilers, P.: Smooth random effects distribution in a linear mixed model. Biometrics **60**(4), 945–953 (2004)
11. Jara, A., Quintana, F., Martín, E.S.: Linear mixed models with skew-elliptical distributions: A Bayesian approach. Comput. Stat. Data Anal. **52**(11), 5033–5045 (2008)
12. Kyung, M., Gill, J., Casella, G.: Characterizing the variance improvement in linear Dirichlet random effects models. Stat. Probab. Lett. **79**, 2343–2350 (2009)
13. Liu, J.S.: Nonparametric hierarchical Bayes via sequential imputations. Ann. Stat. **24**(3), 911–930 (1996)
14. Lo, A.Y.: On a class of Bayesian nonparametric estimates. I. Density estimates. Ann. Stat. **12**(1), 351–357 (1984)
15. Ohlssen, D.I., Sharples, L.D., Spiegelhalter, D.J.: Flexible random-effects models using Bayesian semi-parametric models: Applications to institutional comparisons. Stat. Med. **26**, 2088–2112 (2007)
16. Rao, J.N.K.: Small Area Estimation. Wiley Series in Survey Methodology. Wiley, New York (2003)

17. Sinha, S.K., Rao, J.N.K.: Robust small area estimation. Can. J. Stat. **37**(3), 381–399 (2009)
18. West, M., Müller, P., Escobar, M.D.: Hierarchical priors and mixture models, with application in regression and density estimation. In: Freeman, P.R., Smith, A.F.M. (eds.) Aspects of Uncertainty. A Tribute to D. V. Lindley, pp. 363–386. Wiley, New York (1994)

Chapter 25
Bootstrap Confidence Intervals in Nonparametric Regression Without an Additive Model

Dimitris N. Politis

Abstract The problem of confidence interval construction in nonparametric regression via the bootstrap is revisited. When an additive model holds true, the usual residual bootstrap is available but it often leads to confidence interval under-coverage; the case is made that this under-coverage can be partially corrected using predictive—as opposed to fitted—residuals for resampling. Furthermore, it has been unclear to date if a bootstrap approach is feasible in the absence of an additive model. The main thrust of this paper is to show how the transformation approach put forth by Politis (Test 22(2):183–221, 2013) in the related setting of prediction intervals can be found useful in order to construct bootstrap confidence intervals *without* an additive model.

Keywords Model-free inference • Resampling • Nonparametric function estimation

25.1 Introduction

Consider regression data of the type $\{(Y_t, x_t), \ t = 1, \ldots, n\}$. For simplicity of presentation, the regressor x_t is assumed univariate and deterministic; the case of a multivariate regressor is handled similarly. As usual, it will be assumed that $Y_1, \ldots, Y_n$ are independent but not identically distributed. Attention focuses primarily on the first two moments of the response Y_t, namely

$$\mu(x_t) = E(Y_t|x_t) \ \text{and} \ \sigma^2(x_t) = \text{Var}(Y_t|x_t). \tag{25.1}$$

In the nonparametric setting, the functions $\mu(\cdot)$ and $\sigma(\cdot)$ are considered unknown but assumed to possess some degree of smoothness (differentiability, etc.). There are many approaches towards nonparametric estimation of the functions μ and σ, e.g.,

D.N. Politis (✉)
University of California at San Diego, La Jolla, CA 92093-0112, USA
e-mail: dpolitis@ucsd.edu

© Springer Science+Business Media New York 2014   271
M.G. Akritas et al. (eds.), *Topics in Nonparametric Statistics*, Springer Proceedings
in Mathematics & Statistics 74, DOI 10.1007/978-1-4939-0569-0_25

wavelets and orthogonal series, smoothing splines, local polynomials, and kernel smoothers. For concreteness, this paper will focus on one of the oldest methods, namely the Nadaraya–Watson (N-W) kernel estimators; see Li and Racine [7] and the references therein.

Beyond point estimates of the functions μ and σ, it is important to be able to additionally provide interval estimates in order to have a measure of their statistical accuracy. Suppose, for example, that a practitioner is interested in the expected response to be observed at a future point x_f. A confidence interval for $\mu(x_f)$ is then desirable. Under regularity conditions, such a confidence interval can be given either via a large-sample normal approximation, or via a resampling approach; see, e.g., Freedman [3], Härdle and Bowman [5], Härdle and Marron [6], Hall [4], or Neumann and Polzehl [9].

Typical regularity conditions for the above bootstrap approaches involve the assumption of an additive model with respect to independent and identically distributed (i.i.d.) errors. In Sect. 25.2, we revisit the usual model-based bootstrap for regression adding the dimension of employing predictive as opposed to fitted residuals as advocated by Politis [10, 11] in a related context. More importantly, in Sect. 25.3 we address the problem of constructing a bootstrap confidence interval for $\mu(x_f)$ *without* an underlying additive model.

The *model-free* approach developed in this paper is totally automatic, relieving the practitioner from the need to find an optimal transformation towards additivity and variance stabilization; this is a significant practical advantage because of the multitude of such proposed transformations, e.g. the Box/Cox power family, ACE, AVAS, etc.—see Linton et al. [8] and the references therein. The finite-sample simulations provided in Sect. 25.4 confirm the viability and good performance of the model-free confidence intervals.

25.2 Model-Based Nonparametric Regression

25.2.1 Nonparametric Regression with an Additive Model

An additive model for nonparametric regression is given by the equation

$$Y_t = \mu(x_t) + \sigma(x_t)\,\varepsilon_t \,, \quad t = 1, \ldots, n, \tag{25.2}$$

with $\varepsilon_t \sim$ i.i.d. (0,1) from an (unknown) distribution F. The N-W estimator of $\mu(x)$ is defined as

$$m_x = \sum_{i=1}^{n} Y_i \tilde{K}\left(\frac{x - x_i}{h}\right) \quad \text{with} \quad \tilde{K}\left(\frac{x - x_i}{h}\right) = \frac{K\left(\frac{x-x_i}{h}\right)}{\sum_{k=1}^{n} K\left(\frac{x-x_k}{h}\right)} \tag{25.3}$$

where h is the bandwidth, and $K(x)$ is a symmetric kernel function with $\int K(x)dx = 1$. Similarly, the N-W estimator of $\sigma^2(x)$ is given by $s_x^2 = M_x - m_x^2$ where $M_x = \sum_{i=1}^{n} Y_i^2 \tilde{K} \left(\frac{x-x_i}{h} \right)$.

For $t = 1, \ldots, n$, let $e_t = (Y_t - m_{x_t})/s_{x_t}$ denote the *fitted* residuals, and $\tilde{e}_t = (Y_t - m_{x_t}^{(t)})/s_{x_t}^{(t)}$ the *predictive* residuals. Here, $m_x^{(t)}$ and $M_x^{(t)}$ denote the estimators m_x and M_x, respectively, computed from the *delete-Y_t* dataset: $\{(Y_i, x_i), i = 1, \ldots, t-1 \text{ and } i = t+1, \ldots, n\}$. As before, define $s_{x_t}^{(t)} = \sqrt{M_{x_t}^{(t)} - (m_{x_t}^{(t)})^2}$. Choosing the bandwidth h is often done by *cross-validation*, i.e., picking h to minimize $\sum_{t=1}^{n} \tilde{e}_t^2$, or its L_1 analog: $\sum_{t=1}^{n} |\tilde{e}_t|$.

25.2.2 Model-Based Confidence Intervals

Consider the problem of constructing a confidence interval for the regression function $\mu(x_f)$ at a point of interest x_f. A normal approximation to the distribution of the estimator m_{x_f} implies an approximate $(1 - \alpha)100\%$ equal-tailed, confidence interval for $\mu(x_f)$ given by:

$$[m_{x_f} + v_{x_f} \cdot z(\alpha/2), \ m_{x_f} + v_{x_f} \cdot z(1 - \alpha/2)] \tag{25.4}$$

where $v_{x_f}^2 = s_{x_f}^2 \sum_{i=1}^{n} \tilde{K}^2 (\frac{x_f - x_i}{h})$ with $\tilde{K}$ defined in (25.3), and $z(\alpha)$ being the α-quantile of the standard normal. If the "density" (e.g., histogram) of the design points $x_1, \ldots, x_n$ can be thought to approximate a given functional shape (say, $f(\cdot)$) for large n, then the large-sample approximation

$$\sum_{i=1}^{n} \tilde{K}^2 \left(\frac{x_f - x_i}{h} \right) \sim \frac{\int K^2(x)dx}{nh \, f(x_f)} \tag{25.5}$$

can be used which relies on the assumption that $\int K(x)dx = 1$; see, e.g., Li and Racine [7].

Interval (25.4) may be problematic in two respects: (a) it ignores the bias of m_x, so it must be either explicitly bias-corrected, or a suboptimal bandwidth must be used to ensure undersmoothing; and (b) it is based on a Central Limit Theorem which may not be a good finite-sample approximation if the errors are skewed and/or leptokurtic, or when the sample size is not large enough. For both above reasons, practitioners often prefer bootstrap methods over the normal approximation interval (25.4). When using fitted residuals, the following algorithm is the well-known residual bootstrap pioneered by Freedman [3] in a linear regression setting, and extended to nonparametric regression by Härdle and Bowman [5], and other authors. As an alternative, we also propose the use of predictive residuals for resampling as advocated by Politis [10, 11] in a related context. The predictive residuals have an empirical distribution that has similar shape as that of the fitted residuals but it has larger scale. This is a finite-sample phenomenon only but it may help alleviate the well-known phenomenon of under-coverage of bootstrap confidence intervals.

Our goal is to approximate the distribution of the confidence *root*: $\mu(x_f) - m_{x_f}$ by that of its bootstrap counterpart.

Resampling Algorithm for Model-Based Confidence Intervals for $\mu(x_f)$

1. Based on the $\{(Y_t, x_t), t = 1, \ldots, n\}$ data, construct the estimates m_x and s_x from which the fitted residuals e_i, and predictive residuals $\tilde{e}_i$ are computed for $i = 1, \ldots, n$.
2. For the traditional model-based bootstrap approach (MB), let $r_i = e_i - n^{-1} \sum_j e_j$, for $i = 1, \ldots, n$. For the predictive residual approach (PRMB) as in Politis [10], let $r_i = \tilde{e}_i - n^{-1} \sum_j \tilde{e}_j$, for $i = 1, \ldots, n$.

   (a) Sample randomly (with replacement) the residuals $r_1, \ldots, r_n$ to create the bootstrap pseudo-residuals $r_1^*, \ldots, r_n^*$ whose empirical distribution is denoted by $\hat{F}_n^*$.
   (b) Create pseudo-data in the Y domain by letting $Y_i^* = m_{x_i} + s_{x_i} r_i^*$, for $i = 1, \ldots, n$.
   (c) Based on the pseudo-data $\{(Y_t^*, x_t), t = 1, \ldots, n\}$, re-estimate the functions $\mu(x)$ and $\sigma(x)$ by the kernel estimators m_x^* and s_x^* (with same kernel and bandwidths as the original estimators m_x and s_x).
   (d) Calculate a replicate of the *bootstrap confidence root*: $m_{x_f} - m_{x_f}^*$.

3. Steps (a)–(d) in the above are repeated B times, and the B bootstrap root replicates are collected in the form of an empirical distribution with α-quantile denoted by $q(\alpha)$.
4. Then, a $(1 - \alpha)100\%$ equal-tailed confidence interval for $\mu(x_f)$ is given by:

$$[m_{x_f} + q(\alpha/2), m_{x_f} + q(1 - \alpha/2)]. \tag{25.6}$$

Remark 2.1. As in all nonparametric smoothing problems, choosing the bandwidth is often a key issue due to the ever-looming problem of bias; the addition of a bootstrap algorithm as above further complicates things. Different authors have used various tricks to account for the bias. For example, Härdle and Bowman [5] construct a kernel estimate for the second derivative $\mu''(x)$, and use this estimate to explicitly correct for the bias; the estimate of the second derivative is known to be consistent but it is difficult to choose its bandwidth. Härdle and Marron [6] estimate the (fitted) residuals using the optimal bandwidth but the resampled residuals are then added to an oversmoothed estimate of μ; the bootstrapped data are then smoothed using the optimal bandwidth. Neumann and Polzehl [9] use only one bandwidth but it is of smaller order than the mean square error optimal rate; this *undersmoothing* of curve estimates was first proposed by Hall [4] and is perhaps the easiest theoretical solution towards confidence band construction although the recommended degree of undersmoothing for practical purposes is not obvious.

Remark 2.2. An important feature of all bootstrap procedures is that they can handle *joint* confidence intervals, i.e., confidence *regions*, with the same ease as the univariate ones. This is especially true in regression where simultaneous confidence intervals are typically constructed in the form of confidence *bands*; the details are well known in the literature and are omitted due to lack of space.

25.3 Model-Free Nonparametric Regression

25.3.1 Nonparametric Regression Without an Additive Model

We now revisit the nonparametric regression setup but in a situation where a model such as (25.2) cannot be considered to hold true (not even approximately). As an example of model (25.2) not being valid, consider the setup where the skewness and/or kurtosis of Y_t depends on x_t, and thus centering and studentization will not result in "i.i.d.–ness." The dataset is still $\{(Y_t, x_t), \ t = 1, \ldots, n\}$ where the regressor x_t is univariate and deterministic, and the variables $Y_1, Y_2, \ldots$ are *independent* but not identically distributed. Define the conditional distribution $D_x(y) = P\{Y_f \leq y | x_f = x\}$ where (Y_f, x_f) represents the random response Y_f associated with regressor x_f. Attention still focuses on constructing an interval estimate of $\mu(x_f) = E(Y_f | x_f) = \int y \, D_{x_f}(dy)$.

Throughout this section, we will assume that the function $D_x(y)$ is continuous in both x and y. Consequently, we can estimate $D_x(y)$ by the local (weighted) empirical distribution

$$\hat{D}_x(y) = \sum_{i=1}^{n} \mathbf{1}\{Y_i \leq y\} \tilde{K}\left(\frac{x - x_i}{h}\right); \tag{25.7}$$

this is just an N-W smoother of the variables $\mathbf{1}\{Y_t \leq y\}$, $t = 1, \ldots, n$. Estimator $\hat{D}_x(y)$ enjoys many desirable properties, including asymptotic consistency, but is discontinuous as a function of y. To construct a continuous (and differentiable) estimator, let b be a positive bandwidth parameter and $\Lambda(y)$ be a (differentiable) distribution function that is strictly increasing, and define

$$\bar{D}_x(y) = \sum_{i=1}^{n} \Lambda\left(\frac{y - Y_i}{b}\right) \tilde{K}\left(\frac{x - x_i}{h}\right). \tag{25.8}$$

Under regularity conditions, Li and Racine [7, Theorem 6.2] show that

$$\text{Var}(\bar{D}_x(y)) = O\left(\frac{1}{hn}\right) \quad \text{and} \quad \text{Bias}(\bar{D}_x(y)) = O(h^2 + b^2) \tag{25.9}$$

assuming that $h \to 0, b \to 0, hn \to \infty$ and $\sqrt{hn}(h^3 + b^3) = o(1)$; to minimize the asymptotic Mean Squared Error of $\bar{D}_x(y)$, the optimal bandwidths are $h \sim c_h n^{-1/5}$ and $b \sim c_b n^{-2/5}$ for some positive constants c_h, c_b.

Recall that the Y_ts are non-i.i.d. only because they do not have identical distributions. Since they are continuous random variables, the *probability integral transform* is applicable. If we let $\eta_i = D_{x_i}(Y_i)$ for $i = 1, \ldots, n$, then $\eta_1, \ldots, \eta_n$ are i.i.d. Uniform(0,1). Of course, $D_x(\cdot)$ is not known but we can define

$$u_i = \bar{D}_{x_i}(Y_i) \quad \text{for } i = 1, \ldots, n; \tag{25.10}$$

by the consistency of $\bar{D}_x(\cdot)$, we can now claim that $u_1, \ldots, u_n$ are approximately i.i.d. Uniform(0,1).

Using (25.10) and following the *Model-free Prediction Principle* of Politis [10], the quantity

$$\Pi_{x_f} = n^{-1} \sum_{i=1}^{n} \hat{D}_{x_f}^{-1}(u_i) \tag{25.11}$$

was proposed as an L_2-optimal predictor of Y_f, i.e., an approximation to the conditional expectation $\mu(x_f) = E(Y_f|x_f)$. Note that $\hat{D}_{x_f}(y)$ is a step function in y, and thus not invertible; the notation $\hat{D}_{x_f}^{-1}$ denotes the quantile inverse. Alternatively, one could propose the quantity $n^{-1} \sum_{i=1}^{n} \bar{D}_{x_f}^{-1}(u_i)$ where a true inverse is used; the difference between the two is negligible, and definition (25.11) is straightforward.

Note that Π_{x_f} is defined as a function of the approximately i.i.d. variables $u_1, \ldots, u_n$; as such, it may be amenable to the original i.i.d. bootstrap of Efron [2]. Two questions arise: (a) is the estimator Π_{x_f} quite different from the standard N-W estimator m_{x_f}? and (b) could m_{x_f} itself be bootstrapped using i.i.d. resampling? The answers to these questions are NO and YES, respectively, due to the following fact. To motivate it, recall that the N-W estimator m_x can be expressed alternatively as

$$m_x = \sum_{i=1}^{n} Y_i \tilde{K}\left(\frac{x - x_i}{h}\right) = \int y \, \hat{D}_x(dy) = \int_0^1 \hat{D}_x^{-1}(u) du. \tag{25.12}$$

The last equality in (25.12) is the identity $\int y \, F(dy) = \int_0^1 F^{-1}(u) du$ that holds true for any distribution F.

Fact 3.1. *Assume that $D_x(y)$ is continuous in x, and differentiable in y with derivative that is everywhere positive on its support. Then, Π_{x_f} and m_{x_f} are asymptotically equivalent, i.e., $\sqrt{nh}\,(\Pi_{x_f} - m_{x_f}) = o_p(1)$ for any x_f that is not a boundary point.*

One way to prove the above is to show that the average appearing in (25.11) is close to a Riemann sum approximation to the integral at the RHS of (25.12) based on a grid of n points. The law of the iterated logarithm for order statistics of uniform spacings can be useful here; see Devroye [1] and the references therein.

Remark 3.1. The above line of arguments indicates that there is a variety of estimators that are asymptotically equivalent to m_{x_f} in the sense of Fact 3.1. For example, the Riemann sum $M^{-1} \sum_{k=1}^{M} \hat{D}_{x_f}^{-1}(k/M)$ is such an approximation as long as $M \geq n$. A stochastic approximation can also be concocted as $M^{-1} \sum_{i=1}^{M} \hat{D}_{x_f}^{-1}(W_i)$ where $W_1, \ldots, W_M$ are i.i.d. generated from a Uniform(0,1) distribution and $M \geq n$.

25.3.2 Bootstrap Algorithm for Model-free Confidence Intervals

Let $\hat{\mu}(x_f)$ denote our chosen estimator of $\mu(x_f) = E(Y_f|x_f)$, i.e., either m_{x_f} or Π_{x_f}, or even one of the other asymptotically equivalent estimators discussed in Remark 3.1. Our goal is to approximate the distribution of the confidence *root*: $\mu(x_f) - \hat{\mu}(x_f)$ by that of its bootstrap counterpart. The algorithm reads as follows.

Resampling Algorithm for Model-Free Confidence Intervals for $\mu(x_f)$

1. Based on the $\{(Y_t, x_t), t = 1, \ldots, n\}$ data, construct the estimates $\hat{D}_x(\cdot)$ and $\bar{D}_x(\cdot)$, and use (25.10) to obtain the transformed data $u_1, \ldots, u_n$ that are approximately i.i.d. Uniform $(0,1)$.

   (a) Sample randomly (with replacement) the transformed data $u_1, \ldots, u_n$ to create bootstrap pseudo-data $u_1^*, \ldots, u_n^*$.
   (b) Use the quantile inverse transformation $\hat{D}_x^{-1}$ to create bootstrap pseudo-data in the Y domain, i.e., let $\underline{Y}_n^* = (Y_1^*, \ldots, Y_n^*)$ where $Y_t^* = \hat{D}_{x_t}^{-1}(u_t^*)$. Note that Y_t^* is paired with the original x_t design point; hence, the bootstrap dataset is $\{(Y_t^*, x_t), t = 1, \ldots, n\}$.
   (c) Based on the pseudo-data $\{(Y_t^*, x_t), t = 1, \ldots, n\}$, re-estimate the conditional distribution $D_x(\cdot)$; denote the bootstrap estimates by $\hat{D}_x^*(\cdot)$ and $\bar{D}_x^*(\cdot)$.
   (d) Calculate a replicate of the *bootstrap confidence root*: $\hat{\mu}(x_f) - \hat{\mu}^*(x_f)$ where $\hat{\mu}^*(x_f)$ equals either $\int y \, \hat{D}_{x_f}^*(dy) = \int_0^1 \hat{D}_{x_f}^{*-1}(u)du$ or $n^{-1} \sum_{i=1}^{n} \hat{D}_{x_f}^{*-1}(u_i^*)$ according to whether $\hat{\mu}(x_f)$ was chosen as m_{x_f} or Π_{x_f}.

2. Steps (a)–(d) in the above are repeated B times, and the B bootstrap root replicates are collected in the form of an empirical distribution with α-quantile denoted by $q(\alpha)$.
3. Then, the *Model-Free* (MF) $(1 - \alpha)100\%$ equal-tailed, confidence interval for $\mu(x_f)$ is[(a)]

$$[\hat{\mu}(x_f) + q(\alpha/2), \hat{\mu}(x_f) + q(1 - \alpha/2)].[(a)] \qquad (25.13)$$

Remark 3.2. An alternative way to implement step 1(a) of the above algorithm is:

a'. Generate bootstrap pseudo-data $u_1^*, \ldots, u_n^*$ i.i.d. from an exact Uniform $(0, 1)$ distribution.

If the above choice is made, then there is no need to use (25.10) to obtain the transformed data $u_1, \ldots, u_n$; in this sense, the smooth estimator $\bar{D}_x(\cdot)$ is not needed, and the step function $\hat{D}_x(\cdot)$ suffices for the algorithm.

The downside to the above proposal is that the option to use "predictive" u-data is unavailable. To elaborate, recall that Politis [10] defined the model-free "predictive" u-data as follows. Let $\bar{D}_{x_t}^{(t)}$ denote the estimator $\bar{D}_{x_t}$ as computed from the delete-Y_t dataset, i.e., $\{(Y_i, x_i), i = 1, \ldots, t-1 \text{ and } i = t+1, \ldots, n\}$. Now let

$$u_t^{(t)} = \bar{D}_{x_t}^{(t)}(Y_t) \quad \text{for } t = 1, \ldots, n. \tag{25.14}$$

The $u_t^{(t)}$ variables are the model-free analogs of the predictive residuals $\tilde{e}_t$ of Sect. 25.2.

Remark 3.3. We can now define *Predictive Model-Free* (PMF) confidence intervals for $\mu(x_f)$. The PMF Resampling Algorithm is identical to the above with one exception; replace step 1(a) with the following:

a''. Sample randomly (with replacement) the predictive u-data $u_1^{(1)}, \ldots, u_n^{(n)}$ to create bootstrap pseudo-data $u_1^*, \ldots, u_n^*$.

Remark 3.4. Recall that the model-free L_1-optimal predictor of Y_f is given by the median$\{\bar{D}_{x_f}^{-1}(u_i)\}$; see Politis [10, 11]. Therefore, by analogy to Fact 3.1, we have: median$\{\bar{D}_{x_f}^{-1}(u_i)\} = \bar{D}_{x_f}^{-1}(\text{median}\{u_i\}) \simeq \bar{D}_{x_f}^{-1}(1/2)$ since the u_is are approximately Uniform (0,1). Hence, if the practitioner wanted to estimate the median (as opposed to the mean) of the conditional distribution of Y_f given x_f, then the local median $\bar{D}_{x_f}^{-1}(1/2)$, could be bootstrapped using i.i.d. resampling in the same manner that median$\{\bar{D}_{x_f}^{-1}(u_i)\}$ can be bootstrapped.

25.4 Simulations

25.4.1 When a Nonparametric Regression Model Is True

The building block for the simulation in Sect. 25.4.1 is model (25.2) with $\mu(x) = \sin(x)$, $\sigma(x) = 1/2$, and errors ε_t i.i.d. N(0,1) or two-sided exponential (Laplace) rescaled to unit variance. Knowledge that the variance $\sigma(x)$ is constant was not used in the estimation, i.e., $\sigma(x)$ was estimated from the data. For each distribution, 500 datasets each of size $n = 100$ were created with the design points $x_1, \ldots, x_n$ being equi-spaced on $(0, 2\pi)$, and N-W estimates of $\mu(x) = E(Y|x)$ and $\sigma^2(x) = \text{Var}(Y|x)$ were computed using a normal kernel in R.

Confidence intervals with nominal level $\alpha = 0.90$ were constructed using the two methods presented in Sect. 25.2.2: Traditional Model-Based (MB) and Predictive Residual Model-Based (PRMB); the two methods presented in Sect. 25.3.2: Model-Free (MF) of (25.13), and Predictive Model-Free (PMF) from Remark 3.3; and the

Table 25.1 Empirical coverage levels (CVR) of confidence intervals according to different methods at several x_f points spanning the interval $(0, 2\pi)$

$x_f/\pi =$	0.15	0.3	0.5	0.75	1	1.25	1.5	1.7	1.85
MB	0.802	0.735	0.725	0.756	0.811	0.760	0.736	0.738	0.765
PRMB	0.860	0.821	0.813	0.826	0.878	0.840	0.806	0.812	0.837
MF	0.843	0.796	0.780	0.798	0.853	0.821	0.815	0.811	0.831
PMF	0.925	0.856	0.851	0.859	0.891	0.875	0.853	0.858	0.878
Normal	0.829	0.836	0.773	0.805	0.860	0.827	0.774	0.820	0.845

Nominal coverage was 0.90, and sample size $n = 100$; error distribution: i.i.d. Normal

Table 25.2 (Average) lengths (LEN)—with standard errors below them—of the confidence intervals reported in Table 25.1

$x_f/\pi =$	0.15	0.3	0.5	0.75	1l	1.25	1.5	1.7	1.85
MB	0.380	0.346	0.332	0.358	0.380	0.359	0.334	0.345	0.377
	0.003	0.003	0.003	0.003	0.003	0.003	0.003	0.003	0.004
PRMB	0.466	0.432	0.418	0.441	0.473	0.451	0.418	0.427	0.466
	0.008	0.009	0.010	0.007	0.010	0.011	0.009	0.008	0.009
MF	0.448	0.418	0.398	0.424	0.455	0.428	0.399	0.420	0.455
	0.005	0.005	0.005	0.004	0.005	0.005	0.005	0.005	0.005
PMF	0.518	0.487	0.468	0.490	0.513	0.492	0.470	0.487	0.517
	0.005	0.005	0.005	0.004	0.005	0.005	0.006	0.005	0.005
Normal	0.382	0.368	0.368	0.369	0.368	0.371	0.367	0.367	0.378
	0.002	0.002	0.002	0.002	0.002	0.002	0.002	0.002	0.002

Table 25.3 As in Table 25.1 but with error distribution: i.i.d. Laplace

$x_f/\pi =$	0.15	0.3	0.5	0.75	1	1.25	1.5	1.7	1.85
MB	0.789	0.739	0.752	0.782	0.816	0.770	0.741	0.730	0.801
PRMB	0.860	0.858	0.852	0.878	0.905	0.865	0.852	0.857	0.886
MF	0.853	0.835	0.837	0.877	0.870	0.843	0.831	0.835	0.872
PMF	0.924	0.906	0.915	0.930	0.930	0.897	0.913	0.908	0.929
Normal	0.810	0.836	0.820	0.849	0.877	0.843	0.817	0.844	0.852

NORMAL approximation interval (25.4). The smoothing kernel Λ in (25.8) was taken to be the standard normal density. All required bandwidths were computed by L_1 cross-validation. For each type of interval, the corresponding empirical coverage level (CVR) and average length (LEN) were recorded together with the (empirical) standard error associated with each average length.

Tables 25.1, 25.2, 25.3, and 25.4 summarize our findings and contain a number of important features.

- The standard error of the reported coverage levels over the 500 replications is 0.013.

Table 25.4 As in Table 25.2 but with error distribution: i.i.d. Laplace

$x_f/\pi =$	0.15	0.3	0.5	0.75	1	1.25	1.5	1.7	1.85
MB	0.362	0.332	0.316	0.348	0.370	0.340	0.304	0.323	0.354
	0.004	0.003	0.003	0.003	0.003	0.003	0.003	0.003	0.004
PRMB	0.455	0.417	0.398	0.436	0.465	0.424	0.382	0.407	0.444
	0.007	0.006	0.007	0.006	0.006	0.005	0.005	0.005	0.006
MF	0.424	0.394	0.372	0.411	0.441	0.405	0.359	0.388	0.426
	0.004	0.004	0.004	0.004	0.004	0.004	0.004	0.004	0.005
PMF	0.494	0.461	0.440	0.473	0.500	0.468	0.431	0.455	0.489
	0.004	0.004	0.004	0.004	0.004	0.004	0.003	0.004	0.004
Normal	0.368	0.362	0.362	0.358	0.362	0.361	0.362	0.362	0.366
	0.002	0.002	0.002	0.002	0.002	0.002	0.002	0.002	0.002

- By construction, this simulation problem has some symmetry that helps us further appreciate the variability of the CVRs. To elaborate, note that for any $x \in [0, \pi]$ we have $|\mu(x)| = |\mu(2\pi - x)|$ and the same symmetry holds for the derivatives of $\mu(x)$ as well due to the sinusoidal structure. Hence, the expected CVRs should be the same for $x_f = 0.15\pi$ and 1.85π in all methods. So for the NORMAL case of Table 25.1, the CVR would be better estimated by the average of 0.829 and 0.845, i.e., closer to 0.837; similarly, the PMF CVR for the same points could be better estimated by the average of 0.925 and 0.878, i.e., 0.902.

- The NORMAL intervals are characterized by under-coverage even when the true distribution is Normal. This under-coverage is more pronounced when $x_f = \pi/2$ or $3\pi/2$ due to the high bias of the kernel estimator at the points of a "peak" or "valley" that the normal interval (25.4) "sweeps under the carpet".

- The length of the NORMAL intervals is quite less variable than those based on bootstrap; this is not surprising since the extra randomization from the bootstrap is expected to inflate the overall variances.

- Although regression model (25.2) holds true here, the MB intervals show pronounced under-coverage; this is a phenomenon well known in the bootstrap literature. As previously mentioned, the predictive residuals have generally larger scale than the fitted ones. Consequently, the PRMB intervals are wider, and manage to partially correct the under-coverage of the MB intervals.

- The performance of MF intervals is better than that of MB intervals despite the fact that the former are constructed without making use of (25.2). However, as with the MB intervals, the MF intervals also show a tendency towards under-coverage.

- The PMF intervals appear to nicely correct the MF under-coverage in the Normal case although in the Laplace case they yield an over-correction. However, even with this over-correction, the PMF coverages are closer to the nominal in most entries of Tables 25.1 and 25.3 with only a few exceptions in Table 25.3 where the PRMB intervals are more accurate.

25.4.2   When a Nonparametric Regression Model Is Not True

In this subsection, we investigate the performance of different confidence intervals in the absence of model (25.2). For easy comparison with Sect. 25.4.1, we will keep the same (conditional) mean and variance, i.e., we will generate independent Y data such that $E(Y|x) = \sin(x)$, $\text{Var}(Y|x) = 1/2$, and design points $x_1, \ldots, x_{100}$ equi-spaced on $(0, 2\pi)$. However, the error structure $\varepsilon_x = (Y - E(Y|x))/\sqrt{\text{Var}(Y|x)}$ has skewness and/or kurtosis that depends on x, thereby violating the i.i.d. assumption. For our simulation, we considered

$$\varepsilon_x = \frac{c_x Z + (1 - c_x) W}{\sqrt{c_x^2 + (1 - c_x)^2}} \tag{25.15}$$

where $c_x = x/(2\pi)$ for $x \in [0, 2\pi]$, and $Z \sim N(0, 1)$ independent of W that will either be distributed as $\frac{1}{2}\chi_2^2 - 1$ to capture a changing *skewness*, or as $\sqrt{\frac{3}{5}}\, t_5$, to capture a changing *kurtosis*; note that $EW = 0$ and $EW^2 = 1$.

Our results are summarized in Tables 25.5, 25.6, 25.7, and 25.8. The findings are qualitatively similar to those in Sect. 25.4.1. The PMF intervals are the undisputed winners here in terms of coverage accuracy. By contrast, the NORMAL and the MB bootstrap intervals show pronounced under-coverage; interestingly, these are the two methods that most practitioners use at the moment.

Table 25.5   As in Table 25.1 but with error distribution (25.15): non-i.i.d. skewed

$x_f/\pi =$	0.15	0.3	0.5	0.75	1	1.25	1.5	1.7	1.85
MB	0.741	0.743	0.731	0.766	0.820	0.775	0.732	0.750	0.778
PRMB	0.813	0.817	0.815	0.844	0.892	0.865	0.825	0.845	0.862
MF	0.799	0.775	0.753	0.820	0.877	0.839	0.783	0.804	0.839
PMF	0.881	0.853	0.851	0.891	0.907	0.884	0.837	0.858	0.882
Normal	0.811	0.826	0.758	0.793	0.879	0.813	0.764	0.817	0.814

Table 25.6   As in Table 25.2 but with error distribution (25.15): non-i.i.d. skewed

$x_f/\pi =$	0.15	0.3	0.5	0.75	1	1.25	1.5	1.7	1.85
MB	0.361	0.332	0.314	0.343	0.370	0.349	0.322	0.335	0.366
	0.005	0.004	0.004	0.003	0.003	0.003	0.003	0.003	0.003
PRMB	0.456	0.417	0.395	0.430	0.463	0.442	0.408	0.420	0.460
	0.009	0.008	0.007	0.006	0.006	0.007	0.008	0.006	0.007
MF	0.422	0.390	0.369	0.405	0.441	0.419	0.388	0.410	0.446
	0.005	0.005	0.004	0.004	0.004	0.004	0.004	0.004	0.005
PMF	0.492	0.460	0.437	0.467	0.499	0.479	0.452	0.472	0.506
	0.006	0.005	0.004	0.004	0.004	0.005	0.004	0.004	0.004
Normal	0.372	0.358	0.358	0.357	0.358	0.359	0.358	0.358	0.367
	0.002	0.002	0.002	0.002	0.002	0.002	0.002	0.002	0.002

Table 25.7 As in Table 25.1 but with error distribution (25.15): non-i.i.d. kurtotic

$x_f/\pi =$	0.15	0.3	0.5	0.75	1	1.25	1.5	1.7	1.85
MB	0.795	0.754	0.730	0.747	0.792	0.782	0.750	0.756	0.770
PRMB	0.868	0.840	0.809	0.843	0.870	0.852	0.834	0.837	0.845
MF	0.863	0.820	0.804	0.811	0.838	0.833	0.814	0.808	0.834
PMF	0.918	0.896	0.889	0.880	0.890	0.886	0.868	0.869	0.877
Normal	0.808	0.815	0.788	0.800	0.861	0.821	0.781	0.828	0.812

Table 25.8 As in Table 25.2 but with error distribution (25.15): non-i.i.d. kurtotic

$x_f/\pi =$	0.15	0.3	0.5	0.75	1	1.25	1.5	1.7	1.85
MB	0.363	0.339	0.327	0.353	0.380	0.360	0.332	0.341	0.372
	0.004	0.004	0.004	0.003	0.003	0.003	0.003	0.003	0.003
PRMB	0.446	0.416	0.402	0.433	0.465	0.443	0.408	0.420	0.455
	0.007	0.007	0.007	0.007	0.006	0.006	0.007	0.005	0.006
MF	0.423	0.402	0.382	0.416	0.449	0.429	0.397	0.410	0.451
	0.004	0.004	0.004	0.004	0.004	0.004	0.004	0.004	0.004
PMF	0.496	0.469	0.452	0.480	0.508	0.489	0.461	0.477	0.507
	0.004	0.005	0.004	0.004	0.004	0.004	0.004	0.004	0.004
Normal	0.377	0.365	0.365	0.365	0.365	0.367	0.365	0.365	0.373
	0.002	0.002	0.002	0.002	0.002	0.002	0.002	0.002	0.002

References

1. Devroye, L.: Laws of the iterated logarithm for order statistics of uniform spacings. Ann. Probab. **9**(6), 860–867 (1981)
2. Efron, B.: Bootstrap methods: another look at the jackknife. Ann. Stat. **7**, 1–26 (1979)
3. Freedman, D.A.: Bootstrapping regression models. Ann. Stat. **9**, 1218–1228 (1981)
4. Hall, P.: On Edgeworth expansion and bootstrap confidence bands in nonparametric curve estimation. J. R. Stat. Soc. Ser. B **55**, 291–304 (1993)
5. Härdle, W., Bowman, A.W.: Bootstrapping in nonparametric regression: local adaptive smoothing and confidence bands. J. Am. Stat. Assoc. **83**, 102–110 (1988)
6. Härdle, W., Marron, J.S.: Bootstrap simultaneous error bars for nonparametric regression. Ann. Stat. **19**, 778–796 (1991)
7. Li, Q., Racine, J.S.: Nonparametric Econometrics. Princeton University Press, Princeton (2007)
8. Linton, O.B., Chen, R., Wang, N., Härdle, W.: An analysis of transformations for additive nonparametric regression. J. Am. Stat. Assoc. **92**, 1512–1521 (1997)
9. Neumann, M., Polzehl, J.: Simultaneous bootstrap confidence bands in nonparametric regression. J. Nonparametr. Stat. **9**, 307–333 (1998)
10. Politis, D.N.: Model-free model-fitting and predictive distributions. Discussion Paper, Department of Economics, Univ. of California, San Diego (2010). Retrievable from http://escholarship.org/uc/item/67j6s174
11. Politis, D.N.: Model-free model-fitting and predictive distributions. Test **22**(2), 183–221 (2013)

Chapter 26
Heteroskedastic Linear Regression: Steps Towards Adaptivity, Efficiency, and Robustness

Dimitris N. Politis and Stefanos Poulis

Abstract In linear regression with heteroscedastic errors, the Generalized Least Squares (GLS) estimator is optimal, i.e., it is the Best Linear Unbiased Estimator (BLUE). The Ordinary Least Squares (OLS) estimator is suboptimal but still valid, i.e., unbiased and consistent. White, in his seminal paper (White, Econometrica 48:817–838, 1980) used the OLS residuals in order to obtain an estimate of the standard error of the OLS estimator under an unknown structure of the underlying heteroscedasticity. The GLS estimator similarly depends on the unknown heteroscedasticity, and is thus intractable. In this paper, we introduce two different approximations to the optimal GLS estimator; the starting point for both approaches is in the spirit of White's correction, i.e., using the OLS residuals to get a rough estimate of the underlying heteroscedasticity. We show how the new estimators can benefit from the Wild Bootstrap both in terms of optimising them, and in terms of providing valid standard errors for them despite their complicated construction. The performance of the new estimators is compared via simulations to the OLS and to the exact (but intractable) GLS.

Keywords BLUE • Least squares estimation • Minimum variance

26.1 Introduction

Standard regression methods rely on the assumption that the regression errors are either independent, identically distributed (i.i.d.), or at least being uncorrelated having the same variance; this latter property is called *homoscedasticity*. The Generalized Least Squares (GLS) estimator is Best Linear Unbiased Estimator (BLUE) but its computation depends on the structure of the underlying heteroscedasticity. Typically, this structure is unknown, and the GLS estimator is intractable; in this case, practitioners may be forced to use the traditional Ordinary Least Squares (OLS) estimator which will still be valid, i.e., unbiased and consistent, under general conditions.

D.N. Politis (✉) • S. Poulis
University of California, La Jolla, San Diego, CA 92093-0112, USA
e-mail: dpolitis@ucsd.edu; spoulis@ucsd.edu

© Springer Science+Business Media New York 2014   283
M.G. Akritas et al. (eds.), *Topics in Nonparametric Statistics*, Springer Proceedings
in Mathematics & Statistics 74, DOI 10.1007/978-1-4939-0569-0_26

Under the assumption that the error variance is an unknown but smooth function of the regressors it is possible to give an approximation to the GLS estimator. For example, Caroll [2] showed that one can construct estimates of regression parameters that are asymptotically equivalent to the weighted least squares estimates. Chatterjee and Mächler [1] proposed an iterative weighted least squares algorithm to approximate the optimal GLS. In a different spirit, Yuan and Whaba [12] introduced a penalized likelihood procedure to estimate the conditional mean and variance simultaneously. In the same framework, Le, Smola, and Canu [4], using Gaussian process regression, estimate the conditional mean and variance by estimating the natural parameters of the exponential family representation of the Gaussian distribution.

What happens if no smoothness assumption on the error variance can be made? In his seminal paper, White [10] used the OLS residuals in order to get an estimate of the standard error of the OLS estimator that is valid, i.e., consistent, under an unknown structure of the underlying heteroscedasticity. In the paper at hand, we introduce two new approximations to the optimal GLS estimator; the starting point for both approaches is in the spirit of White's [10] correction, i.e., using the OLS residuals to get a rough estimate of the underlying heteroscedasticity. The paper is structured as follows; in Sect. 26.2 we formally state the problem, introduce our first estimator, and show how a convex combination of the *under-correcting* OLS and the *over-correcting* approximate GLS estimators can yield improved results. In Sect. 26.3, in our effort to approximate the quantities needed for the GLS, we introduce a second estimator. In Sect. 26.4 we present a series of simulation experiments to compare performance of the new estimators to the OLS and to the exact (but intractable) GLS.

26.2 Adaptive Estimation: A First Attempt

Consider the general linear regression setup where the data vector $Y = (Y_1, \ldots, Y_n)'$ satisfies:

$$Y = X\beta + \epsilon. \tag{26.1}$$

As usual, β is a $p \times 1$ unknown parameter vector, X is an $n \times p$ matrix of observable regressors, and $\epsilon = (\epsilon_1, \ldots, \epsilon_n)'$ is an unobservable error vector. The regressors may be fixed or random variables (r.v.); in either case, it will be assumed that the regressor matrix X is independent from the error vector ϵ, and that the $n \times p$ matrix is of full rank almost surely.

Letting x_i denote the ith row of the matrix X, we will further assume that

$$\{(x_i, \epsilon_i) \text{ for } i = 1, \ldots, n\} \text{ is a sequence of independent r.v.'s} \tag{26.2}$$

and that the first two moments of ϵ are finite and satisfy:

$$E(\epsilon) = 0 \text{ and } V(\epsilon) = \Sigma \text{ where } \Sigma = \text{diag}(\sigma_1^2, \ldots, \sigma_n^2), \tag{26.3}$$

i.e., the ϵ_i errors are mean zero, uncorrelated but with heteroskedasticity of arbitrary form. The OLS estimator of β is $\hat{\beta}_{\text{LS}} = (X'X)^{-1}X'Y$. Its variance–covariance matrix, conditional on X, is given by

$$V(\hat{\beta}_{\text{LS}}) = (X'X)^{-1}X'\Sigma X(X'X)^{-1}$$

that can be estimated consistently by White's (1980)[10] heteroskedasticity consistent estimator (HCE):

$$(X'X)^{-1}X'\hat{\Sigma}X(X'X)^{-1}$$

where

$$\hat{\Sigma} = \text{diag}(\hat{\sigma}_1^2, \ldots, \hat{\sigma}_n^2) \text{ with } \hat{\sigma}_i^2 = r_i^2/(1 - h_i)^{\gamma}. \tag{26.4}$$

In the above, r_i is the ith element of the residual vector $r = Y - X\hat{\beta}_{\text{LS}}$, and h_i is the "leverage" of point x_i, i.e., the ith diagonal element of the projection ("hat") matrix $H = X(X'X)^{-1}X'$. White's [10] original proposal had $\gamma = 0$ in Eq. (26.4); later proposals recommended $\gamma = 1$, i.e., studentized residuals, or $\gamma = 2$, i.e., delete-1 jackknife residuals; see MacKinnon [7] for a review.

Nevertheless, in the presence of heteroskedasticity, $\hat{\beta}_{\text{LS}}$ is not optimal. Efficiency in this case is attained by the GLS estimator $\hat{\beta}_{\text{GLS}}$ which is the solution of

$$(X'\Sigma^{-1}X)\,\hat{\beta}_{\text{GLS}} = X'\Sigma^{-1}Y. \tag{26.5}$$

Under the stated assumptions, the variance–covariance matrix of $\hat{\beta}_{\text{GLS}}$, conditional on X, is given by

$$V(\hat{\beta}_{\text{GLS}}) = (X'\Sigma^{-1}X)^{-1}. \tag{26.6}$$

The problem, of course, is that Σ is unknown, so $\hat{\beta}_{\text{GLS}}$ is unobtainable. However, despite the fact that $\hat{\Sigma}$ is an inconsistent estimator of Σ, we may still construct estimators of the matrices $X'\Sigma^{-1}X$ and $X'\Sigma^{-1}$ that are needed to compute $\hat{\beta}_{\text{GLS}}$; this is done in the spirit of White's [10] HCE.

Before doing that though it is important to consider the possibility that $\hat{\beta}_{\text{GLS}}$ is not well defined due to the fact that Σ might be non-invertible. We can define a small perturbation of Σ that is invertible; to do this, let $\delta > 0$ and define $\Sigma_\delta = \text{diag}(\sigma_1^2 + \delta, \ldots, \sigma_n^2 + \delta)$. We now define $\hat{\beta}_{\text{GLS},\delta}$ as the solution of

$$(X'\Sigma_\delta^{-1}X)\,\hat{\beta}_{\text{GLS},\delta} = X'\Sigma_\delta^{-1}Y.$$

Note that $\hat{\beta}_{\text{GLS},\delta}$ is always well defined, and—for δ small enough—is close to $\hat{\beta}_{\text{GLS}}$ when it is well defined.

In the same spirit, let $\delta > 0$ and define

$$\tilde{\Sigma}_\delta = \mathrm{diag}(\tilde{\sigma}_1^2, \ldots, \tilde{\sigma}_n^2) \quad \text{with} \quad \tilde{\sigma}_i^2 = \frac{r_i^2 + \delta}{(1 - h_i)^\gamma} \tag{26.7}$$

where r and γ are as in Eq. (26.4). Note that $\tilde{\Sigma}_\delta$ reduces to $\hat{\Sigma}$ when $\delta = 0$; the advantage of $\tilde{\Sigma}_\delta$, however, is that it is always invertible with $\tilde{\Sigma}_\delta^{-1} = \mathrm{diag}(\tilde{\sigma}_1^{-2}, \ldots, \tilde{\sigma}_n^{-2})$.

Remark 2.1. In practice, δ could/should be taken to be a fraction of the residual sample variance $S^2 = (n - p)^{-1} \sum_{i=1}^n r_i^2$, e.g., $\delta = 0.01\, S^2$ or $\delta = 0.001\, S^2$; among other things, this ensures equivariance of $\hat{\Sigma}_\delta$.

We now define a preliminary estimator $\tilde{\beta}_\delta$ as the solution of

$$(X' \tilde{\Sigma}_\delta^{-1} X)\, \tilde{\beta}_\delta = X' \tilde{\Sigma}_\delta^{-1} Y. \tag{26.8}$$

To investigate the behavior of the preliminary estimator $\tilde{\beta}_\delta$ we need to define the diagonal matrix W_δ that has as ith element the quantity $E[(r_i^2 + \delta)^{-1}]$. Now let $\tilde{\beta}_{W,\delta}$ as the solution of

$$(X' W_\delta X)\, \tilde{\beta}_{W,\delta} = X' W_\delta Y. \tag{26.9}$$

Under conditions similar to the assumptions of Theorem 1 of White [10], we can now claim the following *large-sample* approximations:

$$X' \tilde{\Sigma}_\delta^{-1} X \approx_c X' W_\delta X \quad \text{and} \quad X' \tilde{\Sigma}_\delta^{-1} \approx_c X' W_\delta \tag{26.10}$$

where symbol $A \approx_c B$ for matrices $A = (a_{ij})$ and $B = (b_{ij})$ is short-hand to denote that $a_{ij} \approx b_{ij}$ for all i, j, i.e., coordinate-wise approximation. As a consequence, it follows that $\tilde{\beta}_\delta \approx \tilde{\beta}_{W,\delta}$ for large enough samples.

Now if we could claim that $W_\delta \approx_c \tilde{\Sigma}_\delta^{-1}$, then we would have that $\tilde{\beta}_{W,\delta} \approx \hat{\beta}_{\mathrm{GLS},\delta}$, and thus $\tilde{\beta}_\delta$ would be a consistent approximation to the GLS estimator. However, the approximation $W_\delta \approx_c \tilde{\Sigma}_\delta^{-1}$ is not a good one. In fact, by Jensen's inequality we have $E r_i^{-2} \geq 1/E r_i^2 = 1/\sigma_i^2$; hence, it follows that W_δ will be biased upward as an estimator of Σ_δ^{-1}. In this sense, $\tilde{\beta}_\delta$ is an *over-correction* in trying to take account of the covariance of the errors. Since the OLS estimator $\hat{\beta}_{\mathrm{LS}}$ is an *under-correction*, it is interesting to explore the hybrid estimator $\tilde{\beta}_{\delta,\lambda}$ whose kth coordinate is given by

$$\tilde{\beta}_{\delta,\lambda;k} = \lambda_k \tilde{\beta}_{\delta;k} + (1 - \lambda_k)\hat{\beta}_{\mathrm{LS};k} \tag{26.11}$$

where $\lambda_k, \tilde{\beta}_{\delta;k}, \hat{\beta}_{\mathrm{LS};k}$ denote the kth coordinates of $\lambda, \tilde{\beta}_\delta, \hat{\beta}_{\mathrm{LS}}$, respectively, and λ is a *mixing* weight vector of the practitioner's choice. By choosing the tuning parameter λ to be multi-dimensional, we allow each coordinate to receive a different weight; this might be especially helpful if/when heteroscedasticity is due to only a subset of the predictors. In this case, Eq. (26.11) becomes

$$\tilde{\beta}_{\delta,\lambda} = \lambda \circ \tilde{\beta}_\delta + (1 - \lambda) \circ \hat{\beta}_{LS} \qquad (26.12)$$

where ($\circ$) denotes the *Hadamard* (point-wise) vector product.

Even if efficiency is not attained, $\tilde{\beta}_{\delta,\lambda}$ is still a step in the right direction towards the GLS estimator. In addition, it satisfies the two robustness criteria of Chatterjee and Mächler [1] as it gives reduced weight to (Y_i, x_i) points that involve either an outlier in the error ϵ_i, or an outlier in the regressor x_i, i.e., a high "leverage" point.

Remark 2.2. To fine-tune the choice of the weight vector λ, the *wild bootstrap* of Wu [11] is found useful. See Mammen [6] for a detailed study. Here we will use the simplest version proposed by Liu [5] and further discussed by Davidson and Flachaire [3], namely bootstrapping the signs of the residuals. Specifically, define the bootstrap residuals

$$r_i^* = s_i\, r_i / (1 - h_i)^{\gamma/2} \text{ for } i = 1, \dots, n$$

where $s_1, \dots, s_n$ are i.i.d. sign changes with Prob$\{s_i = 1\} =$ Prob$\{s_i = -1\} = 1/2$, and $\gamma = 1$ or 2, i.e., studentized or jackknife residuals, respectively, as in Eq. (26.4). The use of jackknife residuals—also known as predictive residuals—in bootstrapping has also been recommended by Politis [8, 9].

The wild bootstrap datasets are generated via Eq. (26.1) using the same X matrix, and β replaced by $\hat{\beta}_{LS}$. Letting $r^* = (r_1^*, \dots, r_n^*)'$, the bootstrap data vector Y^* satisfies the equation:

$$Y^* = X\hat{\beta}_{LS} + r^*. \qquad (26.13)$$

Then the pseudo-data Y^* can be treated as (approximate) replicates of the original data. Of course, this is justified under the additional assumption that the errors ϵ_i are (approximately) symmetrically distributed around zero; if this is not true, then the error skewness can be incorporated in the wild bootstrap by using a non-symmetric auxiliary distribution for the s_i random variables used above.

The bootstrap procedure for choosing the weight vector λ is summarized in Algorithm 1 below that works as follows. Firstly, d denotes the number of candidate λ vectors that the bootstrap procedure should choose from. Typically, a grid will be formed spanning the allowed [0,1] interval for each coordinate of λ; a large d corresponds to a fine grid. Then, B bootstrap datasets are generated based on Eq. (26.13). For a particular choice of the vector λ, $\tilde{\beta}_{\delta,\lambda}$ is computed as in Eq. (26.12) on every bootstrap dataset. The algorithm finds λ_{opt} that minimizes the empirical MSE with respect to the $\hat{\beta}_{LS}$, as estimated in the original dataset. In the algorithm, k denotes the kth parameter vector, and j denotes the jth bootstrap dataset.

Algorithm 1 Find optimal tuning parameter vector λ for $\tilde{\beta}_{\delta,\lambda}$ via Wild Bootstrap

---

1: Choose d candidate tuning parameter vectors $\lambda^{(1)}, \ldots, \lambda^{(d)}$, where $\lambda^{(k)} \in (0, 1)^p$
2: Create B wild bootstrap datasets by Eq. (26.13)
3: **for** k in 1 to d **do**
4:     **for** j in 1 to B **do**
5:         $\tilde{\beta}_\delta^{(j)} = (X'\tilde{\Sigma}_\delta^{-1(j)}X)^{-1}X'\tilde{\Sigma}_\delta^{-1(j)}Y^{(j)}$
6:         $\hat{\beta}_{LS}^{(j)} = (X'X)^{-1}X'Y^{(j)}$
7:         $\tilde{\beta}_{\delta,\lambda}^{(kj)} = \lambda^{(k)} \circ \tilde{\beta}_\delta^{(j)} + (1 - \lambda^{(k)}) \circ \hat{\beta}_{LS}^{(j)}$
8:     **end for**
9: **end for**
10: $\lambda_{opt} = \underset{1 \leq k \leq d}{\text{argmin}} \frac{1}{B} \sum_{j=1}^{B} (\tilde{\beta}_{\delta,\lambda}^{(kj)} - \hat{\beta}_{LS})^2$
11: output $\tilde{\beta}_{\delta,\lambda} = \lambda_{opt} \circ \hat{\beta}_\delta + (1 - \lambda_{opt}) \circ \hat{\beta}_{LS}$

26.3 Adaptive Estimation: A Second Attempt

One may consider linearization as an alternative approximation to adaptive GLS. To do this, consider linearizing the $1/x$ function around a point x_0, i.e.,

$$1/x \approx 1/x_0 - (1/x_0^2) * (x - x_0) = 2/x_0 - (1/x_0^2) * x. \tag{26.14}$$

Using the above approximation on each of the elements of the diagonal matrix Σ, it follows that we may be able to approximate the $X'\Sigma^{-1}X$ needed for GLS by: $2x_0^{-1}X'X - x_0^{-2}X'\Sigma X$. But the last expression is estimable by a usual HCE procedure, i.e., by:

$$2x_0^{-1}X'X - x_0^{-2}X'\hat{\Sigma}X \tag{26.15}$$

where $\hat{\Sigma}$ is defined as in Eq. (26.4). Similarly, $X'\Sigma^{-1} \approx 2x_0^{-1}X' - x_0^{-2}X'\Sigma$ which is estimable by:

$$2x_0^{-1}X' - x_0^{-2}X'\hat{\Sigma}. \tag{26.16}$$

Because of the convexity of the $1/x$ function the linear approximation (Eq. 26.14) becomes a supporting line, i.e.,

$$1/x > 2/x_0 - (1/x_0^2) * x$$

for all positive $x \neq x_0$ (and becomes an approximation for x close to x_0). Hence, the quantity in Eq. (26.15) *under*-estimates $X'\Sigma^{-1}X$ in some sense.

Note, however, that the RHS of Eq. (26.14) goes negative when $x > 2x_0$ in which case it is not only useless, but may also cause positive definiteness problems if the approximation is applied to $1/\sigma_i^2$ or $1/r_i^2$ type quantities. To address this, we propose a two-step solution:

(a) Truncate (rather Winsorize) the r_i^2 by letting $\tilde{r}_i = r_i$ when $|r_i| < \zeta$, and $\tilde{r}_i = \zeta \cdot sign(r_i)$ when $|r_i| \geq \zeta$ for some $\zeta > 0$. This implies that the influence of too large residuals will be bounded. The number ζ can be picked in a data-based manner, e.g. we can take ζ^2 as the λ quantile of the empirical distribution of the r_i^2 for some $\lambda \in (0, 1)$ but close to 1; for example, $\lambda = 0.9$ may be reasonable in which case ζ^2 is the upper decile.

(b) Choose a big enough linearization point x_0 to ensure positive definiteness of the quantity in Eq. (26.15). To do this, let $x_0 = (\zeta^2 - \delta)/2$ where δ is a small, positive number; as in Remark 2.1, δ can be chosen to be a fraction of S^2 or—in this case—a fraction of ζ^2, e.g. $\delta = 0.001\zeta^2$.

With choices of x_0 and λ as in parts (a) and (b) above, we now define $\check{\beta}_{\delta,\lambda}$ as the solution of

$$(2x_0^{-1} X'X - x_0^{-2} X' \hat{\Sigma} X) \check{\beta}_{\delta,\lambda} = 2x_0^{-1} X'Y - x_0^{-2} X' \hat{\Sigma} Y$$

which is equivalent to

$$(2X'X - x_0^{-1} X' \hat{\Sigma} X) \check{\beta}_{\delta,\lambda} = 2X'Y - x_0^{-1} X' \hat{\Sigma} Y. \tag{26.17}$$

The construction of estimator $\check{\beta}_{\delta,\lambda}$ is described in full detail in Algorithm 2. The procedure for the choice of the optimal linearization point (or truncation point) is similar to that of Algorithm 1. B bootstrap datasets are generated based on Eq. (26.13). For a particular choice of truncation point, $\check{\beta}_{\delta,\lambda}$ is computed as in Eq. (26.17) on every bootstrap dataset. The algorithm finds ζ_{opt} that minimizes the empirical MSE with respect to the $\hat{\beta}_{LS}$, as estimated in the original dataset.

In Algorithm 2, d denotes the total number of possible truncation points to be examined and denotes how many quantiles of the OLS squared residuals are considered as candidates; for example, one may consider the 70 %, 80 %, and 90 % quantiles, i.e., $d = 3$. In the Algorithm, k denotes the kth candidate truncation point and j denotes the jth bootstrap dataset.

Remark 3.1. Computing the distribution of $\tilde{\beta}_\delta$, $\tilde{\beta}_{\delta,\lambda}$, and $\check{\beta}_{\delta,\lambda}$ analytically seems a daunting task even under simplifying assumptions such as normality; the same is true regarding their asymptotic distribution. In order to estimate/approximate the intractable distribution of our estimators, we propose the simple *wild bootstrap* procedure mentioned in Sect. 26.2, i.e., changing the signs of the ith residual with probability $1/2$.

Luckily, it seems that the bias of the above estimators—although not identically zero—is negligible; see the empirical results of Sect. 26.4.4. Hence, the main concern is estimating the variance of our estimators. Section 26.4.3 provides a simulation experiment showing that estimation of variance can be successfully performed via the wild bootstrap.

Algorithm 2 Find optimal linearization point x_0 for $\check{\beta}_{\delta,\lambda}$ via Wild Bootstrap

---

1: Choose d candidate truncation parameters $\zeta^{(1)}, \ldots, \zeta^{(d)}$.
2: Create B wild bootstrap datasets by Eq. (26.13)
3: **for** k in 1 to d **do**
4:     Set $\tilde{r}_i \leftarrow r_i$ when $|r_i| < \zeta^{(k)}$ and $\tilde{r}_i \leftarrow \zeta^{(k)} sign(r_i)$ otherwise
5:     $x_0^{(k)} = \frac{(\zeta^{(k)})^2 - \delta}{2}$, for some small δ
6:     **for** j in 1 to B **do**
7:         $\check{\beta}_{\delta,\lambda}^{(kj)} = (X'X - \frac{1}{x_0^{(k)}} X' \hat{\Sigma}^{(j)} X)^{-1}(X'Y^{(j)} - \frac{1}{x_0^{(k)}} X' \hat{\Sigma}^{(j)} Y^{(j)})$
8:     **end for**
9: **end for**
10: $\zeta_{opt} = \underset{1 \le k \le d}{\arg\min} \frac{1}{B} \sum_{j=1}^{B} (\check{\beta}_{\delta,\lambda}^{(kj)} - \hat{\beta}_{LS})^2$
11: $x_0 \leftarrow \frac{\zeta_{opt}^2 - \delta}{2}$
12: $\check{\beta}_{\delta,\lambda} = (X'X - \frac{1}{x_0} X' \hat{\Sigma} X)^{-1}(X'Y - \frac{1}{x_0} X' \hat{\Sigma} Y)$

26.4 Simulation Experiments

To empirically study the performance of the proposed estimators, extensive finite-sample simulations were carried through. The setup for the simulation experiments was the same as in MacKinnon [7]. In particular, the model employed was

$$y_i = \beta_1 + \sum_{k=2}^{p} (\beta_k X_{ik}) + u_i, \tag{26.18}$$

where $X_k \sim \text{lognormal}(0,1)$, and $u_i = \sigma_i \epsilon_i$ where $\sigma_i = |1 + \beta_k X_{ik}|^\eta$ and $\epsilon_i \sim N(0, 1)$. Experiments were conducted for $p = 2$. The heteroscedasticity inducing factor $\eta \in [0, 2]$ was chosen in increments of 0.2.

26.4.1 Choosing the Tuning Parameter Vector λ

In this experiment, we explore the empirical behavior of the tuning parameter vector λ for the estimator of Sect. 26.2. We first choose λ to be simply a binary vector, that is, its kth coordinate is either 0 or 1. Here, we simply count the number of *correctly* captured λ's. *Correctly*, means closeness to the true parameter as measured by the squared difference (L_2 distance). Specifically, when $\hat{\beta}_{LS;k}$ is closer to the true parameter in L_2 than what $\tilde{\beta}_{\delta;k}$ is, λ_k should be equal to 0. When the opposite is true, λ_k should be equal to 1. The above strategy is equivalent to a test procedure, in the sense that only the optimal parameter stays in the final model. We carry out the experiment for three cases; Case I, $\beta_{\text{TRUE}} = (1, 0.5)'$; Case II, $\beta_{\text{TRUE}} = (1, 1)'$; Case III, $\beta_{\text{TRUE}} = (1, 1.5)'$.

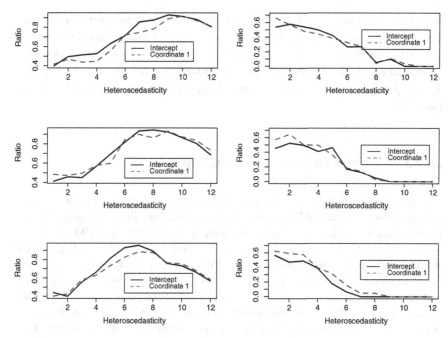

Fig. 26.1 Ratio of counts when $\lambda = 1$ and when $\lambda = 0$ to the counts of times when the corresponding coefficient was closer to the true parameter in L_2. Plots on the left show the ratio when $\tilde{\beta}_\delta$ was optimal, while plots on the right show the ratio when $\hat{\beta}_{LS}$ was optimal. Each row in the figure corresponds to an experiment

Equation (26.18) was used for the generation of the response variable Y. The experiment was carried out with $n = 200$ and it was repeated 199 times while the number of bootstrap samples was 201. For all the experiments, we choose the perturbation on the diagonal of $\hat{\Sigma}$ to be $\delta = 0.001\ S^2$. Results for the first part of the experiment are shown in Fig. 26.1. We report the ratio of counts of times when $\lambda = 1$ and when $\lambda = 0$ to the counts of times when the corresponding estimate was closer to the true parameter in L_2. As expected, as heteroscedasticity increases, $\tilde{\beta}_{\delta;k}$ is closer to the kth coordinate of the true parameter than what $\hat{\beta}_{LS;k}$ is. As it can be seen, the ratio of counts of times when $\lambda_k = 1$ to the counts of times when $\tilde{\beta}_{\delta;k}$ is optimal increases as a function of heteroscedasticity. On the other hand, the ratio of counts of times when $\lambda_k = 0$ to the counts of times when $\hat{\beta}_{LS;k}$ is optimal decreases as a function of heteroscedasticity. This suggested that our procedure for estimating $\tilde{\beta}_{\delta,\lambda}$ is functioning properly.

26.4.2   Comparing the Performance of the New Estimators

We now compare the performance of our estimators in terms of their average square difference to the true parameter, i.e., their Mean Squared Error (MSE); our empirical results are summarized in Table 26.1. For $\tilde{\beta}_{\delta,\lambda}$, the *mixing* parameter vector was the same as above. For $\check{\beta}_{\delta,\lambda}$, the search for the optimal truncation point begun at the 70 % quantile of the EDF of the squared OLS residuals. In several cases where heteroscedasticity is mild, $\tilde{\beta}_{\delta,\lambda}$ and $\check{\beta}_{\delta,\lambda}$ outperform $\hat{\beta}_{LS}$ and they are closer to the optimal $\hat{\beta}_{GLS}$. In cases of heavy heteroscedasticity, $\tilde{\beta}_{\delta}$ is always closer to the true parameter.

26.4.3 Variance Estimation via Wild Bootstrap

As suggested in Remark 3.1, it may be possible to estimate the variance of our estimators via the wild bootstrap. In this simulation, data were generated as before using Case II. The sample sizes were $n = 50, 100$, and 200, while the heteroscedasticity inducing factor was $\eta = 0, 1$, and 2. It is important to note that we modified our data generating model in Eq. (26.18) by changing the parameters of the lognormal distribution from $(0, 1)$ to $(0, .25)$. In our experiments we found that the former parameters generate a heavy-tailed distribution, therefore possible outliers in the predictor variables could show up. This might cause problems in the variance estimation procedure via the wild bootstrap that are magnified in the presence of heteroscedasticity. By modifying the parameters and controlling for heavy tails, we alleviate this problem. We compare the true variances of our estimators to their bootstrap counterparts. Specifically, we evaluate the true variance of our estimators, denoted by $\mathrm{var}\beta$ by a Monte Carlo simulation based on 100 datasets. Secondly, to obtain estimates of $\mathrm{var}\beta$ we use the wild bootstrap. Our bootstrapped estimates are denoted by $\mathrm{var}\beta^*$ and were averaged over 100 bootstrap samples. Our results are shown in Table 26.2. As it can be seen, the ratio $\frac{\mathrm{var}\beta^*}{\mathrm{var}\beta}$ is nearly always, very close to 1. This suggests that the wild bootstrap can be safely used to estimate standard errors. For this experiment, given that our estimators are already computationally intensive, we did not do any fine-tuning.

26.4.4 The Bias of the Proposed Estimators

If the bias of the proposed estimators were appreciable, then we would need to also estimate it via bootstrap, otherwise the standard errors developed in Sect. 26.4.3 would be to no avail. Luckily, as hinted at in Remark 3.1, the bias of the proposed estimators appears to be negligible; this implies that a practitioner need not worry about the bias (or estimation thereof) when conducting inference using the new estimators. To empirically validate this claim, we performed the same simulation experiment as in Sect. 26.4.3 but with a larger number of Monte Carlo samples (1,000) in order to accurately gauge the bias. As can be seen in Table 26.3, all estimators have a bias that may be considered negligible. Also reported are the

Table 26.1 Comparison of MSEs of various estimators. Highlighted numbers indicate the estimator that had the least MSE in each case

MSE		$\eta = 0$					$\eta = 0.2$			
	$\hat{\beta}_{\delta,\lambda}$	$\hat{\beta}_{LS}$	$\tilde{\beta}_{\delta}$	$\tilde{\beta}_{\delta,\lambda}$	$\hat{\beta}_{GLS}$	$\hat{\beta}_{\delta,\lambda}$	$\hat{\beta}_{LS}$	$\tilde{\beta}_{\delta}$	$\tilde{\beta}_{\delta,\lambda}$	$\hat{\beta}_{GLS}$
Case I	0.05378	0.04339	0.04570	0.04319	0.04339	0.05417	0.05447	0.05474	0.05399	0.05366
	0.01531	0.00835	0.00899	0.00832	0.00835	0.01334	0.01161	0.01187	0.01148	0.01121
Case II	0.04263	0.04344	0.04323	0.04359	0.04344	0.06832	0.051	0.05021	0.05107	0.04852
	0.00763	0.00761	0.00769	0.00762	0.007615	0.01966	0.01147	0.01204	0.01154	0.010781
Case III	0.05459	0.04348	0.04329	0.04318	0.04348	0.05838	0.05778	0.06082	0.05887	0.05542
	0.01609	0.00788	0.00788	0.00771	0.007888	0.02273	0.01841	0.01931	0.01889	0.01705

MSE		$\eta = 0.6$					$\eta = 0.8$			
	$\hat{\beta}_{\delta,\lambda}$	$\hat{\beta}_{LS}$	$\tilde{\beta}_{\delta}$	$\tilde{\beta}_{\delta,\lambda}$	$\hat{\beta}_{GLS}$	$\hat{\beta}_{\delta,\lambda}$	$\hat{\beta}_{LS}$	$\tilde{\beta}_{\delta}$	$\tilde{\beta}_{\delta,\lambda}$	$\hat{\beta}_{GLS}$
Case I	0.37866	0.12574	0.12319	0.12078	0.06432	0.18478	0.17819	0.1645	0.17111	0.07202
	0.21391	0.06465	0.05943	0.06233	0.03657	0.10541	0.09371	0.08488	0.08947	0.04991
Case II	0.36578	0.27534	0.24874	0.26025	0.13435	0.75059	0.61153	0.55498	0.57962	0.1176
	0.2253	0.15584	0.15005	0.15085	0.0819	0.49334	0.31246	0.29469	0.29966	0.12589
Case III	0.70688	0.78015	0.73473	0.74693	0.19797	2.29265	2.44659	1.89875	2.21672	0.1991
	0.44353	0.48215	0.44644	0.45875	0.17111	1.79619	1.88462	1.59625	1.81567	0.40225

MSE	$\eta = 1.0$					$\eta = 1.2$				
	$\check{\beta}_{\delta,\lambda}$	$\hat{\beta}_{LS}$	$\tilde{\beta}_{\delta}$	$\tilde{\beta}_{\delta,\lambda}$	$\hat{\beta}_{GLS}$	$\check{\beta}_{\delta,\lambda}$	$\hat{\beta}_{LS}$	$\tilde{\beta}_{\delta}$	$\tilde{\beta}_{\delta,\lambda}$	$\hat{\beta}_{GLS}$
Case I	0.59474	0.60417	0.51433	0.56085	0.10105	9.79127	0.82338	0.73297	0.7738	0.15385
	0.32983	0.30275	0.26862	0.29096	0.08065	12.49075	0.53116	0.48447	0.5065	0.13754
Case II	1.26204	1.36173	1.01231	1.24952	0.14238	4.47709	4.60854	3.35955	4.02106	0.25482
	0.72283	0.75983	0.64193	0.75699	0.16291	3.32433	3.40715	2.67437	3.18252	0.45618
Case III	18.09569	18.2068	9.06757	17.16014	0.40708	22.72211	22.98654	15.51993	21.06294	0.53693
	7.70597	7.76887	5.53547	7.37507	0.76265	14.18275	14.37411	10.60903	12.6128	1.11989

MSE	$\eta = 1.8$					$\eta = 2.0$				
	$\check{\beta}_{\delta,\lambda}$	$\hat{\beta}_{LS}$	$\tilde{\beta}_{\delta}$	$\tilde{\beta}_{\delta,\lambda}$	$\hat{\beta}_{GLS}$	$\check{\beta}_{\delta,\lambda}$	$\hat{\beta}_{LS}$	$\tilde{\beta}_{\delta}$	$\tilde{\beta}_{\delta,\lambda}$	$\hat{\beta}_{GLS}$
Case I	19.833	20.417	5.631	19.372	0.168	26.271	26.914	11.45	25.286	0.196
	12.044	12.364	3.782	11.976	0.248	14.655	15.073	7.516	13.816	0.432
Case II	100.48	100.756	31.418	95.755	0.437	1,213.308	1,213.636	118.394	1,187.726	0.363
	70.156	70.326	24.712	67.256	1.441	520.303	520.536	73.39	506.098	1.259
Case III	5,221.765	5,222.393	1,020.255	5,048.798	1.077	10,365.66	10,366.458	1,639.68	9,932.69	1.26
	2,504.414	2,504.788	598.076	2,403.024	6.202	4,975.22	4,975.733	1,056.367	4,719.818	8.529

Table 26.2 Comparison of the true variance to the variance estimated by the wild bootstrap for the various estimators

$\frac{\text{var}\beta^*}{\text{var}\hat{\beta}}$	$n = 50$			$n = 100$			$n = 200$		
	$\eta = 0$	$\eta = 1$	$\eta = 2$	$\eta = 0$	$\eta = 1$	$\eta = 2$	$\eta = 0$	$\eta = 1$	$\eta = 2$
$\hat{\beta}_{LS}$	0.98452	0.92712	0.90215	0.97599	1.00946	0.98036	1.03372	0.95064	0.97939
	0.94926	0.94358	0.90952	1.00786	1.00354	0.97330	1.02632	0.94079	0.97425
$\tilde{\beta}_{\delta}$	0.97823	0.93632	0.89858	0.96645	1.00206	0.98862	1.01824	0.93914	0.99269
	0.94020	0.94542	0.90791	0.99466	0.99601	0.98258	1.01044	0.92939	0.98661
$\tilde{\beta}_{\delta,\lambda}$	0.91100	0.90898	0.88396	0.69562	0.88212	1.15183	1.07295	0.96827	0.94053
	0.89474	0.88480	0.87706	0.69291	0.89955	1.06307	1.10091	1.01367	0.92281
$\breve{\beta}_{\delta,\lambda}$	0.88959	0.91235	0.89738	0.70893	0.85388	1.17216	1.02872	0.91514	0.94677
	0.87843	0.88626	0.89392	0.70797	0.86385	1.07663	1.02124	0.96083	0.92529

Table 26.3 Bias of the different estimators with inter-quantile ranges divided by $\sqrt{1000}$ (number of Monte Carlo Samples) in parentheses. As it can be seen in most cases the bias is negligible.

	$\eta = 0$	$\eta = 1$	$\eta = 2$
	$n = 50$		
$\hat{\beta}_{LS}$	0.028 (0.024)	−0.031 (0.047)	−0.043 (0.095)
	−0.017 (0.023)	0.018 (0.048)	0.018 (0.103)
$\tilde{\beta}_{\delta}$	0.029 (0.025)	−0.038 (0.046)	−0.031 (0.09)
	−0.017 (0.024)	0.026 (0.049)	0.002 (0.101)
$\tilde{\beta}_{\delta,\lambda}$	0.027 (0.024)	−0.036 (0.047)	−0.038 (0.093)
	−0.017 (0.023)	0.021 (0.047)	0.009 (0.104)
$\breve{\beta}_{\delta,\lambda}$	0.028 (0.024)	−0.031 (0.047)	−0.043 (0.095)
	−0.016 (0.024)	0.017 (0.049)	0.017 (0.103)
	$n = 100$		
$\hat{\beta}_{LS}$	−0.001 (0.018)	0.022 (0.036)	0.016 (0.082)
	−0.003 (0.016)	−0.031 (0.036)	−0.028 (0.087)
$\tilde{\beta}_{\delta}$	0.000 (0.018)	0.014 (0.037)	0.014 (0.082)
	−0.004 (0.017)	−0.025 (0.037)	−0.027 (0.086)
$\tilde{\beta}_{\delta,\lambda}$	0.000 (0.017)	0.018 (0.037)	0.007 (0.082)
	−0.004 (0.017)	−0.027 (0.036)	−0.021 (0.087)
$\breve{\beta}_{\delta,\lambda}$	−0.001 (0.017)	0.021 (0.036)	0.016 (0.082)
−0.002 (0.017)	−0.031 (0.036)	−0.027 (0.087)	
	$n = 200$		
$\hat{\beta}_{LS}$	0.012 (0.012)	0.003 (0.025)	−0.04 (0.055)
	−0.013 (0.011)	−0.005 (0.024)	0.032 (0.056)
$\tilde{\beta}_{\delta}$	0.013 (0.012)	0.000 (0.025)	−0.039 (0.054)
	−0.014 (0.011)	−0.002 (0.024)	0.03 (0.055)
$\tilde{\beta}_{\delta,\lambda}$	0.012 (0.012)	0.001 (0.025)	−0.038 (0.054)
	−0.013 (0.011)	−0.003 (0.024)	0.031 (0.055)
$\breve{\beta}_{\delta,\lambda}$	0.012 (0.012)	0.003 (0.025)	−0.039 (0.055)
	−0.014 (0.011)	−0.005 (0.024)	0.032 (0.056)

inter-quantile range associated with the 1,000 replications; they are small enough to ensure that if the true bias is different from zero, it cannot be too far away.

26.5   Concluding Remarks

In our attempt to approximate the intractable GLS estimator, we presented several different estimators. The advantage of our estimators lies on the fact that they do not rely on smoothness assumptions regarding the error variance. Our simulation results show that our estimators largely outperform the traditional OLS in the presence of heteroscedasticity of unknown structure. In particular, our preliminary analysis shows that *mixing* different estimators can yield good results. However, a limitation of this approach is computational time, since the optimal grid of the *mixing* parameter is unknown and an extensive search is needed. Nevertheless, the direct estimator $\tilde{\beta}_\delta$ is not computer-intensive, and can always be used since it has been empirically shown to be quite efficient in our simulation. Also, improved optimization techniques can be used for estimating the tuning parameters of both our estimators, i.e. an adaptive grid, as well as parallelization for speed-up.

Further work may include comparisons to techniques that rely on smoothness assumptions, as well as extending these ideas to the time series case where the underlying covariance structure is unknown. Finally, in addition to adapting to heteroscedasticity, any of our procedures can be easily modified to perform model selection by introducing L_2 or L_1 regularizers on our objective function. The latter could be particularly interesting in high-dimensional scenarios, where the rank of the design matrix is less than p and the data are contaminated with outliers. All the above are open directions for future research.

References

1. Chatterjee, S., Mächler, M.: Robust regression: a weighted least squares approach, Commun. Stat. Theory Methods **26**(6), 1381–1394 (1997)
2. Carol, J.R.: Adapting for heteroscedasticity in linear models. Ann. Stat. **10**(4), 1224–1233 (1982)
3. Davidson, R., Flachaire, E.: The wild bootstrap, tamed at last. J. Econometrics **146**, 162–169 (2008)
4. Le, Q.V., Smola, A.J., Canu, S.: Heteroscedastic gaussian process regression. In: Proceedings of the 22nd International Conference on Machine Learning, Bonn (2005)
5. Liu, R.Y.: Bootstrap procedures under some non-iid models. Ann. Stat. **16**, 1696–1708 (1988)
6. Mammen, E.: When Does Bootstrap Work? Asymptotic Results and Simulations. Springer Lecture Notes in Statistics. Springer, New York (1992)
7. MacKinnon, J.G.: Thirty years of heteroskedasticity-robust inference. In: Chen, X., Swanson, N.R. (eds.) Recent Advances and Future Directions in Causality, Prediction, and Specification Analysis, pp. 437–462. Springer, New York (2012)

8. Politis, D.N.: Model-free model-fitting and predictive distributions. Discussion Paper, Department of Economics, Univ. of California—San Diego. Retrievable from http://escholarship.org/uc/item/67j6s174 (2010)
9. Politis, D.N.: Model-free model-fitting and predictive distributions. Invited Discuss. Pap. J. TEST 22(2), 183–221 (2013)
10. White, H.: A heteroskedasticity-consistent covariance matrix estimator and a direct test for heteroskedasticity. Econometrica 48, 817–838 (1980)
11. Wu, C.F.J.: Jackknife, bootstrap and other resampling methods in regression analysis. Ann. Stat. 14, 1261–1295 (1986)
12. Yuan, M., Wahba, G.: Doubly penalized likelihood estimator in heteroscedastic regression. Stat. Probab. Lett. 34, 603–617 (2004)

Chapter 27
Level Set Estimation

P. Saavedra-Nieves, W. González-Manteiga, and A. Rodríguez-Casal

Abstract A density level set can be estimated using three different methodologies: Plug-in methods, excess mass methods, and hybrid methods. The three groups of algorithms to estimate level sets are reviewed in this work. In addition, two new hybrid methods are proposed. Finally, all of them are compared through an extensive simulation study and the results obtained are shown.

Keywords Level set estimation • Excess mass • Plug-in • Hybrid • Shape restrictions

27.1 Introduction

Level set estimation theory deals with the problem of reconstructing an unknown set of type $L(\tau) = \{f \geq f_\tau\}$ from a random sample of points $\mathcal{X}_n = \{X_1, \ldots, X_n\}$, where f stands for the density which generates the sample $\mathcal{X}_n$, $\tau \in (0, 1)$ is a probability, fixed by the practitioner, and $f_\tau > 0$ denotes the biggest threshold such that the level set $L(\tau)$ has a probability at least $1 - \tau$ with respect to the distribution induced by f. Figure 27.1 shows the level sets for three different values of the parameter τ. The problem of estimating $L(\tau)$ has been analyzed using three different methodologies in the literature: Plug-in methods, excess mass methods, and hybrid methods. We will present these three groups of automatic methods to reconstruct level sets and we will compare them through a detailed simulation study for dimension 1. We have restricted ourselves to the one-dimensional case because some of these methods have not yet been extended for higher dimension (see [8] or [6] for example). In Sect. 27.2, we will present and compare the plug-in methods. In Sects. 27.3 and 27.4, we will study the behavior of excess mass methods and hybrids methods, respectively. Finally, we will compare the most competitive methods in each group in Sect. 27.5.

P. Saavedra-Nieves (✉) • W. González-Manteiga • A. Rodríguez-Casal
Universidad de Santiago de Compostela, Santiago de Compostela, Spain
e-mail: paula.saavedra@usc.es; wenceslao.gonzalez@usc.es; alberto.rodriguez.casal@usc.es

© Springer Science+Business Media New York 2014   299
M.G. Akritas et al. (eds.), *Topics in Nonparametric Statistics*, Springer Proceedings
in Mathematics & Statistics 74, DOI 10.1007/978-1-4939-0569-0_27

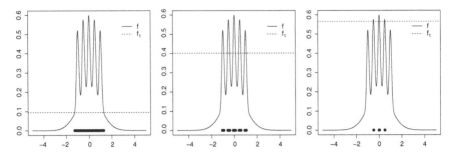

Fig. 27.1 Level sets for a one-dimensional density with $\tau = 0.1$ (*first column*), $\tau = 0.5$ (*second column*) y $\tau = 0.9$ (*third column*)

27.2 Plug-in Methods and Simulations Results

The simplest option to estimate level sets is the so-called plug-in methodology. It is based on replacing the unknown density f by a suitable nonparametric estimator f_n, usually the kernel density estimator. So, this group of methods proposes $\hat{L}(\tau) = \{f_n \geq \hat{f}_\tau\}$ as an estimator, where $\hat{f}_\tau$ denotes an estimator of the threshold. This is the most common approach but its performance is heavily dependent on the choice of the bandwidth parameter for estimating f. Baíllo and Cuevas were interested in choosing the best smoothing parameter to reconstruct a level set in the context of quality control. It was obtained by minimizing a cross-validation estimate of the probability of a false alarm, see [1]. Samworth and Wand proposed an automatic rule to select the smoothing parameter for dimension 1, see [8]. They derived a uniform-in-bandwidth asymptotic approximation of a specific set estimation risk function, $E\{d_{\mu_f}(L(\tau), \hat{L}(\tau))\}$, where $d_{\mu_f}(L(\tau), \hat{L}(\tau)) = \int_{L(\tau)\triangle\hat{L}(\tau)} f(t)dt$ and $\triangle$ denotes the usual difference given by $L(\tau)\triangle\hat{L}(\tau) = (L(\tau) \setminus \hat{L}(\tau)) \cup (\hat{L}(\tau) \setminus L(\tau))$. Of course, it is also possible to consider classical methods such as Seather and Jones or cross validation to select the bandwidth parameter although they are not specific to estimate level sets.

27.2.1 Simulations Results for Plug-in Methods

In this section, we will compare Baíllo and Cuevas' (BC), Samworth and Wand's (SW), Sheather and Jones' (SJ), and cross validation (CV) methods. The first two one are specific bandwidth selectors to estimate level sets. The last two algorithms are general selectors to estimate density functions.

We have generated $1,000$ samples of size $n = 1,600$ for the 15 Marron and Wand's density functions (see [5]) and we have considered three values for the parameter τ: $\tau = 0.2$, $\tau = 0.5$, and $\tau = 0.8$. Although there are several ways to estimate the threshold, we have estimated it by using Hyndman's method, see [3]. This algorithm estimates the threshold by calculating the τ-quantile of the empirical distribution of $f_n(X_1), \ldots, f_n(X_n)$. We have considered the Sheather and Jones selector to calculate f_n. For each fixed random sample and each method, we have estimated the level set $L(\tau)$ and we computed the error of the estimation by calculating $d_{\mu_f}(L(\tau), \hat{L}(\tau))$. So, for a given model and a value of τ we have calculated 1,000 errors for each method.

To facilitate the presentation of the results, we use some figures described below. Each figure is divided into rectangles that are painted with different colors according to the method (vertical axis) and the density model (horizontal axis). Colors are assigned as follows: light colors correspond to low errors and vice versa. So, this representation allows to detect the most or less competitive algorithm fixed the value of τ. Given a density, we have ordered the means of the 1,000 errors calculated by testing if they are equal previously. If we reject the null hypothesis of equality between two means for the same model, then each method will be painted using a different color (darker or lighter according to the mean of the errors is higher or lower). In another case, both algorithms are represented using the same color. We will use this approach in the following sections to compare the methods of the two remaining groups of algorithms.

Figures 27.2 and 27.3 show the plug-in methods comparison for $\tau = 0.5$ and $\tau = 0.8$, respectively. For $\tau = 0.5$, the best results are provided by Sheather and Jones and cross validation selectors. If $\tau = 0.8$, then specific selectors for level sets have better results for the models 1, 2, 3, and 5. All of these densities have an only mode. They are very simple level sets. However, classical selectors are the most competitive for more sophisticated models such as 6, 8, 9, 10, 11, 12, 13, 14, or 15.

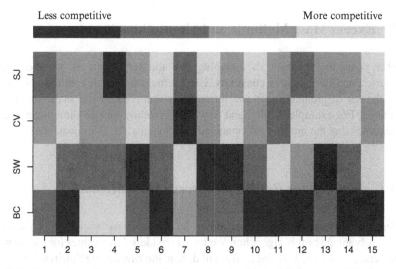

Fig. 27.2 Comparison of plug-in methods (*vertical axis*) with the 15 Marron and Wand's density models (*horizontal axis*), $\tau = 0.5$ and $n = 1,600$. The error criteria is d_{μ_f}

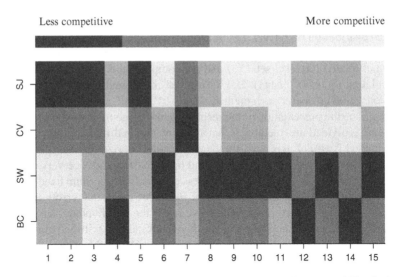

Fig. 27.3 Comparison of plug-in methods (*vertical axis*) with the 15 Marron and Wand's density models (*horizontal axis*), $\tau = 0.8$ and $n = 1,600$. The error criteria is d_{μ_f}.

As a conclusion, specific methods to estimate level sets do not improve the results of the classic bandwidth selection rules. In addition, cross validation and Sheather and Jones methods often provide similar results and they present the best global behavior.

27.3 Excess Mass Methods and Simulations Results

Another possibility consists of assuming that the set of interest satisfies some geometric condition such as convexity. Excess mass approach estimates the level set as the set of greatest mass and minimum volume under the shape restriction considered. For example, Müller and Sawitzki's method for one dimensional level sets assumes that the number of connected components, M, is known, see [6].

27.3.1 Simulations Results for Excess Mass Methods

Müller and Sawitzki's method depends on an unknown parameter M. This is the main disadvantage of this algorithm. We have considered five values for the number of clusters, $M = 1, 2, 3, 4$, and 5. We will denote the Müller and Sawitzki's method with M modes by MS_M.

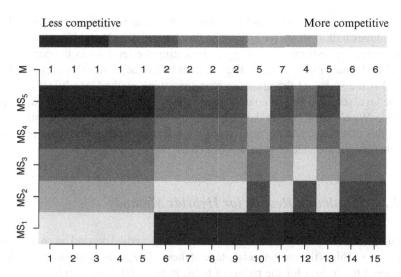

Fig. 27.4 Comparison of Müller and Sawitzki's method for different values of M (*vertical axis*) with the 15 Marron and Wand's density models (*horizontal axis*), $\tau = 0.5$ and $n = 1,600$. The error criteria is d_{μ_f}

To analyze the influence of the parameter M for Müller and Sawitzki's method, we will use Fig. 27.4. In this case, we have written the real number of modes for each density and $\tau = 0.5$ on the vertical axis too.

From Fig. 27.4, it is clear that Müller and Sawitzki's method is very sensitive to the parameter M. For $\tau = 0.5$, densities 1, 2, 3, 4, and 5 are unimodal and $M = 1$ provides the best results. Densities 6, 7, 8, or 9 have two modes and, in this case, the best value of M is $M = 2$. Model 10 has five modes for $\tau = 0.5$ and again $M = 5$ provides the best estimations. However, the best value of M for the Müller and Sawitzki's method is not equal to the real value of M for the models 11, 12, and 13 because some of their modes are not significant. In addition, if misspecification of M occurs, it can be seen that big values of M are better than a small values because the means of errors are lower.

27.4   Hybrid Methods and Simulations Results

As the name suggests, hybrid methods assume geometric restrictions and they use a pilot nonparametric density estimator to decide which sample points can be in the level set, $\mathscr{X}_n^+ = \{f_n \geq \hat{f}_\tau\}$. In this work we proposed two new hybrid methods to estimate convex and r-convex sets with $r > 0$. The last one is a shape condition more general than convexity. In fact, a closed set A is said r-convex with $r > 0$ if $A = C_r(A)$ where $C_r(A) = \bigcap_{\{B_r(x):B_r(x)\cap A=\emptyset\}} (B_r(x))^c$ denotes the r-convex hull of A, $B_r(x)$ denotes the open ball with center in x and radius r and

$(B_r(x))^c$, its complementary. Our two new proposals are based on the convex hull and r-convex hull methods for estimating the support, see [4] and [7], respectively. Under convexity restriction, we suggest estimating the level set as the convex hull of $\mathscr{X}_n^+$ and, under r-convexity, as the r-convex hull of $\mathscr{X}_n^+$. Another classic hybrid method is the so-called the granulometric smoothing method, see [9]. It assumes that the level set $L(\tau)$ and its complementary are r-convex. This method adapts the Devroye–Wise's estimator for the support to the context of level set estimation, see [2]. In this case, the estimator consists of the union of balls around those points in $\mathscr{X}_n^+$ that have a distance of at least r from each point in $\mathscr{X}_n \setminus \mathscr{X}_n^+$.

27.4.1 Simulation Results for Hybrids Methods

Granulometric smoothing method and r-convex hull method depend on an unknown parameter r. This is the main disadvantage of these algorithms. In this work, we have considered five values for the radius of balls, r: $r_1 = 0.01$, $r_2 = 0.05$, $r_3 = 0.1$, $r_4 = 0.2$, and $r_5 = 0.3$. We will denote the methods as follows: Convex hull method by CH, r-convex hull method by CH_r, and granulometric smoothing method with radius r by W_r.

Although these results are not shown here, we have studied the influence of the parameter r for r-convex hull method and granulometric smoothing method. In general, r-convex hull method is less sensitive to the selection to the parameter r. We have compared the three hybrids methods by fixing an intermediate value for r because it is unknown. We have considered $r = r_3$ and use Fig. 27.5 to show the

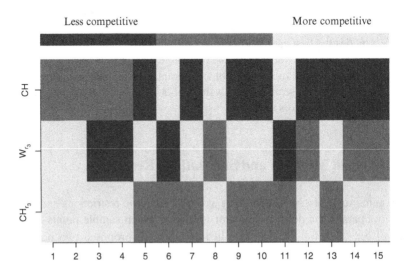

Fig. 27.5 Comparison of hybrid methods (*vertical axis*) with the 15 Marron and Wand's density models (*horizontal axis*), $\tau = 0.2$ and $n = 1,600$. The error criteria is d_{μ_f}

results obtained for $\tau = 0.2$. Each method is represented on the vertical axis and each density model on the horizontal axis.

Some of the density models present convex level sets for $\tau = 0.2$ or $\tau = 0.8$ although they are not unimodal (see, for example, densities 6, 8, or 11 in Fig. 27.5). In this case, when the convexity assumption is true, convex hull method can be very competitive. However, models 1, 2, 3, and 4 have convex level sets for some value of τ and r_3-convex hull method is the most competitive for them. In addition, sometimes convexity hypothesis can be very restrictive (see models 7 or 10, for example) and then, r_3-convex hull or granulometric smoothing methods provide better and similar results although the first one is most competitive for high values of τ.

27.5   Final Conclusions

Finally, we will compare the most competitive methods in each group. So, we will consider cross validation method, Müller and Sawitzki's method, granulometric smoothing method, r-convex hull method, and convex hull method. It is necessary to specify a value for the parameters M and r for Müller and Sawitzki's method and granulometric smoothing method or r-convex hull method. We have fixed $M = 3$ and $r = r_3$ again.

Figures 27.6 and 27.7 show the results for $\tau = 0.2$ and $\tau = 0.5$. Müller and Sawitzki's method with $M = 3$ is not very competitive because most of the models are not trimodal. For low values of τ, cross validation does not present bad results

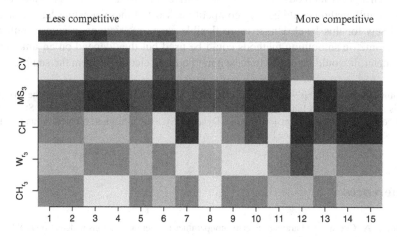

Fig. 27.6 Final comparison of the most competitive methods in each group (*vertical axis*) with the 15 Marron and Wand's density models (*horizontal axis*), $\tau = 0.2$ and $n = 1,600$. The error criteria is d_{μ_f}

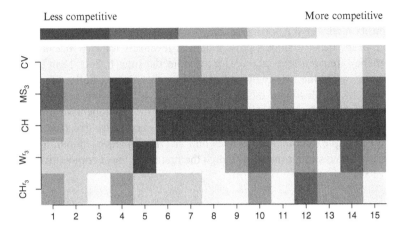

Fig. 27.7 Final comparison of the most competitive methods in each group (*vertical axis*) with the 15 Marron and Wand's density models (*horizontal axis*), $\tau = 0.5$ and $n = 1,600$. The error criteria is d_{μ_f}

but granulometric smoothing or r_3-convex hull methods have a better behavior (see models 3, 4, 6, or 11). But these two methods present a big disadvantage because both depend on an unknown parameter. Convex hull gets worse its results for $\tau = 0.5$ (see models 6, 8, or 11). The rest of the hybrid methods have good results for this value of τ.

In general, if no assumption is made on the shape of the level set, cross validation is a good option. But, if we have some information about the shape of the level set, then hybrid methods can be an alternative. For instance, if τ is small, then convex hull method could be very competitive. Most of these densities have convex level sets for this level. Under more flexible shape restrictions, r-convex hull or granulometric smoothing methods could be used but they depend on an unknown parameter. It would be useful to have a method for selecting it from the sample.

Acknowledgements This work has been supported by Project MTM2008-03010 from the Spanish Ministry of Science and Innovation and the IAP network StUDyS (Developing crucial Statistical methods for Understanding major complex Dynamic Systems in natural, biomedical and social sciences) from Belgian Science Policy.

References

1. Baíllo, A., Cuevas, A.: Parametric versus nonparametric tolerance regions in detection problems. Comput. Stat. **21**, 527–536 (2006)
2. Devroye, L., Wise, G.L.: Detection of abnormal behavior via nonparametric estimation of the support. SIAM J. Appl. Math. **38**, 480–488 (1980)
3. Hyndman, R.J.: Computing and graphing highest density regions. Am. Stat. **50**, 120–126 (1996)

4. Korostelëv, A., Tsybakov, A.: Minimax Theory of Image Reconstruction. Lecture Notes in Statistics, vol. 82. Springer, New York (1993)
5. Marron, J., Wand, M. Exact mean integrated squared error. Ann. Stat. **20**, 712–736 (1992)
6. Müller, D.W., Sawitzki, G.: Excess mass estimates and tests of multimodality. J. Am. Stat. Assoc. **86**, 738–746 (1991)
7. Rodríguez-Casal, A.: Set estimation under convexity type assumptions. Annales de l'I.H.P.-Probabilités Statistiques **43**, 763–774 (2007)
8. Samworth, R.J., Wand , M.P.: Asymptotics and optimal bandwidth selection for highest density region estimation. Ann. Stat. **38**, 1767–1792 (2010)
9. Walther, G.: Granulometric smoothing. Ann. Stat. **25**, 2273–2299 (1997)

Chapter 28
Nonparametric Permutation-Based Control Charts for Ordinal Data

Livio Corain and Luigi Salmaso

Abstract In the literature of statistical process control (SPC), design and implementation of traditional Shewart-based control charts requires the assumption that the process response distribution follows a parametric form (e.g., normal). However, since in practice, ordinal observations may not follow the pre-specified parametric distribution these charts may not be reliable. In this connection, this work aims at providing a contribution to the nonparametric SPC literature, proposing univariate and multivariate nonparametric permutation-based control charts for ordinal response variables which are not only interesting as methodological solution but they have a very practical value particularly within the context of monitoring some measure of user's satisfaction, loyalty, etc. related to use of a given service. As confirmed by the simulation study and by the application to a real case study in the field of monitoring of customer satisfaction in services, we can state that the proposed NPC chart for ordered categorical response variables is certainly a good alternative with respect to the literature counterparts.

Keywords Nonparametric combination • NPC chart • Permutation tests

28.1 Introduction and Motivation

Recently Corain and Salmaso [5] proposed the application of the Non Parametric Combination (NPC) methodology [13] to develop a novel type of multivariate nonparametric control chart called NPC chart which has proved to be particularly effective as statistical process control (SPC) tool when the underlying data generation mechanism is non-normal in nature. In this paper we extend the proposal of Corain and Salmaso [5] to the case of ordered categorical response variables which is a situation difficult to treat by traditional SPC methods such as the Shewart-type charts. The NPC chart provides a flexible and effective analysis in terms both of the specification of the inferential hypotheses and of the nature of the variables involved

L. Corain (✉) • L. Salmaso
Department of Management and Engineering, University of Padova, Padova, Italy
e-mail: livio.corain@unipd.it; luigi.salmaso@unipd.it

© Springer Science+Business Media New York 2014   309
M.G. Akritas et al. (eds.), *Topics in Nonparametric Statistics*, Springer Proceedings
in Mathematics & Statistics 74, DOI 10.1007/978-1-4939-0569-0__28

without the need of modelling, in case of multivariate response, the dependence among variables.

Design and implementation of traditional Shewart-based control charts requires the assumption that the process response distribution follows a parametric form (e.g., normal). The statistical properties of commonly employed control charts are exact only if this assumption is satisfied; however, the underlying process is not normal in many applications, and as a result the statistical properties of the standard charts can be highly affected in such situations. To develop appropriate control charts that do not require specifying the parametric form of the response distribution beforehand, a number of distribution-free or nonparametric control charts have been proposed in literature (Chakraborti et al. [3]). Despite immense use and acceptability of parametric control charts, non-parametric control charts are an emerging area of recent development in the theory of SPC. The main advantage of non-parametric control charts is that they do not require any knowledge about the underlying distribution of the variable.

Das [6] summarized several non-parametric control charts for controlling location from a literature survey (the sign test, Hodges-Lehmann estimator and Mann-Whitney statistic) and compared their efficiency to detect the shift in location while in out of the control state under different situations and identified the best method under the prevailing situation. Chakraborti and van de Wiel [4] outlined that non-parametric or distribution-free charts can be useful in statistical process control problems when there is limited or lack of knowledge about the underlying process distribution. In their paper, a phase II Shewhart-type chart is considered for location, based on reference data from phase I analysis and the well-known Mann-Whitney statistic.

The implementation of most control charts requires that the performance ratings be quantitative on the interval scale of measurement, which may not be the case in many applications (Bakir [1]). Bakir proposes a quality control chart that is particularly useful in the area of performance appraisal when the workers' ratings are categorical on the ordinal scale of measurement. Moreover in a comparison of three control charts for monitoring data from student evaluations of teaching (SET) with the goal of improving student satisfaction with teaching performance, Ding et al. [7] show that a comparison of three charts (an individuals chart, the modified p chart, and the z-score chart) reveals that the modified p chart is the best approach for analyzing SET data because it utilizes distributions that are appropriate for categorical data, and its interpretation is more straightforward. Samimi et al. [16] give an ordinal scale contribution to control charts; their paper presents several control charts classified in two groups based on the scale used to assess customer loyalty. In the first group of control charts, customer loyalty is considered as a binary random variable modeled by Bernoulli distribution whilst in the second group, an ordinal scale is considered to report loyalty level. Performance comparison of the proposed techniques using ARL criterion indicates that chi-square and likelihood-ratio control charts developed based on Pearson chi-square statistic and ordinal logistic regression model respectively are able to rapidly detect the significant changes in loyalty behavior. When the process measurement is multivariate most existing multivariate SPC procedures assume that the in-control distribution of the multivariate process measurement is known and it is a Gaussian distribution (Qiu [14]); Qiu [14] demonstrates that results from conventional multivariate SPC procedures are usually unreliable when the data are non-Gaussian. He suggests a methodology for estimating the in-control multivariate measurement distribution based on log-linear modeling and a multivariate CUSUM procedure (a nonparametric multivariate Cumulative SUM procedure, based on the antiranks of the measurement components) for detecting shifts in the location parameter vector of the measurement distribution is also suggested for Phase II SPC. Qiu and Li [15], have considered the case of non parametric form of the process distribution and categorical data. The authors try to make two contributions to the non-parametric SPC literature. First, they propose an alternative framework for constructing non-parametric control charts, by first categorizing observed data and then applying categorical data analysis methods to SPC. Under this framework, some new non-parametric control charts are proposed. Second, they compare our proposed control charts with several representative existing control charts in various cases.

28.2 Univariate and Multivariate Permutation Tests and Nonparametric Combination Methodology

The importance of the permutation approach in resolving a large number of infer-ential problems is well documented in the literature, where the relevant theoretical aspects emerge, as well as the effectiveness and flexibility from an applicatory point of view ([2, 8, 9, 13]).

For any general testing problem, in the null hypothesis (H_0), which usually assumes that data come from only one (with respect to groups) unknown population distribution P, the whole set of observed data $\mathbf{x}$ is considered to be a random sample, taking values on sample space Ω^n, where $\mathbf{x}$ is one observation of the n-dimensional sampling variable $\mathbf{X}^{(n)}$ and where this random sample has no necessarily independent and identically distributed (i.i.d.) components. We note that the observed data set $\mathbf{x}$ is always a set of sufficient statistics in H_0 for any underlying distribution [12]. Since, in the null hypothesis and assuming exchangeability, the conditional probability distribution of a generic point $\mathbf{x}' \in \Omega^n$, for any underlying population distribution $P \in \mathscr{P}$, is distribution-independent, permutation inferences are invariant with respect to the underlying distribution in H_0. Some authors, emphasizing this invariance property, prefer to give them the name of invariant tests. However, due to this invariance property, permutation tests are distribution-free and nonparametric. Permutation tests have general good properties such as exactness, unbiasedness and consistency (see [10, 13]).

In order to provide details on the construction of multivariate permutation tests via nonparametric combination approach, let us consider, for instance, two multivariate populations and the related two-sample multivariate hypothesis testing problem where p (possibly dependent) variables are considered. The most simple additive statistical model (with fixed effects) can be represented as follows:

$$\mathbf{Y}_{ij} = \mu_j + \epsilon_{ij}, i = 1, \ldots, n_j, j = 0, 1,$$

where μ_j is the p-dimensional mean effect, $\epsilon_{ij} \sim \text{IID}(0, \Sigma)$ is a p-variate random term of experimental errors with zero mean and variance/covariance matrix Σ. The univariate response Y can be either continuous, binary, ordered categorical or mixed variables (some binary/continuous and some other ordered categorical).

The main difficulties when developing a multivariate hypothesis testing procedure arise because of the underlying dependence structure among variables, which is generally unknown. Moreover, a global answer involving several dependent variables is often required, hence the main question is how to combine the information related to the p variables into one global test. In order to better explain the proposed approach let us denote an $n \times p$, $n = n_0 + n_1$, data set with $\mathbf{Y}$:

$$\mathbf{Y} = [\mathbf{Y}_0, \mathbf{Y}_1] = [Y_1, Y_2, \ldots, Y_p] = \begin{matrix} y_{11} & y_{12} & \cdots & y_{1p} \\ y_{21} & y_{22} & \cdots & y_{2p} \\ \cdots & \cdots & \cdots & \cdots \\ y_{n1} & y_{n2} & \cdots & y_{np} \end{matrix},$$

where $\mathbf{Y}_0$ and $\mathbf{Y}_1$ are the $n_0 \times p$ and the $n_1 \times p$ samples drawn from the first and second populations, respectively. In the framework of Nonparametric Combination (NPC) of Dependent Permutation Tests we suppose that, if the global null hypothesis $H_0 : \mu_0 = \mu_1$ of equality of the two populations is true, the hypothesis of exchangeability of random errors holds. Hence, the following set of mild conditions should be jointly satisfied:

a) we suppose that for $\mathbf{Y} = [Y_0, Y_1]$ an appropriate p-dimensional distribution exists, $P_j \in \mathscr{P}, j = 0, 1$, belonging to a (possibly non-specified) family $\mathscr{P}$ of non-degenerate probability distributions;

b) the null hypothesis H_0 states the equality in distribution of the multivariate distribution of the p variables in the two groups: $H_0 : [P_0 = P_1] = \left[Y_0 \overset{d}{=} Y_1 \right]$. Null hypothesis H_0 implies the exchangeability of the individual data vector with respect to the two groups. Moreover H_0 is supposed to be properly decomposed into p sub-hypotheses $H_{0k}, k = 1, \ldots, p$, each appropriate for partial (univariate) tests, thus H_0 (multivariate) is true if all the H_{0k} (univariate) are jointly true: $H_0 : \left[\bigcap_{k=1}^{p} Y_{1k} \overset{d}{=} Y_{2k} \right] = \left[\bigcap_{k=1}^{p} H_{0k} \right]$. H_0 is called the *global or overall null hypothesis*, and $H_{0k}, k = 1, \ldots, p$, are called the *partial null hypotheses*.

c) The alternative hypothesis H_1 can be represented by the union of partial H_{1k} sub-alternatives: $H_1 : \left[\bigcup_{k=1}^{p} H_{1k} \right]$, hence, H_1 is true if at least one of the sub-alternatives is true.

In this context, H_1 is called the *global* or *overall alternative*, and $H_{1k}, k = 1, \ldots, p$, are called the *partial alternatives*.

d) Let $\mathbf{T} = T(\mathbf{Y})$ represent a p-dimensional vector of test statistics, $p \geq 1$, whose components $Tk, k = 1, \ldots, p$, represent the partial univariate and non-degenerate *partial tests* appropriate for testing the sub-hypothesis H_{0k} against H_{1k}. Without loss of generality, all partial tests are assumed to be marginally unbiased, consistent and significant for large values (for more details we refer to Pesarin and Salmaso [13]).

At this stage, in order to test the global null hypothesis H_0 and the p univariate hypotheses H_{0k}, the key idea comes from the partial (univariate) tests which are focused on the k-th component variable, and then combining them through an appropriate combining function, to test the global (multivariate) test which is referred to as the global null hypothesis H_0.

However, we should observe that in most real problems when the sample sizes are large enough, there is a clash over the problem of computational difficulties in calculating the conditional permutation space. Hence, it is not possible to calculate the exact p-value of observed statistic T_{k0}. This is usually overcome by using the CMCP (Conditional Monte Carlo Procedure). The CMCP on the pooled data set $\mathbf{Y}$ is a random sampling from the set of all possible permutations of the same data under H_0. Hence, in order to obtain an estimate of the permutation distribution under H_0 of all test statistics, a CMCP can be used. Every resampling without replacement $\mathbf{Y}^*$ from the data set $\mathbf{Y}$ actually consists of a random attribution of the individual block data vectors to the two treatments. In every $\mathbf{Y}_b^*$ resampling, $b = 1, \ldots, B$, the k partial tests are calculated to obtain the set of values $\left[\mathbf{T}_b^* = \mathbf{T}(\mathbf{Y}_{bk}^*), k = 1, .., p; b = 1, \ldots, B \right]$, from the B independent random resamplings. It should be emphasized that CMCP only considers permutations of individual data vectors, so that all underlying dependence relations which are present in the component variables are preserved.

Without loss of generality, let us suppose that partial tests are significant for large values. More formally, the steps of the CMC procedure are described as follows:

1. calculate the p-dimensional vectors of statistics, each one related to the corresponding partial tests from the observed data:

$$\mathbf{T}_{p \times 1}^{\text{obs}} = \mathbf{T}(\mathbf{Y}) = \left[T_k^{\text{obs}} = T_k(\mathbf{Y}), k = 1, .., p \right],$$

2. calculate the same vectors of statistics for the permuted data:

$$\mathbf{T}_b^* = \mathbf{T}(\mathbf{Y}_b^*) = \left[T_{bk}^* = T_k(\mathbf{Y}_b^*), k = 1, .., p \right],$$

3. repeat the previous step B times independently. We denote with $\{ \mathbf{T}_b^*, b = 1, \ldots, B \}$ the resulting sets from the B conditional resamplings. Each element represents a random sample from the p-variate permutation c.d.f. $F_T(z | \mathbf{Y})$ of the test vector $\mathbf{T}(\mathbf{Y})$.

The resulting estimates are:

$$\hat{F}_T\left(\mathbf{z}|\mathbf{Y}\right) = \left[\frac{1}{2} + \sum_{b=1}^{B} \mathbf{I}\left(\mathbf{T}_b^* \leqslant \mathbf{z}\right)\right] / (B+1), \forall \mathbf{z} \in \mathbf{R}^p,$$

$$\hat{L}_{T_k}\left(z|\mathbf{Y}\right) = \left[\frac{1}{2} + \sum_{b=1}^{B} \mathbf{I}\left(\mathbf{T}_{bk}^* \geqslant z\right)\right] / (B+1), \forall z \in \mathbf{R}^1,$$

$$\hat{\lambda}_k = \hat{L}_{T_k}\left(T_k^{obs}|\mathbf{Y}\right) = \left[\frac{1}{2} + \sum_{b=1}^{B} \mathbf{I}\left(\mathbf{T}_{bk}^* \geqslant T_k^{obs}\right)\right] / (B+1), k = 1, \ldots, p$$

where $\mathbf{I}(.)$ is the indicating function and where with respect to the traditional EDF estimators, 1/2 and 1 have been added, respectively, to the numerators and denominators in order to obtain estimated values in the open interval (0,1), so that transformations by inverse CDF of continuous distributions are continuous and so are always well defined.

Hence, if $\hat{\lambda}_k < \alpha$, the null hypothesis corresponding to the k-th variable (H_{0k}) is rejected at significance level equal to α.

Let us now consider a suitable continuous non-decreasing real function, ϕ : $(0,1)^p \rightarrow P^1$, that applied to the p-values of partial tests T_k defines the second order global (multivariate) test T'' (for details see [13]).

28.3 NPC Charts for Ordered Categorical Response Variables

The theory of permutation tests and of nonparametric combination represent the methodological background from which a nonparametric multivariate control chart for ordered categorical response variables can be developed. Let be the two p-variate samples $\mathbf{Y}_0$ and $\mathbf{Y}_1$ related to the so-called control chart phase I and II, more specifically, denoting with n ($n > 1$) the sample size of the rational subgroup, Y_0 has to be considered as the $n_0 \times p$ pooled sample, $n_0 = n \times m$, of the m in-control samples, used to retrospectively testing whether the process was under control and where the first m subgroups have been drawn. Note that without loss of generality we are assuming that all m initial subgroups are actually in-control samples. The sample Y_1 has to be considered as one of the actual subgroups of size $n_1 \times p$, possibly out-of-control samples ($n = n_0$), used for testing whether the process remains under control when further subgroups will be drawn. Since our reference response is a multivariate ordered categorical variables, for each univariate component we consider suitable permutation test statistics:

- Multi-focus statistic [13]: this approach suggests to decompose the categorical response variable of interest into s binary variables each one related to one category of the response; in this way, it is possible to refer to a further decomposition of the null univariate sub-hypothesis H_{0ks} into s additional sub-hypothesis each one suitable for testing the equality in distribution of each one of the s category of the ordered categorical response variable; this is done by taking into account a set of s Chi-squared based tests calculated from s 2×2 contingency sub-tables to be then combined into a final statistic;

- Anderson–Darling statistic [13]:

$$T_{AD}^* = \sum_{r=1}^{s-1} \left(N_{0r}^* - N_{1r}^* \right) \left[2 \frac{N_{\cdot r}}{n} \left(\frac{2n - N_{\cdot r}}{2n} \right) \frac{n^2}{2n - 1} \right]^{-\frac{1}{2}},$$

- where $N_{\cdot r} = N_{0r} + N_{1r} = N_{0r}^* + N_{1r}^*$ are the observed and the permutation cumulative frequencies in which $N_{sr}^* = \sum_{q \le r} f_{sq}^*$, $r = 1, \ldots, s - 1$, $j = 0, 1$, and f_{sq} is the frequencies of the j-th treatment for the q-th category of the response.

Within our approach we do not need to estimate any parameters of the in-control process. This is consistent with the usual rationale behind the nonparametric permutation approach, i.e. the in-control pooled sample Y_0 plays the role of reference dataset to be kept out and recursively compared with Y_1, which is one of the subgroups to be tested in the future. This conceptual framework leads to a procedure where we consider a sequence of independent multivariate two-sample hypotheses testing problems in case of unbalanced designs ($n_0 > n_1$). Some important remarks have to be pointed out: the control limit for the multivariate control chart at the desired α-level has to be simply calculated as the $(1 - \alpha)$ quantile of the null permutation distribution of the multivariate combined test statistic T'' (for details on how to perform a multivariate permutation test via nonparametric combination methodology see [13]); noting that necessarily the limits will differ for any given subgroup under testing. Finally, we are implicitly assuming that the process parameters are unknown but the extension to the case of known parameters is straightforward; in fact, this case may reduce to the so-called multivariate one-sample problem where a combination-based solution already does exist [13]. In this work we focus on the case of unknown limits, the most interesting for real applications.

28.4 Simulation Study

In order to validate the proposed NPC Chart and to evaluate its relative performance when compared with a traditional multivariate and control chart such as Hotelling T^2 and X-bar (which are not appropriate in case of ordered categorical data), we carried out a suitable Monte Carlo simulation study. The real context we are referring to is a typical customer satisfaction study where a group of 20 people provides their evaluations by a Likert 1–5 rating ordinal scale, where we suppose that the 0.5 scores are admitted as well. Note that we are actually considering a nine point ordered categorical response variable. Let us consider the following simulation setting:

- number of response variables: $p = 5, 10$; where the number of active variables (under the alternative hypothesis) was 1, 2, and 4 (three setting), more precisely while in all cases $\mu_0 = [3, 3, 3, 3, 3]$, under H_1 the mean values were set for the three settings, respectively, as $\mu_1 = [3, 3, 3, 3, 5]$, $\mu_1 = [3, 3, 3, 4, 5]$ and $\mu_1 = [3, 4, 4, 5, 5]$;
- two types of multivariate distributions for random errors: Normal and a moderate heavy-tailed and skewed distribution (with kurtosis equal to 20 and skewness equal to 3). In order to guarantee an ordered categorical response variable we rounded the continuous error to the nearest 0.5 value;
- two types of variance/covariance matrices: i. I_p (identity matrix, i.e. the case of independence where $\sigma_{jh} = 0, \forall j, h = 1, \dots, p$) and Σ_p is such that each univariate random component has $\sigma_j^2 = 1, \forall j = 1, \dots, p$ and $\sigma_{jh} = 0.4$.

The performance of univariate and multivariate NPC Charts has been evaluated in terms of Average Run Length—ARL, i.e. the average number of samples needed before to get the first out-of-control (reject the null hypothesis that a truly out-of-control process is under control). Simulations are designed so that for each of the 1,000 data generations five independent samples of size $n = 20$ are created and the control charts are applied until the first out of control is reported. The in-control pooled dataset Y_0 was generated by $n_0 = n \times 2$ random values ($m = 2$).

Results of normal errors and skewed and heavy-tailed errors (due to space requirements, tables only for the case of dependent random errors), for $p = 5$ and $p = 10$, respectively, are shown in the tables below (Tables 28.1, 28.2, 28.3, 28.4).

First of all, let us consider results under the null hypothesis. We estimated the type one error considering a 0.5 % α-level. As we can see, all NPC charts properly respect the nominal level, while X-bar and Hotelling (which are not appropriate for ordinal data) appear strongly biased (the first) and quite conservative (the second).

Under the alternative hypothesis, the Anderson–Darling with Tippet and Multi-Focus with iterated combining functions seem to perform better than other NPC solutions [13].

28.5 Conclusion

As confirmed by the simulation study and by the real case study, we can state that the proposed NPC chart for ordered categorical response variables may represent a good alternative with respect to existing techniques. Furthermore, control charts based on NPC can manage with any dependence relation among variables and any kind of variable (even mixed variables with possible presence of missing observations). An important property which can be very useful in real applications is represented

Table 28.1 Rejection rates: Normal errors, $p = 5$

Type of chart	Statistic	Shift			
		0	0.5	1	2
Univar. control chart	Anderson–Darling	0.006	0.238	0.841	1.00
	Multi-focus	0.005	0.193	0.719	1.00
	X-bar	0.015	0.253	0.878	1.00

Type of chart	Statistic	Combining function	H1						H0	
			Sett_1	Sett_2	Sett_3	Sett_4	Sett_5	Sett_6	Sett_1-3	Sett_4-6
Multivar. control chart	Anderson–Darling	Fisher	0.277	0.614	0.994	0.034	0.158	0.778	0.006	0.008
	Multi-focus		0.991	0.709	0.983	0.985	0.531	0.817	0.006	0.006
	Anderson–Darling	Tippet	1.00	0.670	0.884	1.00	0.624	0.805	0.006	0.005
	Multi-focus		0.989	0.603	0.828	0.987	0.590	0.769	0.005	0.005
	Anderson–Darling	Liptak	0.113	0.305	0.980	0.018	0.065	0.593	0.004	0.007
	Multi-focus		0.990	0.645	0.971	0.985	0.525	0.777	0.006	0.005
	Anderson–Darling	Iterated	0.313	0.593	0.994	0.157	0.217	0.778	0.007	0.007
	Multi-focus		0.990	0.705	0.983	0.986	0.537	0.817	0.007	0.006
	Hotelling		0.999	0.613	0.982	1.00	0.720	0.876	0.000	0.000

Table 28.2 Rejection rates: G-H, $p = 5$

Type of chart	Statistic	Shift		
		0	0.5	1
Univar. control chart	Anderson–Darling	0.006	0.189	0.585
	Multi-focus	0.006	0.137	0.362
	X-bar	0.038	0.363	0.803

Type of chart	Statistic	Combing function	H1						H0	
			Sett_3	Sett_2	Sett_1	Sett_6	Sett_5	Sett_4	Sett_1-3	Sett_4-6
Multivar. control chart	Anderson–Darling	Fisher	0.146	0.419	0.924	0.025	0.131	0.572	0.004	0.004
	Multi-focus		0.283	0.390	0.738	0.232	0.297	0.537	0.004	0.005
	Anderson–Darling	Tippet	0.386	0.439	0.690	0.359	0.415	0.251	0.007	0.008
	Multi-focus		0.292	0.344	0.530	0.270	0.341	0.529	0.006	0.005
	Anderson–Darling	Liptak	0.063	0.225	0.895	0.017	0.076	0.461	0.005	0.006
	Multi-focus		0.264	0.351	0.706	0.229	0.299	0.514	0.004	0.006
	Anderson–Darling	Iterated	0.172	0.440	0.931	0.103	0.205	0.593	0.007	0.006
	Multi-focus		0.284	0.393	0.735	0.242	0.312	0.546	0.005	0.006
	Hotelling		0.098	0.223	0.882	0.242	0.312	0.586	0.001	0.001

Table 28.3 Rejection rates: Normal errors, $p = 10$

Type of chart	Statistic	Shift		
		0	0.25	0.5
Univar. control chart	Anderson–Darling	0.007	0.042	0.194
	Multi-focus	0.005	0.039	0.165
	X-bar	0.017	0.071	0.268

Type of chart	Combining function	Statistic	H1						H0	
			Sett_3	Sett_2	Sett_1	Sett_6	Sett_5	Sett_4	Sett_1-3	Sett_4-6
Multivar. control chart	Fisher	Anderson–Darling	0.130	0.207	0.733	0.047	0.063	0.210	0.006	0.008
		Multi-focus	0.197	0.277	0.713	0.137	0.127	0.297	0.006	0.006
	Tippet	Anderson–Darling	0.095	0.170	0.723	0.037	0.053	0.202	0.006	0.005
		Multi-focus	0.147	0.113	0.227	0.127	0.100	0.223	0.005	0.005
	Liptak	Anderson–Darling	0.060	0.133	0.713	0.027	0.043	0.193	0.004	0.007
		Multi-focus	0.183	0.230	0.700	0.130	0.117	0.280	0.006	0.005
	Iterated	Anderson–Darling	0.113	0.190	0.747	0.040	0.060	0.203	0.006	0.007
		Multi-focus	0.203	0.287	0.740	0.137	0.117	0.303	0.007	0.006
		Hotelling	0.000	0.000	0.003	0.000	0.000	0.000	0.000	0.000

Table 28.4 Rejection rates: G-H, $p = 10$

Type of chart	Statistic	Shift 0	0.25	0.5
Univar. control chart	Anderson–Darling	0.006	0.037	0.117
	Multi-focus	0.008	0.041	0.131
	X-bar	0.050	0.123	0.359

Type of chart	Statistic	Combing function	H1						H0	
			Sett_3	Sett_2	Sett_1	Sett_6	Sett_5	Sett_4	Sett_1-3	Sett_4-6
Multivar. control chart	Anderson–Darling	Fisher	0.093	0.220	0.693	0.023	0.040	0.110	0.004	0.004
	Multi-focus		0.167	0.243	0.570	0.110	0.163	0.180	0.004	0.005
	Anderson–Darling	Tippet	0.073	0.187	0.690	0.017	0.033	0.102	0.007	0.008
	Multi-focus		0.123	0.150	0.260	0.107	0.157	0.160	0.006	0.005
	Anderson–Darling	Liptak	0.052	0.153	0.687	0.010	0.027	0.093	0.005	0.006
	Multi-focus		0.140	0.200	0.623	0.107	0.163	0.170	0.004	0.006
	Anderson–Darling	Iterated	0.101	0.213	0.727	0.030	0.037	0.113	0.007	0.006
	Multi-focus		0.170	0.240	0.627	0.110	0.163	0.187	0.005	0.006
	Hotelling		0.000	0.006	0.038	0.003	0.100	0.003	0.000	0.001

by the finite sample consistency of NPC-based tests (see [13]) which can help in gain power keeping fixed the sample size and increasing the number of informative variables. Regarding some directions for future research, an effective way to gain additional power could be the inclusion in the same problem of more than one aspect (so-called multi-aspect strategy). Finally, additional research is needed to study the effect of heteroscedastic random errors and extensions to single case data could be of interest for future research.

References

1. Bakir Saad, T.: A quality control chart for work performance appraisal. Qual. Eng. **17**, 429–434 (2005)
2. Basso, D., Pesarin, F., Salmaso, L., Solari, A.: Permutation Tests for Stochastic Ordering and ANOVA: Theory and Applications with R. Lecture Notes in Statistics. Springer, New York (2009)
3. Chakraborti, S., Van Der Laan, P., Bakir, S.T.: Nonparametric control charts: an overview and some results. J. Qual. Technol. **33**(3), 304–315 (2001)
4. Chakraborti, S., van de Wiel, M.A.: A nonparametric control chart based on the Mann-Whitney statistic. Inst. Math. Stat. Collect. **1**, 156–172 (2008)
5. Corain, L., Salmaso, L.: Nonparametric permutation and combination-based multivariate control charts with applications in microelectronics. Appl. Stoch. Model Bus. Ind. **29**, 334–349 (2013)
6. Das, N.: A comparison study of three non-parametric control charts to detect shift in location parameters. Int. J Adv. Manuf. Technol. **41**, 799–807 (2009)
7. Ding, X.: An assessment of statistical process control-based approaches for charting student evaluation scores. Decis. Sci. J. Innov. Educ. **4**(2), 259–272 (2006)
8. Edgington, E.S., Onghena, P.: Randomization Tests, 4th edn. Chapman and Hall, London (2007)
9. Good, P.: Permutation, Parametric, and Bootstrap Tests of Hypotheses, 3rd edn. Springer Series in Statistics, New York (2010)
10. Hoeffding, W.: The large-sample power of tests based on permutations of observations. Ann. Math. Stat. **23**, 169–192 (1952)
11. Ion, R.A., Klaassen, C.A.J.: Non-parametric Shewhart control charts. Nonparametr. Stat. **17**(8), 971–988 (2005)
12. Pesarin, F., Salmaso, L.: Finite-sample consistency of combination-based permutation tests with application to repeated measures designs. J. Nonparametric Stat. **22**(5), 669–684 (2010)
13. Pesarin, F., Salmaso, L.: Permutation Tests for Complex Data: Theory, Applications and Software. Wiley, Chichester (2010)
14. Qiu, P.: Distribution-free multivariate process control based on log-linear modeling. IIE Trans. **40**, 664–677 (2008)
15. Qiu, P., Li, Z.: On nonparametric statistical process control of univariate processes. Technometrics **53**(4), 390–405 (2011)
16. Samimi, Y., Aghaie, A., Tarokh, M.J.: Analysis of ordered categorical data to develop control charts for monitoring customer loyalty. Appl. Stoch. Models Bus. Ind. **26**, 668–688 (2010)

Chapter 29
The Latest Advances on the Hill Estimator and Its Modifications

M. Ivette Gomes and Milan Stehlík

Abstract Recent developments on the Hill and related estimators of the extreme value index are provided. We also discuss their properties, like mean square error, efficiency and robustness. We further discuss the introduction of underlying score functions related to modified Hill estimators.

Keywords Extreme value index • Efficiency • Heavy right tails • Hill estimator • Modified Hill estimators • Mean square error • Robustness • Score function

29.1 Estimators Under Study and Scope of the Paper

Let us consider a sample of size n of independent, identically distributed random variables (r.v.'s), $X_1, \ldots, X_n$, with a common distribution function (d.f.) F. Let us denote by $X_{1,n} \leq \cdots \leq X_{n,n}$ the associated ascending order statistics (o.s.) and let us assume that there exist sequences of real constants $\{a_n > 0\}$ and $\{b_n \in \Re\}$ such that the maximum, linearly normalized, i.e. $(X_{n,n} - b_n)/a_n$, converges in distribution to a non-degenerate random variable. Then, the limit distribution is necessarily of the type of the general *extreme value* (EV) d.f., given by

$$G_\gamma(x) = \begin{cases} \exp(-(1 + \gamma x)^{-1/\gamma}), & 1 + \gamma x > 0, \text{ if } \gamma \neq 0 \\ \exp(-\exp(-x)), & x \in \Re, \quad \text{if } \gamma = 0. \end{cases} \tag{29.1}$$

F is then said to belong to the *max-domain of attraction* of G_γ and we use the notation $F \in \mathscr{D}_{\mathscr{M}}(G_\gamma)$. The parameter γ, in (29.1), is the *extreme value index*

M.I. Gomes (✉)
Centro de Estatística e Aplicações, Faculdade de Ciências, Universidade de Lisboa
e-mail: ivette.gomes@fc.ul.pt

M. Stehlík
Department of Applied Statistics, Johannes Kepler University, Linz, Austria

Departamento de Matemática, Universidad Técnica Federico Santa María, Casilla 110-V, Valparaíso, Chile
e-mail: Milan.Stehlik@jku.at

© Springer Science+Business Media New York 2014
M.G. Akritas et al. (eds.), *Topics in Nonparametric Statistics*, Springer Proceedings in Mathematics & Statistics 74, DOI 10.1007/978-1-4939-0569-0_29

(EVI), the primary parameter of extreme events. We further use the notation $\mathscr{D}_M^+ :=$ $\mathscr{D}_M (G_{\gamma>0})$.

Let us denote RV_a the class of regularly varying functions at infinity, with an index of regular variation equal to $a \in \Re$, i.e. positive measurable functions $g(\cdot)$ such that for all $x > 0$, $g(tx)/g(t) \to x^a$, as $t \to \infty$ (see [6], for details on regular variation). The EVI measures the heaviness of the right *tail function* $\overline{F}(x) := 1 - F(x)$, and the heavier the right tail, the larger γ is. In this paper we shall consider Pareto-type underlying d.f.'s, with a positive EVI, or equivalently, models such that $\overline{F}(x) = x^{-1/\gamma} L(x), \gamma > 0$, with $L \in RV_0$, a slowly varying function at infinity, i.e. a regularly varying function with an index of regular variation equal to zero. These heavy-tailed models are quite common in the most diverse areas of application.

In Sect. 29.2, we provide a few properties of the Hill EVI-estimator and a few related estimators, like the mean of order p of those same statistics, the generalized Hill and corrected-Hill EVI-estimators. We compare asymptotically at optimal levels, in the sense of minimum mean square error (MSE) some of those EVI-estimators. Finally, in Sect. 29.3, we approach the use of score functions for the H-estimator to obtain some desirable properties, e.g. robustness.

29.2 The Hill (H) and a Few Related EVI-Estimators

For heavy-tailed models, i.e. if $F \in \mathscr{D}_M^+$, the simplest class of EVI-estimators is proposed in Hill [28]. The H-estimator, also denoted $H_{k,n}$, is the average of the log-excesses as well as of the scaled log-spacings, given by

$$V_{ik} := \ln X_{n-i+1,n}/X_{n-k,n} \quad \text{and} \quad W_i := i \{\ln X_{n-i+1,n}/X_{n-i,n}\}, \ 1 \leq i \leq k < n,$$
$$(29.2)$$

respectively. We thus have

$$H_{k,n} := \frac{1}{k} \sum_{i=1}^{k} V_{ik} = \frac{1}{k} \sum_{i=1}^{k} W_i, \quad 1 \leq k < n. \tag{29.3}$$

The asymptotic properties of $H_{k,n}$ have been thoroughly studied by several authors. See the recent reviews in [4] and [21]. Note that with $F^{\leftarrow}(x) := \inf\{y : F(y) \geq x\}$ denoting the generalized inverse function of F, and

$$U(t) := F^{\leftarrow}(1 - 1/t), \ t \geq 1, \tag{29.4}$$

the reciprocal quantile function, we can write the distributional identity $X \overset{d}{=} U(Y)$, with Y a unit Pareto r.v., i.e. an r.v. with d.f. $F_Y(y) = 1 - 1/y, \ y \geq 1$. For the o.s. associated with a random Pareto sample $(Y_1, \ldots, Y_n)$, we have the distributional identity $Y_{n-i+1,n}/Y_{n-k,n} \overset{d}{=} Y_{k-i+1,k}, \ 1 \leq i \leq k$. Moreover, $k Y_{n-k,n}/n \xrightarrow[n\to\infty]{p} 1$, i.e. $Y_{n-k,n} \overset{p}{\sim} n/k$. Consequently, and provided that $k = k_n$, $1 \le k < n$, is an intermediate sequence of integers, i.e. if

$$k = k_n \to \infty \quad \text{and} \quad k_n = o(n), \text{ as } n \to \infty, \tag{29.5}$$

we get

$$U_{ik} := \frac{X_{n-i+1,n}}{X_{n-k,n}} \overset{d}{=} \frac{U(Y_{n-i+1,n})}{U(Y_{n-k,n})} \overset{d}{=} \frac{U(Y_{n-k,n}Y_{k-i+1,k})}{U(Y_{n-k,n})} \overset{d}{=} Y_{k-i+1,k}^{\gamma}(1+o_p(1)), \tag{29.6}$$

i.e. $U_{ik} \overset{p}{\sim} Y_{k-i+1,k}^{\gamma}$. Hence, $\ln U_{ik} \approx \gamma \ln Y_{k-i+1,k} = \gamma E_{k-i+1,k}$, $1 \le i \le k$, with E denoting a standard exponential r.v. The log-excesses, $V_{ik} = \ln U_{ik}$, $1 \le i \le k$, in (29.2), are thus approximately the k top o.s. of a sample of size k from an exponential parent with mean value γ. This justifies the H-estimator, in (29.3). For first and second-order conditions see [27] and the references therein.

29.2.1 A Mean of Order p (MOP) and the Generalized Hill (GH) EVI-Estimators

Note that we can write

$$H_{k,n} = \sum_{i=1}^{k} \ln(X_{n-i+1,n}/X_{n-k,n})^{1/k} = \ln(\prod_{i=1}^{k} X_{n-i+1,n}/X_{n-k,n})^{1/k}, \quad 1 \le i \le k < n,$$

the logarithm of the *geometric mean* of the statistics U_{ik}, given in (29.6). More generally, as done in [9], we can consider as basic statistics for the EVI-estimation, the MOP of U_{ik}, i.e. the class of statistics

$$A_p(k) = \begin{cases} \left(\frac{1}{k}\sum_{i=1}^{k} U_{ik}^p\right)^{1/p}, & \text{if } p > 0 \\ \\ \left(\prod_{i=1}^{k} U_{ik}\right)^{1/k}, & \text{if } p = 0. \end{cases} \tag{29.7}$$

From (29.6), we can write $U_{ik}^p = Y_{k-i+1,k}^{\gamma p}(1 + o_p(1))$. Since $E(Y^a) = 1/(1-a)$ if $a < 1$, the law of large numbers enables us to say that if $p < 1/\gamma$, $A_p(k) \overset{p}{\underset{n\to\infty}{\longrightarrow}} (1/(1-\gamma p))^{1/p}$. Hence the reason for the class of MOP EVI-estimators,

$$
\mathrm{H}_{k,n}^{(p)} := \begin{cases} \left(1 - A_p^{-p}(k)\right)/p, \text{ if } p > 0 \\[2mm] \ln A_0(k) = H(k), \text{ if } p = 0, \end{cases} \tag{29.8}
$$

with $A_p(k)$ given in (29.7), and with $\mathrm{H}_{k,n}^{(0)} \equiv \mathrm{H}_{k,n}$, given in (29.3). This class of MOP EVI-estimators depends on this tuning parameter $p \geq 0$, which makes it very flexible, and even able to overpass for finite samples one of the simplest and one of the most efficient EVI-estimators in the literature, the corrected-Hill (CH) EVI-estimator in [10].

The slope of a generalized quantile plot led to the GH EVI-estimator in [2], further studied in [3], with the functional form,

$$
\mathrm{GH}_{k,n} = \mathrm{H}_{k,n} + \frac{1}{k} \sum_{i=1}^{k} \{\ln \mathrm{H}_{i,n} - \ln \mathrm{H}_{k,n}\} .
$$

29.2.2 Asymptotic Properties of the Hill and Related EVI-Estimators

Weak consistency of any of the aforementioned EVI-estimators is achieved whenever $F \in \mathscr{D}_M^+$ and k is intermediate, i.e. (29.5) holds. Under the validity of an adequate second-order condition (see [27]), it is possible to guarantee their asymptotic normality. More precisely, denoting T any of the aforementioned EVI-estimators, and with $A(t)$ a function converging to zero as $t \to \infty$, measuring the rate of convergence of the sequence of maximum values to its non-degenerate limit and such that $|A| \in \mathrm{RV}_\rho$, $\rho \leq 0$ a shape second-order parameter, it is possible to guarantee the existence of $(b_T, \sigma_T) \in \mathfrak{R} \times \mathfrak{R}^+$, such that

$$
T_{k,n} \overset{d}{=} \gamma + \sigma_T P_k^T / \sqrt{k} + b_T A(n/k) + o_p(A(n/k)), \tag{29.9}
$$

with P_k^T an asymptotically standard normal r.v. Details on the values of (b_T, σ_T) in (29.9) are given in the aforementioned papers associated with the T-estimators.

29.2.3 Asymptotic Comparison at Optimal Levels of the H, MOP and GH EVI-Estimators

We shall next proceed to the comparison of the aforementioned EVI-estimators under study at their optimal levels, in a way similar to the one used in several articles on the topic. See [25] and the references therein. Let us assume that $T_{k,n}$ denote any arbitrary semi-parametric EVI-estimator for which (29.9) holds for any intermediate sequence of integers $k = k_n$, and where P_k^T is an asymptotically standard normal r.v. Then, we have $\sqrt{k}\left(T_{k,n} - \gamma\right) \xrightarrow{d} \text{Normal}(\lambda b_T, \sigma_T^2)$ as $n \to \infty$, provided that k is such that $\sqrt{k}\, A(n/k) \to \lambda$, finite, as $n \to \infty$. We then write $\text{Bias}_\infty(T_{k,n}) := b_T\, A(n/k)$ and $\text{Var}_\infty(T_{k,n}) := \sigma_T^2/k$. The so-called *asymptotic mean square error* (AMSE) is then given by $\text{AMSE}(T_{k,n}) := \sigma_T^2/k + b_T^2\, A^2(n/k)$. Regular variation theory [6] enables us to show that, whenever $b_T \neq 0$, there exists a function $\varphi(n) = \varphi(n, \gamma, \rho)$, such that $\lim_{n \to \infty} \varphi(n)\, \text{AMSE}(T_{0n}) = \left(\sigma_T^2\right)^{-\frac{2\rho}{1-2\rho}} \left(b_T^2\right)^{\frac{1}{1-2\rho}} =: \text{LMSE}(T_{0n})$, where $T_{0n} := T_{k_0^T(n),n}$ and $k_0^T(n) := \arg\min_k \text{AMSE}(T_{k,n})$. Moreover, if we more restrictively assume that $A(t) = \gamma \beta t^\rho$, $\rho < 0$,

$$k_0^T(n) := \arg\min_k \text{MSE}\,(T_{k,n}) = \left(\sigma_T^2\, n^{-2\rho} / \left(b_T^2 \gamma^2 \beta^2(-2\rho)\right)\right)^{1/(1-2\rho)} (1 + o(1)).$$

It is then usual to consider the following definition, where we introduce an AREFF-indicator conceived so that the highest the AREFF is, the better is the first estimator.

Definition 1. Given two biased estimators $T_k^{(1)}$ and $T_k^{(2)}$, for which a distributional representation of the type of the one in (29.9) holds, with constants (σ_1, b_1) and (σ_2, b_2), $b_1, b_2 \neq 0$, respectively, both computed at their optimal levels, the *Asymptotic Root Efficiency* (AREFF) of $T_{0n}^{(1)}$ relatively to $T_{0n}^{(2)}$ is

$$\text{AREFF}_{1|2} \equiv \text{AREFF}_{T_{0n}^{(1)}|T_{0n}^{(2)}} := \sqrt{\frac{\text{LMSE}(T_{0n}^{(2)})}{\text{LMSE}(T_{0n}^{(1)})}} = \left(\left(\frac{\sigma_2}{\sigma_1}\right)^{-2\rho} \left|\frac{b_2}{b_1}\right|\right)^{\frac{1}{1-2\rho}}.$$

As detected in [9], at optimal levels, the MOP EVI-estimator, with p optimally chosen, i.e. $p := \arg\inf_{p<1/(2\gamma)} \text{AMSE}(H_{0n}^{(p)})$, denoted merely MOP in Fig. 29.1, can beat the H-estimator in the whole (γ, ρ)-plane. But it can be beaten by the GH-estimator, unless γ is small. In Fig. 29.1 we exhibit the comparative behaviour of the aforementioned EVI-estimators.

The MOP EVI-estimators have thus a nice performance in a region of the (γ, ρ)-plane quite usual in practice. But if we are in a region of (γ, ρ) around the line providing $b_{\text{GH}} = 0$, the GH outperforms the MOP at optimal levels, unless $\rho = 0$.

29.2.4 Location-Invariant H-Estimators and Corrected Hill EVI-Estimators

As mentioned by several authors, the inadequate use of the H-estimator for shifted data can lead to drastic systematic errors. Such a non-invariance for shifts of the H-estimator led the authors in [17] to the study of the location invariant Hill-type estimator

ρ \ γ	0.00	0.10	0.20	0.30	0.40	0.50	0.60	0.70	0.80	0.90	1.00	1.10	1.20	1.30	1.40	1.50	1.60	1.70	1.80	1.90	2.00
0.00	H	MOP	MOP	MOP	MOP	MOP	MOP	MOP	MOP	MOP	MOP	MOP	MOP	MOP	MOP	MOP	MOP	MOP	MOP	MOP	MOP
-0.10	MOP	GH	GH	GH	GH	GH	GH	GH	GH	GH	GH	GH	GH	GH	GH	GH	GH	GH	GH	GH	GH
-0.20	MOP	MOP	GH	GH	GH	GH	GH	GH	GH	GH	GH	GH	GH	GH	GH	GH	GH	GH	GH	GH	GH
-0.30	MOP	MOP	MOP	GH	GH	GH	GH	GH	GH	GH	GH	GH	GH	GH	GH	GH	GH	GH	GH	GH	GH
-0.40	MOP	MOP	MOP	GH	GH	GH	GH	GH	GH	GH	GH	GH	GH	GH	GH	GH	GH	GH	GH	GH	GH
-0.50	MOP	MOP	MOP	GH	GH	GH	GH	GH	GH	GH	GH	GH	GH	GH	GH	GH	GH	GH	GH	GH	GH
-0.60	MOP	MOP	MOP	MOP	GH	GH	GH	GH	GH	GH	GH	GH	GH	GH	GH	GH	GH	GH	GH	GH	GH
-0.70	MOP	MOP	MOP	MOP	GH	GH	GH	GH	GH	MOP	MOP	MOP	MOP	MOP	GH	GH	GH	GH	GH	GH	GH
-0.80	MOP	MOP	MOP	MOP	GH	GH	GH	GH	GH	MOP	MOP	MOP	MOP	MOP	MOP	MOP	MOP	GH	GH	GH	GH
-0.90	MOP	MOP	MOP	MOP	MOP	GH	GH	GH	MOP	MOP	MOP	MOP	MOP	MOP	MOP	MOP	MOP	MOP	GH	GH	GH
-1.00	MOP	MOP	MOP	MOP	MOP	GH	GH	GH	MOP	MOP	MOP	MOP	MOP	MOP	MOP	MOP	MOP	MOP	MOP	MOP	GH
-1.10	MOP	MOP	MOP	MOP	MOP	GH	GH	GH	MOP	MOP	MOP	MOP	MOP	MOP	MOP	MOP	MOP	MOP	MOP	MOP	MOP
-1.20	MOP	MOP	MOP	MOP	MOP	GH	GH	GH	MOP	MOP	MOP	MOP	MOP	MOP	MOP	MOP	MOP	MOP	MOP	MOP	MOP
-1.30	MOP	MOP	MOP	MOP	MOP	MOP	GH	GH	MOP	MOP	MOP	MOP	MOP	MOP	MOP	MOP	MOP	MOP	MOP	MOP	MOP
-1.40	MOP	MOP	MOP	MOP	MOP	MOP	GH	GH	MOP	MOP	MOP	MOP	MOP	MOP	MOP	MOP	MOP	MOP	MOP	MOP	MOP
-1.50	MOP	MOP	MOP	MOP	MOP	MOP	GH	GH	MOP	MOP	MOP	MOP	MOP	MOP	MOP	MOP	MOP	MOP	MOP	MOP	MOP
-1.60	MOP	MOP	MOP	MOP	MOP	MOP	GH	GH	MOP	MOP	MOP	MOP	MOP	MOP	MOP	MOP	MOP	MOP	MOP	MOP	MOP
-1.70	MOP	MOP	MOP	MOP	MOP	MOP	GH	GH	MOP	MOP	MOP	MOP	MOP	MOP	MOP	MOP	MOP	MOP	MOP	MOP	MOP
-1.80	MOP	MOP	MOP	MOP	MOP	MOP	GH	GH	MOP	MOP	MOP	MOP	MOP	MOP	MOP	MOP	MOP	MOP	MOP	MOP	MOP
-1.90	MOP	MOP	MOP	MOP	MOP	MOP	MOP	GH	MOP	MOP	MOP	MOP	MOP	MOP	MOP	MOP	MOP	MOP	MOP	MOP	MOP
-2.00	MOP	MOP	MOP	MOP	MOP	MOP	MOP	GH		MOP	MOP	MOP	MOP	MOP	MOP	MOP	MOP	MOP	MOP	MOP	MOP

Fig. 29.1 Comparative overall behaviour of the classical EVI-estimators under consideration

$$H_{k,n}(k_0) := \frac{1}{k_0} \sum_{i=1}^{k} \ln \left(\left(X_{n-i+1,n} - X_{n-k,n} \right) / \left(X_{n-k_0+1,n} - X_{n-k,n} \right) \right),$$

with $k_0 < k$ adequately chosen (the rate of convergence is of the order of $\sqrt{k_0} < \sqrt{k}$, but the asymptotic bias can be drastically reduced). An algorithm for the choice of k_0 and k is provided in [17], as well as full information on the asymptotic bias and variance of these EVI-estimators. For a discussion on location-invariant H-estimators, we recommend the reading of [1], where the H-estimators are transformed into scale/location invariant estimators through the use of the transformed sample $\underline{X}_n^{(q)} := \left(X_{n:n} - X_{n_q:n}, X_{n-1:n} - X_{n_q:n}, \ldots, X_{n_q+1:n} - X_{n_q:n} \right)$, where $n_q = \lfloor nq \rfloor + 1$, with $\lfloor x \rfloor$ denoting the integer part of x. We can have $0 < q < 1$, for any $F \in D_{\mathscr{M}}(G_{\gamma>0})$ (the random threshold, $X_{n_q:n}$, is an empirical quantile), and $q = 0$, for d.f.'s with a finite left endpoint x_F (the random threshold is the minimum, $X_{1:n}$). Any statistical inference methodology based on the sample of excesses $\underline{X}_n^{(q)}$ is called a PORT-methodology, with PORT standing for peaks over random thresholds, a term coined by the authors in [1]. This methodology enabled the introduction and study of classical location/scale invariant EVI-estimators, like the PORT-Hill estimators, among others, studied for finite-samples in [22]. See also [18, 24] and [26].

Let us next consider any "classical" semi-parametric EVI-estimator, $T_{k,n}$, valid for $\gamma > 0$. Let us also assume that a distributional representation similar to the one in (29.9) holds for $T_{k,n}$, with (b_T, σ_T) replaced by (b, σ) for the sake of simplicity. The pattern of these estimators exhibits thus the same type of peculiarities:

- high variance for high thresholds $X_{n-k,n}$, i.e., for small values of k;
- high bias for low thresholds, i.e., for large values of k;
- a small region of stability of the sample path (plot of the estimates versus k), as a function of k, making problematic the adaptive choice of the threshold, on the basis of any sample paths' stability criterion;
- a "very peaked" MSE, making difficult the choice of $k_0 := \arg \inf_k \text{MSE}(T_{k,n})$.

The preceding peculiarities have led researchers to consider the possibility of reducing the bias term, building new estimators $T_{k,n}^R$, the so-called second-order reduced-bias (SORB) EVI-estimators. Particularly, for heavy tails, i.e., $\gamma > 0$, the reduction of bias is a very important problem involved in the estimation of γ or the Pareto index, $\alpha = 1/\gamma$, in case the slowly varying part of the Pareto type model disappears at a very slow rate. Under an adequate second-order condition and for k intermediate, there exist $\sigma_R > 0$ and an asymptotically standard normal r.v. P_k^T, such that for a large class of models in $\mathscr{D}_M^+$, and with $A(.)$ the function in (29.9), $T_{k,n}^R \overset{d}{=} \gamma + \sigma_R P_k^T / \sqrt{k} + o_p(A(n/k))$. Notice that for such reduced-bias estimators we no longer have a dominant component of bias of the order of $A(n/k)$, as in (29.9). Therefore, $\sqrt{k}(T_{k,n}^R - \gamma) \overset{d}{\underset{n \to \infty}{\longrightarrow}} \text{Normal}(0, \sigma_R^2)$ not only when $\sqrt{k}A(n/k) \to 0$, as for the classical estimators, but also when $\sqrt{k}A(n/k) \to \lambda$, finite and non-null. Such a bias reduction provides usually a stable sample path for a wider region of k-values and a reduction of the MSE at the optimal level, in the sense of minimum MSE. These optimal levels should be such that $\sqrt{k}A(n/k) \to \infty$, as $n \to \infty$.

Such an approach has been carried out for heavy tails in the most diverse manners. The key ideas are either to find ways of getting rid of the dominant component $bA(n/k)$ of bias in (29.9), or to go further into the second-order behaviour of the basic statistics used for the EVI-estimation, like the log-excesses or the scaled log-spacings in (29.2). The accommodation of bias in the log-excesses led the authors in [23] to investigate a class of weighted combinations of the log-excesses denoted *weighted Hill* (WH) estimators, more robust than the H-estimators, and given by

$$T_{k,n,\hat{\beta},\hat{\rho}}^{WH} := \frac{1}{k} \sum_{i=1}^{k} p_{ik}(\hat{\beta}, \hat{\rho}) V_{ik}, \quad p_{ik}(\hat{\beta}, \hat{\rho}) := e^{\hat{\beta} (n/k)^{\hat{\rho}}((i/k)^{-\hat{\rho}} - 1)/(\hat{\rho} \ln(i/k))},$$

where $(\hat{\beta}, \hat{\rho})$ are suitable consistent estimators of the second-order parameters (β, ρ), in $A(t) = \gamma \beta t^\rho$. It is a class of minimum-variance reduced-bias (MVRB) estimators, in the sense that these estimators can have an asymptotic variance equal to the one of the H-estimators but an asymptotic bias of smaller order, and therefore seems to open interesting new perspectives in the field. Related work appears in [10] and [20]. The authors in [20] suggest the class of EVI-estimators,

$\overline{M}_{k,n,\hat{\beta},\hat{\rho}} := H_{k,n} - \hat{\beta} (n/k)^{\hat{\rho}} \hat{C}_1, \quad \hat{C}_1 = \frac{1}{k} \sum_{i=1}^{k} (i/k)^{-\hat{\rho}} W_i$, with $W_i, 1 \le i \le k$, given in (29.2). With the same objectives, but with a slightly simpler analytic expression, we also refer the estimator $\overline{H}_{k,n,\hat{\beta},\hat{\rho}} := H_{k,n}(1 - \hat{\beta} (n/k)^{\hat{\rho}} / (1 - \hat{\rho}))$, studied in [10]. Notice that the dominant component of the bias of the H-estimator is estimated in two different ways for $\overline{M}_{k,n,\hat{\beta},\hat{\rho}}$ and $\overline{H}_{k,n,\hat{\beta},\hat{\rho}}$, and directly removed from the H-estimator. The key of success of these MVRB EVI-estimators lies in the estimation of β and ρ at a level k_1, such that $k = o(k_1)$, with k the number of top o.s. used for the EVI-estimation. The level k_1 needs to be such that $(\hat{\beta}, \hat{\rho})$ is consistent for the estimation of (β, ρ) and $\hat{\rho} - \rho = o_p(1/\ln n)$. For more details on the choice of k_1, see [19, 23], and more recently [11].

29.3 Score Functions for the Hill Estimator

In several applications we need some specific properties of estimators, e.g. robustness (see [31] for normality testing against Pareto tails).

A standard score function is typically considered to be a base for statistical inference. Sometimes it is overlooked that pioneering works on score functions, e.g. [13,16,30] considered as a base the normal distribution, which is defined on the whole line. In such a setup, if a parameter is going to be inferred, we have typically a translation group closely related to its score function. This standard score is the gradient wrt θ of the logarithm of the likelihood function, i.e.

$$S(\theta, X) = \frac{\partial}{\partial \theta} \log L(\theta; X).$$

In this paper we consider only a semi-parametric setup. For a nonparametric analogy see [12] where scores $S_n = \frac{B^2}{A} \frac{\partial}{\partial x_n} \log f(x_n|x_{n-1}) + \frac{x_n}{A}$ are defined for a conditionally exponential family in the linear model $X_n = A S_n + B \eta_n$, where A, B are known constants, η_n is Gaussian noise, $(X_n, S_n), n > 1$ is a two-component Markov process, (X_n) is an observable process and (S_n) is an unobservable useful process.

In our setup, let $\mathscr{X}$ be the support of the distribution F with density f, continuously differentiable according to $x \in \mathscr{X}$ and let $\eta : \mathscr{X} \to \mathfrak{R}$ be given by [29] $\eta(x) = x$, if $\mathscr{X} = \mathfrak{R}, \eta(x) = \log(x - a)$, if $\mathscr{X} = (a, \infty)$ and $\eta(x) = \log \frac{x}{1-x}$, if $\mathscr{X} = (0, 1)$. Then the *transformation-based score* or shortly the *t-score* [14] is defined by

$$T(x) = -\frac{1}{f(x)} \frac{d}{dx} \left(\frac{1}{\eta'(x)} f(x) \right),$$

which expresses a relative change of a "basic component of the density", the density divided by the Jacobian of mapping η.

It is clear that for the Normal distribution, which is an archetypical distribution, we have $\eta(x) = x$, $S(x, \theta) = \frac{d}{d\theta} \log f(x, \theta)$ and $\hat{\theta} = $ MLE, with MLE standing for *maximum likelihood estimator*, is the solution of $\sum_{i=1}^{n} S(X_i, \hat{\theta}) = 0$.

However, for the Pareto distribution we can consider at least two recently implemented approaches, MLE, which is related to the "standard score" estimation with $\eta = id$ and $S_F(X, \alpha) = \frac{1}{\alpha} - \log x$ and t-score estimation with $\eta(x) = \log(x - 1)$ (see [31] and [32]). Notice that the MLE is not robust wrt right outliers, i.e. if $X_i \to \inf$, then $\hat{\alpha} \downarrow 0$. For a t-estimation we have

$$S_F(x) = \alpha\left(1 - \frac{\alpha + 1}{\alpha x}\right).$$

Thus standard estimation $\sum S_F(g)(X_i) = 0$ gives us $\hat{\alpha} = \frac{1}{\overline{x}-1}$ (where $\overline{x} = \frac{n}{\sum \frac{1}{x_i}}$ is a harmonic mean) which is an estimator apparently robust against right-outliers. This allows us to define the t-Hill estimator by (see [31])

$$H_{k,n}^* = \frac{1}{\hat{\alpha}_{k,n}} = \left(\frac{1}{k} \sum_{i=1}^{k} \frac{X_{n-k,n}}{X_{n-j+1,n}}\right)^{-1} - 1.$$

Note that if we consider a generalization to $p < 0$ of the MOP functionals in (29.8), we get $H_{k,n}^* = H_{k,n}^{(-1)}$. To justify such a construction we remark that formal heavy-tailed propositions can only be satisfactorily involved for empirical constructs if sample data can be taken as a reasonable representation of the underlying distribution. In practice, distribution data may be contaminated by errors. The point of departure is the recent research which has shown that the Hill estimator is non-robust. This means that small amounts of data contamination in the wrong place can reverse unambiguous conclusions. The "wrong place" usually means in the upper tail of the distribution. As shown in [7], small errors in the estimation of the tail index can bring large errors in the estimation of quantiles. Robust methods for extreme values have been recently addressed in the literature (e.g. [8] considered robust estimation in the strict Pareto model, [33] proposed robust tail index estimation procedure for the semi-parametric setting of Pareto-type distributions). As discussed in the paper [32], t-estimation is at least a competitive estimation technique in the presence of heavy tails. In [15] we have shown that t-estimation is clearly better when contamination is present. The weak consistency of the t-Hill estimator has been proven under standard regularity conditions in [31]. Asymptotics and robustness of the t-Hill estimator is studied in [5].

Acknowledgements M. Ivette Gomes has been partially supported by National Funds through FCT—Fundação para a Ciência e a Tecnologia, projects PEst-OE/MAT/UI0006/2011 (CEAUL) and EXTREMA (PTDC/MAT/101736/2008).

References

1. Araújo Santos, P., Fraga Alves, M.I., Gomes, M.I.: Peaks over random threshold methodology for tail index and quantile estimation. Revstat **4**, 227–247 (2006)
2. Beirlant, J., Vynckier, P., Teugels, J.: Excess functions and estimation of the extreme-value index. Bernoulli **2**, 293–318 (1996)
3. Beirlant, J., Dierckx, G., Guillou, A.: Estimation of the extreme-value index and generalized quantile plots. Bernoulli **11**(6), 949–970 (2005)
4. Beirlant, J., Caeiro, C., Gomes, M.I.: An overview and open research topics in the field of statistics of univariate extremes. Revstat **10**(1), 1–31 (2012)
5. Beran, J., Schell, D., Stehlík, M.: The harmonic moment tail index estimator: asymptotic distribution and robustness. Ann. Inst. Stat. Math. **66**, 193–220 (2014)
6. Bingham, N., Goldie, C.M., Teugels, J.L.: Regular Variation. Cambridge Univ. Press, Cambridge (1987)
7. Brazauskas, V., Serfling, R.: Robust and efficient estimation of the tail index of a single-parameter Pareto distribution. North Am. Actuar. J. **4**, 12–27 (2000)
8. Brazauskas, V., Serfling, R.: Robust estimation of tail parameters for two-parameter Pareto and exponential models via generalized quantile statistics. Extremes **3**(3), 231–249 (2000).
9. Brilhante, M.F., Gomes, M.I., Pestana, D.: A simple generalization of the Hill estimator. Comput. Stat. Data Anal. **57**, 518–535 (2013)
10. Caeiro, F., Gomes, M.I., Pestana, D.: Direct reduction of bias of the classical Hill estimator. Revstat **3**, 113–136 (2005)
11. Caeiro, F., Gomes, M.I., Henriques-Rodrigues, L.: Reduced-bias tail index estimators under a third order framework. Commun. Stat. Theory Methods **38**, 1019–1040 (2009)
12. Dobrovidov, A.V., Koshkin, G.M. and Vasiliev, V.A.: Non-Parametric State Space Models. Kendrick Press, Heber City (2012)
13. Edgeworth, F.Y.: On the probable errors of frequency-constants (contd.). J.R. Stat. Soc. **71**, 499–512 (1908)
14. Fabián Z.: Induced cores and their use in robust parametric estimation. Commun. Stat. Theory Methods **30**, 537–556 (2001)
15. Fabián, Z., Stehlík, M.: A note on favorable estimation when data is contaminated. Commun. Dependability Qual. Manag. **114**, 36–43 (2008)
16. Fisher, R.A.: Theory of statistical estimation. Procs. Camb. Philos. Soc. **22**(700725), doi: 10.1017/S0305004100009580 (1925)
17. Fraga Alves, M.I.: A location invariant Hill-type estimator. Extremes **4**(3), 199–217 (2001)
18. Fraga Alves, M.I., Gomes, M.I., de Haan, L. Neves, C.: Mixed moment estimators and location invariant alternatives. Extremes **12**, 149–185 (2009)
19. Gomes, M.I., Pestana, D.: A sturdy reduced bias extreme quantile (VaR) estimator. J. Am. Stat. Assoc. **102**(477), 280–292 (2007)
20. Gomes, M.I., Martins, M.J., Neves, M.M.: Improving second order reduced bias extreme value index estimation. Revstat **5**(2), 177–207 (2007)
21. Gomes, M.I., Canto e Castro, L., Fraga Alves, M.I., Pestana, D.: Statistics of extremes for iid data and breakthroughs in the estimation of the extreme value index: Laurens de Haan leading contributions. Extremes **11**(1), 3–34 (2008)
22. Gomes, M.I., Fraga Alves, M.I. Araújo Santos, P.: PORT Hill and moment estimators for heavy-tailed models. Commun. Stat. Simul. Comput. **37**, 128–1306 (2008)
23. Gomes, M.I., de Haan, L., Henriques-Rodrigues, L.: Tail Index estimation for heavy-tailed models: accommodation of bias in weighted log-excesses. J. R. Stat. Soc. B **70**(1), 31–52 (2008)
24. Gomes, M.I., Henriques-Rodrigues, L. and Miranda, C.: Reduced-bias location-invariant extreme value index estimation: a simulation study. Commun. Stat. Simul. Comput. **40**(3), 424–447 (2011)

25. Gomes, M.I., Martins, M.J., Neves, M.M.: Generalised jackknife-based estimators for univariate extreme-value modelling. Commun. Stat. Theory Methods **42**(7), 1227–1245 (2013)
26. Gomes, M.I., Henriques-Rodrigues, L., Fraga Alves, M.I. and Manjunath, B.G.: Adaptive PORT-MVRB estimation: an empirical comparison of two heuristic algorithms. J. Stat. Comput. Simul. (2013). doi:10.1080/00949655.2011.652113
27. de Haan, L., Ferreira, A.: Extreme Value Theory: An Introduction. Springer, New York (2006)
28. Hill, B.: A simple general approach to inference about the tail of a distribution. Ann. Stat. **3**(5), 1163–1174 (1975)
29. Johnson, N.L.: Systems of frequency curves generated by methods of translations, Biometrika **36**, 149–176 (1949)
30. Pearson, K., Filon, L.N.G.: Mathematical contributions to the theory of evolution. IV. On the probable errors of frequency constants and on the influence of random selection on variation and correlation. Philos. Trans. R. Soc. Lond. A **191**, 229–311 (1898). Reprinted in Karl Pearsons Early Statistical Papers (1956) 179–261; Cambridge Univ. Press. Abstr. Proc. R. Soc. Lond. **62**, 173–176 (1897)
31. Stehlík, M., Fabián, Z., Střelec, L.: Small sample robust testing for Normality against Pareto tails. Commun. Stat. Simul. Comput. **41**, 1167–1194 (2012)
32. Stehlík, M., Potocký, R., Waldl, H., Fabián, Z.: On the favourable estimation of fitting heavy tailed data. Comput. Stat. **25**, 485–503 (2010)
33. Vandewalle, B., Beirlant, J., Christmann, A., Hubert, M.: A robust estimator for the tail index of Pareto-type distributions, Comput. Stat. Data Anal. **51**, 6252–6268 (2007)

Chapter 30
Multivariate Stochastic Volatility Estimation Using Particle Filters

K. Triantafyllopoulos

Abstract A particle filter algorithm is proposed for sequential estimation of volatility and cross-correlation of multivariate financial time series. The returns of prices of assets, such as shares traded in the stock market, are modelled with a skew-t distribution, which is able to capture the heavy tails and asymmetry of financial returns, and the inverse of the volatility covariance matrix is modelled via a Wishart autoregressive process. Motivated from the conjugacy between the Gaussian and the Wishart distributions, we describe a new choice for the importance density and we modify an existing approach of dealing with the model hyperparameters, using Gaussian mixtures. The proposed methodology is illustrated with data consisting of three constituents of the FTSE-100 stock index.

Keywords Multivariate volatility • Particle filters • Skew-t distribution • Wishart process

30.1 Introduction

Over the last two decades, multivariate volatility models and related computational algorithms have been established to quantify forecast uncertainty and to enable risk management [4]. There are two main categories of models (a) multivariate generalised autoregressive conditional heteroscedastic models (MGARCH) and (b) multivariate stochastic volatility models (MSV). MGARCH, reviewed in [3], model the volatility covariance matrix as a function of past returns and adopt likelihood-based estimation methods for inference and forecasting. MSV, reviewed in [1,11] and in [20], treat the volatility as a stochastic process and adopt simulation-based estimation methods, usually Markov chain Monte Carlo (MCMC). The advantages of Bayesian inference via MCMC, widely reported by many authors [5], offer increased flexibility and adaptability to real data. However, MCMC is aimed at off-line application, because each time out-of-sample estimation or

K. Triantafyllopoulos (✉)
School of Mathematics and Statistics, University of Sheffield,
Hicks Building, Sheffield S3 7RH, UK
e-mail: kostas@sheffield.ac.uk

© Springer Science+Business Media New York 2014   335
M.G. Akritas et al. (eds.), *Topics in Nonparametric Statistics*, Springer Proceedings
in Mathematics & Statistics 74, DOI 10.1007/978-1-4939-0569-0_30

forecasting is required, the Markov chain has to be re-initialised. Together with the fact that MCMC is computationally demanding, off-line application limits its use, in particular with the view of portfolio selection [4]. The nature of sequential application required in industry (e.g. for daily or tick data) call for sequential simulation-based methods. Such methods, including particle filters or sequential Monte Carlo, have been discussed previously [10,12], but the literature is dominated by off-line estimation methods such as MCMC.

In this paper we develop a particle filter procedure, aimed at sequential volatility estimation. Following recent advances in covariance matrix processes [8, 9, 14, 15, 18, 19], we adopt a Wishart autoregressive process for the evolution of the precision covariance matrix (inverse of the volatility covariance matrix). We consider a skew-t distribution in the returns model, which is able to capture heavy tails and asymmetry of the returns. We propose a novel approach for the importance density, based on the conjugacy between the Gaussian and the Wishart distributions. In order to estimate the hyperparameters of the model, such as the skewness parameters, we adopt a modified version of [12], according to which the distribution of the hyperparameters is modelled via a Gaussian mixture. Unlike the previous reference, we do not adopt auxiliary particle filtering; instead, we develop a slightly different simulation approach. We illustrate the methodology by considering data, consisting of returns of three constituents of the FTSE-100 stock index.

The remainder of the paper is organised as follows. Section 30.2 describes the model and the following section discusses inference using the particle filter. Section 30.4 analyses the three-variate financial data mentioned above and finally conclusions are given in Sect. 30.5.

30.2 Model Set-up

Suppose that $y_t = [y_{1t}, \ldots, y_{pt}]^T$ denotes the log-return column vector at time t, i.e. $y_{it} = \log(p_{it}/p_{i,t-1})$, where p_{it} is the price of asset i at time t and T denotes transposition ($i = 1, \ldots, p$). The conditional covariance matrix of y_t is known as volatility matrix and its estimation is the main purpose of this paper. A first model postulates that y_t follows a Gaussian distribution, described by

$$y_t = \mu + \epsilon_t, \quad \epsilon_t \sim N_p(0, \Sigma_t), \tag{30.1}$$

where μ is the mean vector of y_t, ϵ_t is a random vector following the p-dimensional Gaussian distribution with zero mean vector and volatility covariance matrix Σ_t, which is assumed to be symmetric and positive definite. The innovation vector ϵ_t is assumed to be independent of ϵ_s, for $t \neq s$. The mean vector μ may be equal to the zero vector, but usually it is fluctuating around zero, allowing positive or negative returns estimation necessary for portfolio optimisation. We consider that the volatility matrix Σ_t is stochastic and is generated by an inverse Wishart process, described by

$$\Sigma_t^{-1} \mid \Sigma_{t-1} \sim W(k, k^{-1} A \Sigma_{t-1}^{-1} A^T),\tag{30.2}$$

where A is a $p \times p$ autoregressive parameter matrix of full rank and $W(k, S)$ denotes a Wishart distribution with $k > p - 1$ degrees of freedom and scale matrix S. From properties of the Wishart distribution it follows that $E(\Sigma_t^{-1} \mid \Sigma_{t-1}) = kS = A \Sigma_{t-1}^{-1} A^T$, which defines an autoregressive process for the precisions. It is further assumed that Σ_t is independent of ϵ_t and that initially $\Sigma_0^{-1} \sim W(v, V)$, for some degrees of freedom $v > p - 1$ and some scale matrix V. The model consists of (30.1) and (30.2), together with the initial distribution of Σ_0^{-1}; in this model, the hyperparameters are μ, k, A, v, V. From (30.2) it follows that given Σ_{t-1}, Σ_t follows an inverse Wishart distribution with k degrees of freedom and scale matrix S^{-1}, i.e. $\Sigma_t \mid \Sigma_{t-1} \sim IW(k, k(A^T)^{-1} \Sigma_{t-1} A^{-1})$. From this we can write that $E(\Sigma_t \mid \Sigma_{t-1}) = C \Sigma_{t-1} C^T$ and hence Σ_t follows an autoregressive process, where $C = k^{1/2}(k - p - 1)^{-1/2}(A^T)^{-1}$. Similar processes for the volatility are discussed in [19].

The above model relies on the Gaussian distribution assumption of the log-returns y_t. However, in financial time series it is well known that such an assumption is usually wrong or not empirically motivated. This is because financial returns typically exhibit the following characteristics (sometimes referred to as stylised facts):

1. Returns have heavy tails, certainly heavier than the Gaussian distribution. In practical terms this means that tail or extreme events (such as positive high and negative low returns) have more probability to occur than in the Gaussian distribution.
2. Returns are asymmetric, that is a low negative return (linked to loss) does not have the same probability as a high positive return (linked to profit). Typically, there are more frequent positive returns, but their magnitude is lower than that of negative returns (an example of this is shown in Sect. 30.4). Again the Gaussian distribution is inappropriate as it is symmetric and it places same probability to positive and negative returns.

In this paper we propose replacing the Gaussian distribution of the returns with the multivariate skew-t distribution of Azzalini and Capitanio [2]. Thus the returns model is

$$y_t \mid \Sigma_t \sim St_p(0, \Sigma_t, \alpha, v),\tag{30.3}$$

where $\alpha = [\alpha_1, \ldots, \alpha_p]^T$ is the skewness parameter vector and v the degrees of freedom. Some comments are in order. First, note that $\alpha = 0$ reduces the distribution to a multivariate Student t distribution; $\alpha_i > 0$ introduces positive skewness to the distribution of the return y_{it} and $\alpha_i < 0$ negative skewness to y_{it}. We remak that the distribution allows both $\alpha_i \geq 0$ and $\alpha_j \leq 0$, for $i \neq j$ (both positively and negatively skew returns can be incorporated). If v is large, the above distribution is approximated by a skew-Gaussian distribution. In our study we expect to have

$\alpha_i < 0$ (as returns are negatively asymmetric) and v to be small (as we expect the distribution of the returns to exhibit heavy tails).

Secondly, we note that from [2], the mean vector and the covariance matrix of y_t are given by

$$\mu = E(y_t \mid \Sigma_t) = c\text{diag}(\Sigma_t)^{1/2}\delta,$$

and

$$\text{Var}(y_t \mid \Sigma_t) = \frac{v}{v-2}\Sigma_t - c^2\text{diag}(\Sigma_t)^{1/2}\delta\delta^T\text{diag}(\Sigma_t)^{1/2}, \qquad (30.4)$$

where $\text{diag}(\Sigma_t)$ denotes the diagonal matrix with diagonal elements the diagonal elements of Σ_t, $c = v^{1/2}\pi^{-1/2}\Gamma[(v-1)/2]\Gamma(v/2)^{-1}$, $\delta = (1+a^T\Sigma_t^* a)^{-1/2}\Sigma_t^* a$, $\Sigma_t^* = \text{diag}(\Sigma_t)^{-1/2}\Sigma_t\text{diag}(\Sigma_t)^{-1/2}$, $\Gamma(x)$ denotes the gamma function with argument x, and $v > 2$. Thus, the mean return μ in this distribution depends on the skewness parameter vector α and the covariance matrix Σ_t. Interestingly, non-zero μ is implied by a non-zero α; large shifts in Σ_t (which imply large volatility), imply large absolute mean return. Now, Σ_t is not the volatility, as the conditional covariance matrix of y_t, given by (30.4) is not equal to Σ_t.

The working model adopts the skew-t distribution for the returns (30.3) together with the Wishart autoregressive evolution of Σ_t and the initial distribution $\Sigma_0^{-1} \sim W(v, V)$; the hyperparameters are α, v, k, A, v, V.

30.3 Inference Using Particle Filters

Suppose we wish to compute the integral

$$I = \int f(x)p(x)\,dx = E[f(x)],$$

where $p(\cdot)$ is a density function and the integration is performed over the domain of $p(\cdot)$. If we simulate N values $x^{(1)}, \ldots, x^{(N)}$ from $p(x)$, we can approximate I by $N^{-1}\sum_{i=1}^{N} x^{(i)}$.

Usually, it is difficult to sample from $p(\cdot)$. Importance sampling suggests to sample from another density function $g(\cdot)$ and to approximate I by

$$I = \int f(x)\frac{p(x)}{g(x)}g(x)\,dx = E[f(x)w(x)] \approx \frac{1}{N}\sum_{i=1}^{N} f(x^{(i)})w^{(i)},$$

where $x^{(1)}, \ldots, x^{(N)}$ is a sample from $g(\cdot)$ and $w = w(x) = p(x)/g(x)$ is called weight. For this approach to work, the main requirement is that $g(\cdot)$ should have the same domain as $p(\cdot)$ and obviously it should be a convenient density we can sample from.

In many situations, we only know $p(\cdot)$ up to a proportionality constant, e.g. when $p(\cdot)$ is a posterior distribution computed by an application of the Bayes' theorem as posterior $\propto$ likelihood $\times$ prior. It turns out that it suffices to know $p(\cdot)$ up to a proportionality constant, if we operate with the normalised weights defined by $\tilde{w} = w_i / \sum_{i=1}^{N} w_i$; for more details the reader is referred to [13].

This simple idea is applied sequentially over time. First we introduce some notation. For any time $t = 1, 2, \ldots$, let $y_{1:t} = \{y_1, \ldots, y_t\}$ be the collection of observed data up to and including time t; likewise, let $\Sigma_{0:t} = \{\Sigma_0, \Sigma_1, \ldots, \Sigma_t\}$ be the collection of covariance matrices up to and including time t. Let $g_t(\Sigma_{0:t} \mid y_{1:t})$ be an importance function, from which we can simulate from. Then we have

$$w_t = \frac{p(\Sigma_{0:t}|y_{1:t})}{g(\Sigma_{0:t}|y_{1:t})} \propto \frac{p(\Sigma_{0:t}, y_t|y_{1:t-1})}{g(\Sigma_{0:t}|y_{1:t})} \propto \frac{p(\Sigma_t, y_t|\Sigma_{0:t-1}, y_{1:t-1})}{g(\Sigma_t|\Sigma_{0:t-1}, y_{1:t})} \frac{p(\Sigma_{0:t-1}|y_{1:t-1})}{g(\Sigma_{0:t-1}|y_{1:t-1})}$$

$$\propto \frac{p(y_t|\Sigma_t)p(\Sigma_t|\Sigma_{t-1})}{g(\Sigma_t|\Sigma_{t-1}, y_{1:t})} w_{t-1}, \tag{30.5}$$

where it is assumed that the importance function can be decomposed as $g(\Sigma_{0:t} \mid y_{1:t}) = g(\Sigma_t|\Sigma_{0:t-1}, y_{1:t})g(\Sigma_{0:t-1}|y_{1:t-1})$. We note that this decomposition is true if $g(\Sigma_t \mid \Sigma_{t-1}, y_t)$ is equal to $p(\Sigma_t \mid \Sigma_{t-1}, y_t)$ (the density of Σ_t, given Σ_{t-1} and y_t); in general, the above rule enforces that $g(\cdot)$ behaves in a way similar to $p(\Sigma_t \mid \Sigma_{t-1}, y_t)$, which is the optimal importance function (see below).

The particle filter algorithm commences by generating $\Sigma_t^{(1)}, \ldots, \Sigma_t^{(N)}$ from $g(\Sigma_t|\Sigma_{t-1}^{(i)}, y_t)$, then evaluating the non-normalised weights

$$w_t^{(i)} = \frac{p(y_t|\Sigma_t^{(i)})p(\Sigma_t^{(i)}|\Sigma_{t-1}^{(i)})}{g(\Sigma_t^{(i)}|\Sigma_{t-1}^{(i)}, y_t)} \tilde{w}_{t-1}^{(i)}$$

and finally normalise the weights at time t, $\tilde{w}_t^{(i)} = N^{-1} \sum_{i=1}^{N} w_t^{(i)}$, and proceed to time $t + 1$. The posterior distribution of Σ_t at t is approximated as $p(\Sigma_t|y_{1:t}) \approx \sum_{i=1}^{N} \tilde{w}_t^{(i)} \delta(\Sigma_t - \Sigma_t^{(i)})$, where $\delta(\cdot)$ denotes the dirac point mass.

In the application of the above algorithm there are three important issues needed to be dealt with.

1. **Particle degeneration.** It has been observed that the particles degenerate, i.e. very few particles have significant weights, the rest are very close to zero. This results in poor and biased Monte Carlo estimates as very few particles are used. The solution is to resample the particles, if this happens. The problem is detected if the *effective sample size* $N_{eff} = (\sum(\tilde{w}_t^{(i)})^2)^{-1}$ is lower than $N/3$, in which case we resample the particles; there are several algorithms of resampling in this context, but here we use multinomial resampling, see, e.g., [6] and [13].
2. **Choice of the importance function.** Clearly, we should choose an importance function that we can sample from, but a poor choice would lead to poor

performance of the algorithm. An obvious choice is to sample from the prior $p(\Sigma_t \mid \Sigma_{t-1})$, but this is suboptimal, in a sense that we do not take into account the observation vector y_t. On the other extreme the *optimal importance function*, which minimises the variance of the weights, is $p(\Sigma_t \mid \Sigma_{t-1}, y_t)$, but it is usually hard or not possible to sample from.

We propose to sample from $p(\Sigma_t \mid \Sigma_{t-1}, y_t)$, assuming model (30.1) for y_t, and use this sampling distribution for the general model (30.3). In other words, we equate $g(\Sigma_t \mid \Sigma_{t-1}, y_t)$ to the optimal importance function $p(\Sigma_t \mid \Sigma_{t-1}, y_t)$ obtained in the case of model (30.1). From (30.1) and (30.2) the posterior distribution of Σ_t^{-1}, given Σ_{t-1} and y_t is

$$p(\Sigma_t^{-1} \mid \Sigma_{t-1}, y_t) \propto p(\Sigma_t^{-1} \mid \Sigma_{t-1}) p(y_t \mid \Sigma_t) \propto \det(\Sigma_t)^{(k+1-p-1)/2}$$

$$\times \exp\left\{-\frac{1}{2}\mathrm{tr}\left[(y_t-\mu)(y_t-\mu)^T + (A^T)^{-1}\Sigma_{t-1}A^{-1}\right]\Sigma_t^{-1}\right\},$$

where $\det(\cdot)$ denotes determinant and $\mathrm{tr}(\cdot)$ is the trace operator. Thus, $\Sigma_t^{-1} \mid \Sigma_{t-1}, y_t \sim W(k+1, S_t)$, where $S_t^{-1} = (y_t - \mu)(y_t - \mu)^T + (A^T)^{-1}\Sigma_{t-1}A^{-1}$. This implies that $\Sigma_t \mid \Sigma_{t-1}, y_t \sim IW(k+1, S_t^{-1})$.

3. **Hyperparameter estimation.** Model (30.2), (30.3) depends on some hyperparameters, which are subject to estimation. The successful application of the particle filter depends on the estimation of such hyperparameters; in the model of Sect. 30.2 these are $\mu, \alpha, \nu, k, A, v, V$.

A first estimation approach is to augment θ into the state Σ_t, i.e. define the state $x_t = (\Sigma_t, \theta)$ and provide the particle filter approximation to the posterior $p(x_t \mid y_{1:t})$. The problem with this approach is that since θ is time-invariant ($\theta_t = \theta_{t-1}$ for all t), this approach is equivalent to sampling θ from the prior distribution $p(\theta)$; for a detailed discussion see [7] and [13]. To overcome this problem, [7] suggest to build an artificial evolution for θ, i.e. $\theta_t = \theta_{t-1} + \zeta_t$, where ζ_t is some random vector with zero mean. This basically overcomes the problem that $\theta_t = \theta_{t-1}$ is sampled from $p(\theta)$, but it introduces the undesirable feature that now the hyperparameters are time-varying.

This problem has attracted significant interest, see, e.g., [12, 16] and [17]. In this paper we adopt a modified version of [12]. These authors suggest to approximate the posterior distribution of θ by a finite Gaussian mixture with weights derived from the particle filter of the states. Let $\bar{\theta}$ and V_θ be the mean vector and covariance matrix of θ. The joint posterior distribution of Σ_t and θ is approximated by

$$p(\Sigma_t, \theta \mid y_{1:t}) \approx \sum_{i=1}^{N} \tilde{w}_t^{(i)} N_p(\theta; m^{(i)}, h^2 V_\theta) \delta(\Sigma_t - \Sigma_t^{(i)}),$$

where $m^{(i)} = a\theta^{(i)} + (1 - a)\bar{\theta}$, for some a in $(0, 1)$ and $a^2 + h^2 = 1$. This setting is necessary to make sure that the mean vector and covariance matrix of θ implied by the Gaussian mixture match those of $p(\theta)$.

[12] use the above mixture in order to propose a target importance function to sample from. Here we propose an alternative approach. Assume that at time $t - 1$ we have obtained a sample of $\Sigma_{t-1}^{(i)}$ and $\theta^{(i)}$ together with the weights $\tilde{w}_{t-1}$. At time t we simulate N draws from the Gaussian mixture (using a classification variable as described in [12] or [13]), then we simulate from the importance density $g(\Sigma_t \mid \Sigma_{t-1}, y_t, \theta)$ (this is given in (2) above) and finally we calculate the weights $w_t^{(i)}$ according to (30.5), which is conditional on $\theta = \theta^{(i)}$, we standardise the weights and perform the resampling step, if needed. Thus, our approach differs from that of [12] as we avoid the use of auxiliary filtering to sample from the importance function.

Only briefly we outline that model choice and model comparison may be performed via an evaluation of the log-likelihood function. Given a data path $y_{1:n}$, by noting the prediction error decomposition, the log-likelihood of $(\Sigma_1, \ldots, \Sigma_n)$ is $\ell = \sum_{t=2}^n \log p(y_t \mid y_{1:t-1})p(y_1)$. Thus we can approximate ℓ by

$$\hat{\ell} = \sum_{t=1}^n \log \sum_{i=1}^N \tilde{w}_{t-1}^{(i)} p(y_t \mid \Sigma_t^{(i)}),$$

where $p(y_t \mid \Sigma_t^{(i)})$ is the density of a skew-t distribution. More details of likelihood-based model comparison for particle filters are given in [10].

30.4 Illustration

In this section we consider data consisting of share prices of three assets, Cairn Energy (CE), Anglo America (AA) and Associate British Foods (ABF) traded in the FTSE-100 stock index. The data collected on daily frequency, span a four-year period, 2 January 2006 to 24 December 2009. The prices are transformed to log-returns, which are depicted on the left panel of Fig. 30.1. The returns of AA and ABF are somewhat similar: there appears to be increased uncertainty around the end of 2008. Up to 2009 the returns of CE are quite similar to the returns of AA and ABF, but in the end of 2009 there is a very low negative return of CE. A simple descriptive statistical analysis reveals that the skewness coefficients of CE, AA and ABF are $-24.97, -0.15$ and -0.07, respectively. We note that they are all negative, indicating presence of skewness in the distribution of the returns, in particular for CE, which is clearly shown in Fig. 30.1. This and the presence of heavy tails (sample statistics are not given here) motivate the adoption of the skew-t distribution of Sect. 30.2.

We have applied model (30.2) and (30.3). The hyperparameters of this model are $\mu, v, V, \alpha, \nu, k$ and A. Out of those, we propose to use historical data (prior

Log–returns with estimated volatilities

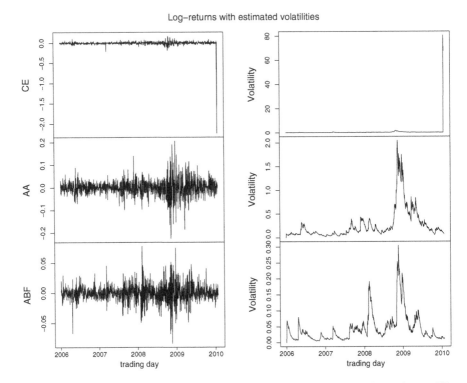

Fig. 30.1 Estimated volatility (*right panel*) against log-returns of the FTSE-100 constituents CE, AA and ABF

to 2 January 2006) to estimate μ, v, V and so reduce the hyperparameters subject to estimation to $\theta = (\alpha, v, k, A)$. μ is the mean of y_t in model (30.1), used for sampling from the importance function, and is estimated as the sample mean using historical data; here $\mu = [-0.0019, 0.0001, -0.00002]^T$. v and V are the degrees of freedom and scale matrix of the initial distribution $\Sigma_0^{-1} \sim W(v, V)$. Since $E(\Sigma_0^{-1}) = vV$, we can choose any arbitrary value $v > p - 1$ (for the distribution to be non-singular) and $V = v^{-1}S$, where S is the sample covariance matrix, calculated from historical data.

For θ, we have adopted the Gaussian mixture approach described in Sect. 30.3, after first transforming v (which must be > 2) to $\log(v - 2)$ and k (which is greater than $p - 1 = 2$) to $\log(k - p + 1)$, so as the Gaussian mixture be valid (the domain of each scalar component of θ to be the real line). Furthermore, we define $\alpha = [\alpha_1, \alpha_2, \alpha_3]^T$ the vector containing the skewness parameters of the return of each asset and allow each α_i to take values in the real line. Finally, we have transformed the 3×3 AR parameter matrix A to its vectorised form (a nine-dimensional column vector) so as to enable θ to be a vector and to avoid working with matrix-variate Gaussian mixtures; in this set-up, θ is 13-dimensional vector. For the simulation of the skew-t distribution, we have used the package sn, available in http://cran.

Estimated dynamic correlations

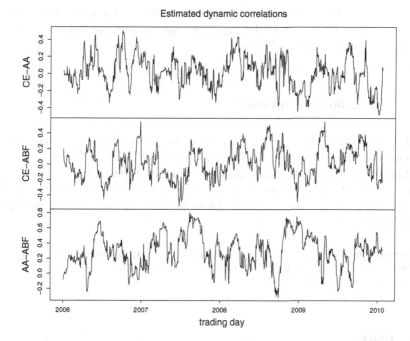

Fig. 30.2 Estimated cross-correlations of the FTSE-100 constituents CE, AA and ABF. Shown are the correlation between CE and AA, CE and ABF and AA and ABF

r-project.org/web/packages/sn/index.html of the programming language R and for the simulation of the Wishart distribution, we have used the package MCMCpack available in http://cran.r-project.org/web/packages/MCMCpack/index.html.

We have used $N = 1,000$ particles for each time t. Figure 30.1 shows the estimated volatilities against the returns. These volatilities are computed by (30.4) by replacing modes of posterior samples of size N at each time point t of Σ_t and θ. As we can see the volatility estimates seem to reflect the dynamics of the returns. Figure 30.2 shows the cross-correlation estimates over time. We observe that these correlations, which again are the modes of each sample correlations, show a clear dynamic nature with AA and ABF having the highest correlations.

At the last time point (24 December 2009) the estimated inverse of A (which drives the AR evolution of Σ_t) is

$$\hat{A}^{-1} = \begin{bmatrix} 0.07179 & 0.00002 & 0.00009 \\ -0.00036 & 0.06199 & 0.00006 \\ -0.00336 & -0.00045 & 0.06150 \end{bmatrix}$$

and those of k, ν and α are $\hat{k} = 8.345$, $\hat{\nu} = 3.537$ and $\hat{\alpha} = [-4.511, -0.558, -0.540]^T$. We note the negative skewness of the returns (as a result of negative skewness parameter), the heavy tails of the returns (as a result of low degrees of freedom $\hat{v}$) and the stationarity of the volatility process (as a result of the matrix $\hat{C}^T = \hat{k}^{1/2}(\hat{k} - p - 1)^{-1/2}\hat{A}^{-1}$ having eigenvalues inside the unit circle, i.e. $\lambda_1 = 0.099$, $\lambda_2 = 0.086$ and $\lambda_3 = 0.085$).

30.5 Concluding Comments

In this paper we describe a particle filter algorithm for sequential estimation of time-varying volatility. We propose a skew-t distribution for the returns, which is capable to deal with the heavy tails and the asymmetry of the returns, and an autoregressive process for the volatility evolution. A suitable important function is put forward in conjunction with a new approach to deal with the estimation of the hyperparameters of the model.

Acknowledgements I should like to thank Dimitris Kugiumtzis for inviting me to present a paper at the first Conference of the International Society for Non-Parametric Statistics. I am also grateful to the editors and an anonymous referee.

References

1. Asai, M., McAleer, M., Yu, J.: Multivariate stochastic volatility: a review. Econ. Rev. **25**, 145–175 (2006)
2. Azzalini, A., Capitanio, A.: Distributions generated by perturbation of symmetry with emphasis on a multivariate skew t distribution. J. R. Stat. Soc. Ser. B **65**, 367–389 (2003)
3. Bauwens, L., Laurent, S., Rombouts, J.V.K.: Multivariate GARCH models: a survey. J. Appl. Econ. **21**, 79–109 (2006)
4. Brandt, M.W., Santa-Clara, P.: Dynamic portfolio selection by augmenting the asset space. J. Fin. **61**, 2187–2217 (2006)
5. Chib, S., Nardari, F., Shephard, N.: Analysis of high dimensional multivariate stochastic volatility models. J. Econ. **134**, 341–371 (2006)
6. Doucet, A., de Freitas, N., Gordon, N.J.: Sequential Monte Carlo Methods in Practice. Springer, New York (2001)
7. Gordon, N.J., Salmond, D.J., Smith, A.F.M.: Novel approach to non-linear/non-Gaussian Bayesian state estimation. IEE Proc. F **140**, 107–113 (1993)
8. Gourieroux, C., Jasiak, J., Sufana, R.: The Wishart autoregressive process of multivariate stochastic volatility. J. Econ. **150**, 167–181 (2009)
9. Jin, X., Maheu, J.M.: Modeling realized covariances and returns. J. Financ. Econ. **11**, 335–369 (2012)
10. Kim, S., Shephard, N., Chib, S.: Stochastic volatility: likelihood inference and comparison with ARCH models. Rev. Econ. Stud. **65**, 361–393 (1998)
11. Liesenfeld, R., Richard, J.-F.: Univariate and multivariate stochastic volatility models: estimation and diagnostics. J. Empir. Financ. **10**, 505–531 (2003)
12. Liu, J., West, M.: Combined parameter and state estimation in simulation-based filtering. In: Sequential Monte Carlo Methods in Practice, pp. 197–223. Springer, New York (2001)
13. Petris, G., Petrone, S., Campagnoli, P.: Dynamic Linear Models with R. Springer, New York (2009)

14. Philipov, A., Glickman, M.E.: Factor multivariate stochastic volatility via Wishart processes. Econ. Rev. **25**, 311–334 (2006)
15. Philipov, A., Glickman, M.E.: Multivariate stochastic volatility via Wishart processes. J. Bus. Econ. Stat. **24**, 313–328 (2006)
16. Polson, N.G., Stroud, J.R., Müller, P.: Practical filtering with sequential parameter learning. J. R. Stat. Soc. Ser. B **70**, 413–428 (2008)
17. Storvik, G.: Particle filters for state-space models with the presence of unknown static parameters. IEEE Trans. Signal Process. **50**, 281–289 (2002)
18. Triantafyllopoulos, K.: Multivariate stochastic volatility with Bayesian dynamic linear models. J. Stat. Plan. Inference **138**, 1021–1037 (2008)
19. Triantafyllopoulos, K.: Multivariate stochastic volatility modelling using Wishart autoregressive processes. J. Time Ser. Anal. **33**, 48–60 (2012)
20. Yu, J., Meyer, R.: Multivariate stochastic volatility models: Bayesian estimation and model comparison. Econ. Rev. **25**, 361–384 (2006)

Chapter 31
Guaranteed Estimation of Logarithmic Density Derivative by Dependent Observations

Vyacheslav Vasiliev

Abstract We provide a truncated estimation method to analyze the non-asymptotic properties of a variety of ratio type functional's estimators when the data may possibly be dependent. As an illustration, the parametric and nonparametric estimation problems on a time interval of a fixed length are considered. In particular, parameters of linear multivariate autoregressive process are estimated. Moreover, the estimation problem of a multivariate logarithmic derivative of a noise density of an autoregressive process with guaranteed accuracy is solved. It is shown that all the truncated estimators have not only guaranteed accuracy in the sense of the L_m-norm, $m \geq 2$, as well as asymptotic properties of basic estimators.

Keywords Ratio estimation • Truncated estimation method • Fixed sample size • Guaranteed accuracy • Multivariate autoregression • Non-parametric logarithmic density derivative estimation

31.1 Introduction

Evolution of mathematical statistics is turned to development of data processing methods by dependent sample of fixed size. One of the possibilities for finding estimators (parametric and nonparametric) with the guaranteed quality of inference using a sample of fixed size is provided by the approach of truncated sequential estimation. The truncated sequential estimation method was developed in [3–5] (among others) for parameter estimation problems in discrete-time dynamic models. Using a sequential approach, estimators of dynamic system parameters with known variance by sample of fixed size were constructed in these papers.

Nonparametric truncated sequential estimators of a regression function were presented in [7, 8] on the basis of Nadaraya–Watson estimators calculated at a special stopping time. These estimators have known mean square errors as well. The duration of observations is also random but bounded from above by a

V. Vasiliev (✉)
Tomsk State University, Lenin Ave. 36, Tomsk, 634050 Russia
e-mail: vas@mail.tsu.ru

© Springer Science+Business Media New York 2014
M.G. Akritas et al. (eds.), *Topics in Nonparametric Statistics*, Springer Proceedings in Mathematics & Statistics 74, DOI 10.1007/978-1-4939-0569-0_31

non-random fixed number. Results in non-asymptotic parametric and nonparametric problem statements can be found in [6, 9–11] among others.

In this paper the truncated estimation method of ratio type functionals by dependent sample of fixed size is presented. This method makes it possible to obtain estimators with guaranteed accuracy in the sense of the L_m-norm, $m \geq 2$. The main purpose of the paper is to obtain an estimator of the multivariate logarithmic derivative of a distribution density of noises of an autoregressive process with unknown parameters. Early similar results for scalar ratio type estimators were published in [13].

31.2   General Problem Statement and Main Result

Let $(\Omega, \mathscr{F}, \mathrm{P})$ be a probability space with a filtration $\{\mathscr{F}_n\}_{n\geq 0}$ and let $(f_n)_{n\geq 1}$ and $(g_n)_{n\geq 1}$ be $\{\mathscr{F}_n\}$-adapted sequences of random $s \times q$ matrices and numbers respectively.

Let

$$\Psi_N = f_N/g_N, \quad N \geq 1 \tag{31.1}$$

be an estimator of a matrix Ψ. For instance, the matrix Ψ can be a ratio

$$\Psi = f/g$$

and f_N and g_N are estimators of some matrix f and number $g \neq 0$ respectively.

Consider the following modification of the estimator Ψ_N :

$$\tilde{\Psi}_N(H) = \Psi_N \cdot \chi(|g_N| \geq H), \quad N \geq 1, \tag{31.2}$$

where H is a positive number or sequence $H = (H_N)$, defined below and the notation $\chi(A)$ means the indicator function of set A.

Our main aim is to formulate general conditions on the sequences (f_N), (g_N) and on the parameter H giving a possibility to estimate Ψ with a guaranteed accuracy in the sense of the L_m-norm, $m \geq 2$.

Define for some $\varphi_N(m)$, $w_N(\mu)$, H and C_V the function

$$V_N(m, \mu, H) = C_V(\varphi_N(m) + w_N(\mu)),$$

as well as for positive integer $p < m$, positive numbers H_N and number $\tilde{C}_V$ the function

$$V_N(p) = \tilde{C}_V[\varphi_N(p) + H_N^{-2p}\varphi_N^{p/m}(m)w_N^{p/\mu}(\mu) + w_N(\mu) + \gamma_N],$$

where $\gamma_N = \chi(H_N > (1 - \beta)|g|)$, $\beta \in (0, 1)$.

Theorem 1. *Assume for some integers $m \geq 1$ and $\mu \geq 1$ there exist sequences of positive numbers $(\varphi_N(m))_{N\geq 1}$ and $(w_N(\mu))_{N\geq 1}$, decreasing to zero, as well as a number $g \neq 0$ such that for every $N \geq 1$ the following assumptions hold*

(1) $E||f_N - \Psi g_N||^{2m} \le \varphi_N(m)$;
(2) $E(g_N - g)^{2\mu} \le w_N(\mu)$.

Then, the estimator $\tilde{\Psi}_N(H)$ defined in (31.2) has the following properties

(i) in the case of known positive lower bound g_* for $|g|$, the parameter H in the definition of the truncated estimator (31.2) should be taken from the interval $(0, g_*)$ and

$$E||\tilde{\Psi}_N(H) - \Psi||^{2m} \le V_N(m, \mu, H); \qquad (31.3)$$

(ii) in the case of unknown g (and g_*) for every (possibly slowly decreasing to zero) sequence $H = (H_N)$ of positive numbers H_N, $N \ge 1$ and every positive integer p, satisfying

$$\frac{mp}{m - p} \le \mu, \quad m > 1 \text{ and } \mu > 1,$$

it holds

$$E||\tilde{\Psi}_N(H_N) - \Psi||^{2p} \le V_N(p). \qquad (31.4)$$

Corollary 1. *Assume that $\Psi = f/g$ for some matrix f, where g is defined in Theorem 1 and, instead of the assumption (1), for some $v \ge 1$ there exists sequence $(v_N(v))_{N \ge 1}$ of non-negative numbers, decreasing to zero, such that*

$$E||f_N - f||^{2v} \le v_N(v), \quad N \ge 1.$$

Then the assumption (1) of Theorem 1 is fulfilled, where the function $\varphi_N(m)$ should be replaced by the following one:

$$\varphi_N(m) = v_N^{m/v}(m) + w_N^{m/\mu}(v), \quad m = min(v, \mu).$$

Remark 1. The functions $V_N(m, \mu, H)$ and $V_N(p)$ may depend on unknown parameters. At the same time the knowledge of the rate of L_m-convergence of proposed estimators can be useful in various adaptive procedures (control, prediction etc.; see Sect. 31.3.2 below as well) and for the construction of pilot estimators (see, e.g., [2, 12, 14]).

Remark 2. The properties of estimators of the often encountered form $G_N^{-1}\Phi_N$ (G_N and Φ_N are random matrices) can be investigated using the presented method (see, e.g., Sect. 31.3.1 below).

31.3   Density Estimation of Noises of a Stable First Order Multivariate Autoregression

31.3.1   Estimation of Parameters of an Autoregression

We show in this section a possibility to apply the presented general truncated method for guaranteed estimation of matrix parameters in multivariate systems.

Consider the s-dimensional process ($s > 1$) satisfying the following equation

$$x(n) = Ax(n-1) + \xi(n), \quad n \geq 1, \tag{31.5}$$

where noises $\xi(n)$, $n \geq 1$ are i.i.d. zero mean random vectors with finite moments of the order $16(s-1)$, as well as $E||x(0)||^{16(s-1)} < \infty$ and the stability condition for the process (31.5) is satisfied, i.e. all the eigenvalues of the matrix A lie within the unit circle, see, e.g., [1]. We suppose that the matrix parameter A to be estimated belongs to a compact set Λ from the stable region.

Consider the estimation problem of A with a guaranteed accuracy. We define the estimator of the type (31.2) on the basis of the LSE of the form (31.1)

$$\hat{A}_N = \overline{\Phi}_N \overline{G}_N^{-1}, \quad N \geq 1,$$

where

$$\overline{G}_N = \frac{1}{N} G_N, \quad G_N = \sum_{n=1}^{N} x(n-1)x'(n-1),$$

$$\overline{\Phi}_N = \frac{1}{N} \Phi_N, \quad \Phi_N = \sum_{n=1}^{N} x(n)x'(n-1), \quad N \geq 1.$$

Define the matrix

$$\overline{G}_N^+ = \overline{\Delta}_N \overline{G}_N^{-1}, \quad \overline{\Delta}_N = \det(\overline{G}_N).$$

According to the general notation, in this case we have

$$\Psi = A, \quad \Psi_N = \hat{A}_N, \quad f_N = \overline{\Phi}_N \overline{G}_N^+, \quad g_N = \overline{\Delta}_N.$$

Using formula (31.5) it is easy to verify that with P_A-probability one it holds

$$\lim_{N \to \infty} \overline{G}_N = F \quad \text{and} \quad \lim_{N \to \infty} \overline{\Delta}_N = \Delta > 0,$$

where F is a positive definite $s \times s$-matrix, such that $\Delta_* = \inf_{A \in \Lambda} \Delta > 0$ (see, e.g., [1]). Then

$$f = A\Delta, \quad g = \Delta$$

and $\tilde{\Psi}_N = \tilde{A}_N$,

$$\tilde{A}_N = \hat{A}_N \cdot \chi(\overline{\Delta}_N \geq H), \tag{31.6}$$

where $H \in (0, \Delta_*)$.

Assumptions of Theorem 1 can be verified for $m = \mu = 4$ similar to, e.g., [2]

$$\sup_{A \in \Lambda} E_A \|f_N - A\overline{\Delta}_N\|^4 \leq \frac{\overline{C}_1}{N^2}; \quad \sup_{A \in \Lambda} E_A (\overline{\Delta}_N - \Delta)^4 \leq \frac{\overline{C}_2}{N^2}.$$

Using the assertion (i) of Theorem 1 it can be shown that there exists a given number C_Λ such that for every $N \geq 1$,

$$\sup_{A \in \Lambda} E_A \|\tilde{A}_N - A\|^4 \leq \frac{C_\Lambda}{N^2}. \tag{31.7}$$

31.3.2 Nonparametric Estimation of a Multivariate Logarithmic Density Derivative

Consider the problem of estimating the logarithmic derivative ($q = 1$ in the general problem statement),

$$\Psi(t) = \nabla f(t)/f(t)$$

($\nabla f(t)$ is a $s \times 1$-vector of the first order partial derivatives of $f(t)$) of a distribution density $f(t)$ of the i.i.d. vector noises $\xi(n) = (\xi_1(n), \ldots, \xi_s(n))'$ in the model (31.5), considered in Sect. 31.3.1. Noises $\xi(n)$, $n \geq 1$ are zero mean random vectors with finite moments of the order 4α, where $\alpha = \max\{4(s - 1), \nu + 1 + \delta\}$ for some $\delta > 0$, as well as $E\|x(0)\|^{4\alpha} < \infty$ (the number ν will be defined below in Assumption (f)) and the stability condition is satisfied. It is supposed that the matrix parameter A to be estimated belongs to a compact set Λ from the stable region (see also Sect. 31.3.1).

ASSUMPTION (f) *We suppose that the function $f(\cdot)$ satisfies the following condition*

$$\sup_{z \in \mathscr{R}^s} f(z) \leq C_f,$$

and, for some even $v \geq 2$, *as well as* $\mathscr{L} > 0$ *and* $\gamma \in (0, 1]$, *for all the partial derivatives of the order* $1 + v$ *the Lipshitz condition*

$$|f^{(1+v)}(x) - f^{(1+v)}(y)| \leq \mathscr{L}||x - y||^\gamma$$

holds.

The knowledge of $\Psi(t)$ is important in various statistical problems, e.g. for constructing the algorithm of optimal control of an autoregressive process; estimating of a regression curve; testing close hypotheses. These problems are of a peculiar interest in the case of dependent observations: for example, where the logarithmic derivative of a density is used for designing the optimal algorithms of nonlinear filtering and adaptive control of random processes (see, e.g., [2] and the references therein).

We will construct estimators of $f(t)$ and $\nabla f(t)$ using the following estimators $\tilde{\xi}(n)$ of noises $\xi(n)$ in (31.5):

$$\tilde{\xi}(n) = x(n) - A_{n-1}^* x(n-1), \quad n = \overline{1, N}, \tag{31.8}$$

where $A_{n-1}^* = \text{proj}_A \tilde{A}_{n-1}$, $\tilde{A}_n$ is the estimator defined in (31.6). By the definition (31.8), the estimators $\tilde{\xi}(n)$ can be represented in the form

$$\tilde{\xi}(n) = \xi(n) + (A - A_{n-1}^*)x(n-1), \quad n = \overline{1, N}.$$

Note that the matrices $A - A_{n-1}^*$ are uniformly bounded

$$\sup_{n, A \in \Lambda} ||A - A_{n-1}^*|| \leq C \tag{31.9}$$

and, according to (31.7), the following properties of the estimator A_{n-1}^* can be obtained,

$$\sup_{A \in \Lambda} E_A ||A - A_{n-1}^*||^4 \leq \sup_{A \in \Lambda} E_A ||A - \tilde{A}_{n-1}||^4 \leq \frac{C_\Lambda}{n^2}, \quad n = \overline{1, N}. \tag{31.10}$$

Using (31.9) and (31.10), we can find the known numbers C_1 and C_m, such that

$$\sup_{A \in \Lambda} \sum_{n=1}^N E_A ||(A - A_{n-1}^*)x(n-1)||^{2m} \leq \begin{cases} C_1 \log N, & m = 1, \\ C_m, & 1 < m \leq v + 1. \end{cases} \tag{31.11}$$

Similar relations were obtained in [2] (see Lemmas 5.1.3 and 5.1.5) for another type of estimators.

As a nonparametric estimator of a density $f(t) = f^{(0)}(t)$ satisfying Assumption (f) and its partial derivative $f^{(1)}(t) = \partial f(t)/\partial t_j$, we use the combined statistic of the form

$$\widehat{f_N^{(r)}}(t) = \frac{1}{N h_{r,N}^{s+r}} \sum_{i=1}^{N} K^{(r)}\left(\frac{t - \tilde{\xi}(i)}{h_{r,N}}\right), \quad r = 0, 1, \tag{31.12}$$

where $K^{(0)}(u) = K(u) = \prod_{k=1}^{s} K(u_k)$ is an s-dimensional multiplicative kernel which, generally speaking, does not necessarily possess the characterizing properties of density (nonnegativity and normalization to 1), $K^{(1)}(u) = \partial K(u)/\partial u_j$, sequences of numbers $h_{r,N} \downarrow 0$, $N \to \infty$.

Using technique of Theorems 4.3.1 and 5.1.3 from [2] and (31.11), by appropriate chosen kernels $K(\cdot)$ in (31.12), we can find known numbers C_i^*, $i = 0; 1$, such that

$$\sup_{A \in \Lambda} E_A(\hat{f}_N(t) - f(t))^4 \le C_0^*\left(\frac{1}{N^2 h_{0,N}^{2s}} + h_{0,N}^{4(\nu+1+\gamma)}\right),$$

$$\sup_{A \in \Lambda} E_A \|\widehat{\nabla f}_N(t) - \nabla f(t)\|^4 \le C_1^*\left(\frac{1}{N^2 h_{1,N}^{2(s+2)}} + h_{1,N}^{4(\nu+\gamma)}\right).$$

Thus, to minimize the obtained upper bounds, it is natural to put

$$h_{0,N} = h_{1,N} = N^{-\frac{1}{2(\nu+1+\gamma)+s}}$$

and for obviously defined numbers $\tilde{C}_0$ and $\tilde{C}_1$, we have

$$\sup_{A \in \Lambda} E_A(\hat{f}_N(t) - f(t))^4 \le \tilde{C}_0 N^{-\frac{4(\nu+1+\gamma)}{2(\nu+1+\gamma)+s}},$$

$$\sup_{A \in \Lambda} E_A \|\widehat{\nabla f}_N(t) - \nabla f(t)\|^4 \le \tilde{C}_1 N^{-\frac{4(\nu+\gamma)}{2(\nu+1+\gamma)+s}}.$$

As an estimator of the ratio $\Psi(t)$ from the observations $(x(n))_{n \ge 1}$, one can use the ratio

$$\hat{\Psi}_N(t) = \widehat{\nabla f}_N(t)/\hat{f}_N(t)$$

of statistics defined in (31.12).

Estimators of type (31.12) of the density and its derivatives from observations (31.8) (with estimators (A_n^*) of an another type) were considered in [2], Chap. 4, where it was established, in particular, their asymptotic normality and convergence with probability one. The results on asymptotic ratio estimation of the partial derivatives of the noise distribution density in multivariate dynamic systems are given in [2], Sect. 5.1.

To obtain estimators of $\Psi(t)$ with a known MSE we apply Theorem 1. Define the estimator

$$\tilde{\Psi}_N(t) = \hat{\Psi}_N \chi(\hat{f}_N(t) \geq H), \quad N \geq 1.$$

Put $H = (\log N)^{-1}$ in the definition of the estimator $\tilde{\Psi}_N(t)$. Then, according to (31.4) with $p = 1$, $m = \mu = 2$, for some number $\tilde{C}$ using Corollary 1 for N large enough (to eliminate γ_N) we have

$$E_A\|\tilde{\Psi}_N(t) - \Psi(t)\|^2 \leq \tilde{C}[N^{-\frac{2(v+\gamma)}{2(v+1+\gamma)+s}} + (\log N)^2 N^{-\frac{4(v+\gamma)+2}{2(v+1+\gamma)+s}} + N^{-\frac{4(v+1+\gamma)}{2(v+1+\gamma)+s}}].$$

It should be noted that the obtained rate of convergency of the estimator $\tilde{\Psi}_N(t)$ is similar to the case of independent observations (see, e.g., [2]).

31.4   Simulation Study

To confirm theoretical results of Sect. 31.3.1 we realize simulations of the truncated estimator of the parameter λ in the software package MATLAB, where λ is the parameter of the scalar autoregressive process $(x_n)_{n \geq 0}$, satisfying the equation

$$x_n = \lambda x_{n-1} + \xi_n, \quad n \geq 1. \tag{31.13}$$

Here x_0 and $(\xi_n)_{n \geq 1}$ are i.i.d. standard Gaussian random variables and parameter λ is assumed to belong to the stable region $(-1,1)$.

Table 31.1 Truncated estimation of the parameter λ

		N=100		N=200		N=500	
	λ	$\tilde{\lambda}(N)$	$S_\lambda^2(N)$	$\tilde{\lambda}(N)$	$S_\lambda^2(N)$	$\tilde{\lambda}(N)$	$S_\lambda^2(N)$
H=0,6							
	0,2	0,201	0,009	0,191	0,005	0,198	0,002
	−0,2	−0,219	0,011	−0,199	0,004	−0,196	0,001
	0,5	0,496	0,007	0,496	0,003	0,502	0,001
	−0,5	−0,489	0,007	−0,493	0,003	−0,501	0,001
	0,9	0,887	0,002	0,892	0,001	0,897	3,9e–004
	−0,9	−0,869	0,005	−0,890	0,001	−0,895	3,9e–004
H=0,8							
	−0,2	−0,189	0,011	−0,190	0,005	−0,198	0,002
	0,5	0,493	0,007	0,486	0,004	0,497	0,001
	−0,9	−0,884	0,003	−0,891	0,002	−0,895	3,6e–004

[a] The presented results of modeling belong T. Dogadova (aurora1900@mail.ru)

In Table 31.1 the average

$$\hat{\lambda}(H, N) = \frac{1}{100} \sum_{k=1}^{100} \tilde{\lambda}^{(k)}(H, N)$$

of the truncated estimators of the type (31.6)

$$\tilde{\lambda}^{(k)}(H, N) = \frac{\sum_{n=1}^{N} x_n^{(k)} x_{n-1}^{(k)}}{\sum_{n=1}^{N} (x_{n-1}^{(k)})^2} \cdot \chi \left[\sum_{n=1}^{N} (x_{n-1}^{(k)})^2 \geq HN \right] \qquad (31.14)$$

for the k-th realization $x^{(k)} = (x_n^{(k)})$, $k = 1 \ldots 100$ of the process (31.13) as well as quality characteristics

$$S_{\tilde{\lambda}}^2(N) = \frac{1}{100} \sum_{k=1}^{100} (\tilde{\lambda}^{(k)}(H, N) - \lambda)^2$$

of estimators (31.14) for different N and H are given.

31.5 Summary

We have presented the truncated estimation method of ratio type multivariate functionals constructed by dependent samples of finite size. This method allows to obtain estimators with a guaranteed accuracy (31.3) on a time interval of a fixed length.

As an illustration, parametric and nonparametric estimation problems are considered. The presented method was applied to estimation of parameters of a linear multivariate autoregressive process and of a multivariate logarithmic derivative of its noise density. Results of simulation confirm the efficiency of the parameter estimation procedure.

The presented method can be similarly applied to samples from continuous-time models.

References

1. Anderson, T.W.: The Statistical Analysis of Time Series. Wiley, New York (1971)
2. Dobrovidov, A.V., Koshkin, G.M., Vasiliev, V.A.: Non-parametric state space models. Kendrick Press, Heber City (2012) http://www.kendrickpress.com/nonpara.html (Russian original: Vasiliev, V.A, Dobrovidov, A.V., Koshkin, G.M.: Nonparametric Estimation of Functionals of Stationary Sequences Distributions. Nauka, Moscow (2004))

3. Fourdrinier, D., Konev, V., Pergamenshchikov, S.: Truncated sequential estimation of the parameter of a first order autoregressive process with dependent noises. Math. Methods Stat. **18**(1), 43–58 (2009)
4. Konev, V.V., Pergamenshchikov, S.M.: Truncated sequential estimation of the parameters in random regression. Seq. Anal. **9**(1), 19–41 (1990)
5. Konev, V.V., Pergamenshchikov, S.M.: On truncated sequential estimation of the drifting parametermeant in the first order autoregressive models. Seq. Anal. **9**(2), 193–216 (1990)
6. Mikulski, P.W., Monsour, M.J.: Optimality of the maximum likelihood estimator in first-order autoregressive processes. J. Time Ser. Anal. **12**(3), 237–253 (1991). doi: 10.1007/s00184-010-0333-5
7. Politis, D.N., Vasiliev, V.A.: Sequential kernel estimation of a multivariate regression function. In: Proceedings of the IX International Conference "System Identification and Control Problems", SICPRO'12, pp. 996–1009, V. A. Tapeznikov Institute of Control Sciences, Moscow, 30 January–2 February 2012
8. Politis, D.N., Vasiliev, V.A.: Non-parametric sequential estimation of a regression function based on dependent observations. Seq. Anal. **32**(3), 243–266 (2013)
9. Roll, J., Nazin, A., Ljung L.: A non-asymptotic approach to local modelling. In: The 41st IEEE CDC, Las Vegas, Nevada, 10–13 December 2002. (Regular paper) (2002). Predocumentation is available at http://www.control.isy.liu.se/research/reports/2003/2482.pdf
10. Roll, J., Nazin, A., Ljung L.: Non-linear system identification via direct weight optimization. Automat. IFAC J. Special Issue on Data-based modelling and system identification **41**(3), 475–490 (2005). Predocumentation is available at http://www.control.isy.liu.se/research/reports/2005/2696.pdf
11. Shiryaev, A.N., Spokoiny, V.G.: Statistical Experiments and Decisions. Asymptotic Theory. World Scientific, Singapore (2000)
12. Vasiliev, V.A.: On identification of dynamic systems of autoregressive type. Automat. Remote Control **12**, 106–118 (1997)
13. Vasiliev, V.A.: One investigation method of ratio type estimators. Preprint 5 of the Math. Inst. of Humboldt University, Berlin, 1–15 (2012). http://www2.mathematik.hu-berlin.de/publ/pre/2012/p-list-12.html
14. Vasiliev, V.A., Koshkin, G.M.: Nonparametric identification of autoregression. Probab. Theory Appl. **43**(3), 577–588 (1998)

Chapter 32
Nonparametric Statistics and High/Infinite Dimensional Data

Frédéric Ferraty and Philippe Vieu

Abstract One of the current challenge proposed to the nonparametric community is to deal with high dimensional (and possibly infinite dimensional) data. In high (but finite) dimensional setting the key question is often to proceed to some variable selection stage. In the infinite framework, which involves the so-called functional data, an usual approach consists in adapting (or trying to adapt) standard methodologies by taking suitably into account the infinite dimensional feature of the problem. While both fields have been widely studied in the few last past years under parametric (mainly linear) modelling, the challenge is now to develop more trustable nonparametric methods. The aim of this paper is to discuss the few recent nonparametric advances made in this direction, as well in functional data analysis as in model selection when considering a large number of variables, with main objective to point some interesting tracks for the future. Even if both fields may look similar for the unadvertised people, we will also emphasize on the structural differences existing between high (but finite) dimensional data setting and functional data one.

Keywords Functional data • Infinite dimensional variable • High dimension statistics

32.1 Introduction

All along their developments most of revolutionary scientific advances had to cope with old ideas. This is exactly the case for nonparametric techniques in Statistics. In a natural way, each new problem in Statistics is firstly taken on by simple linear approach and at each time the main challenge for the nonparametricians has s been to convince people of the interest and of the feasibility of non-linear approaches. Because of increasing sources of applications (being possible by technological progress in collecting, storing and processing data) and because of important open theoretical questions, most of the current problems in Statistics are concerning

F. Ferraty • P. Vieu (✉)

Institut de Mathématiques, Université Paul Sabatier, Toulouse, France

e-mail: ferraty@math.univ-toulouse.fr; vieu@math.univ-toulouse.fr

© Springer Science+Business Media New York 2014   357

M.G. Akritas et al. (eds.), *Topics in Nonparametric Statistics*, Springer Proceedings
in Mathematics & Statistics 74, DOI 10.1007/978-1-4939-0569-0_32

"high" dimensional variables and the challenge for the nonparametricians is to convince the community of the possible using of nonparametric ideas in this situation. The aim of this contribution is to discuss two important kinds of problems involving "high" dimension: functional data analysis and variables selection. These two problems are at the centre of thousands of papers since a few years but once again the scientific community is currently far to be convinced of the interest for developing nonparametric approaches. After a short historical discussion in Sect. 32.2, we discuss in Sect. 32.3 the main ideas making possible the using of nonparametrics in functional data analysis while in Sect. 32.4 we present the few recent advances existing in nonparametric variable selection procedures. The aim of Sect. 32.5 is double: firstly discussing both the similarities and the structural differences between functional data and high dimensional ones, and secondly presenting some situations in which there is some interest for mixing both approaches. One of the main goals of this contribution is to highlight some tracks for the future in both fields.

32.2 Nonparametric Modelling in Low Dimensions: Some Short Historical Background

While the main precursor papers in one-dimensional nonparametric statistics [36, 38, 51] go back to the end of fifties (and even sooner with Tukey's regressogram [50]), it took much more time to convince people of the interest of such flexible modelling approaches. It was not rare for the nonparametricians, even in the 1980s, to receive sceptical comments as well from the statistical community as from unspecialist scientists, like *Nonparametric smoothing methods will never work, they need too much data*... or *Nonparametric regression is great but unuseful to provide some kind of correlation coefficient* ... and so on. Thanks to the efforts of the statistical community and to the increasing capacity of computers, these old ideas have been quickly overpassed and nonparametric smoothing has been popularized to become at the end of the 1980s an incontournable preliminary stage when one has to deal with univariate data. In this moment various general contributions have popularized this field of Statistics [10, 13, 28, 46].

Curiously, in the moment as the nonparametric ideas became popular for univariate problems the same phenomenon occurred when one was thinking in multivariate smoothing! Based on a narrow interpretation of the exponential deterioration (when the dimension of the problem increases) of the rates of convergence of nonparametric estimates [47], it was not unusual to face arguments like *Nonparametric modelling has no interest nor future when the dimension is greater than* 3 *or* 4, ..., going so far as to raise the spectre of the devil through the famous *curse of dimensionality* which was one among the most commonly used expressions in the 1990s in the statistical community. Once again the development of numerous researches has allowed to overpass these ideas and it is now commonly accepted that the dimension is not a curse but a chance for the statisticians (and for the scientists in general). It is also widely admitted now that nonparametric ideas can be very efficiently adapted in multivariate situations. Additive modelling (popularized by Stone [48] and Hastie and Tibshirani [29]) and semiparametric statistics [44,45] are two famous examples of how smoothing ideas can be suitably used in multivariate situations.

32.3 Nonparametric Functional Data Analysis

Functional data analysis consists in extracting statistical informations from a sample $\chi_1, \ldots, \chi_n$ of objects being of infinite dimensional nature. The technological progresses have dramatically increased the number of scientific fields having now to deal with such kind of data, and the need for accurate statistical analyses of them has been popularized by the various monographies by Ramsay and Silverman (see [39–41]). Even if they are mostly oriented towards time series modelling the works by Bosq have also contributed to popularize this field (see [6]). Basically, the χ_i's can be curves, images, continuous periods of a single time series, or even objects having more complicated structure. A typical example is the following spectrometric dataset (see Fig. 32.1), which serves now as usual benchmark for testing the behaviour of any new statistical method. In these data, coming from food quality problems, the χ_i's are the spectra of light absorbance as function of the wavelength for a sample of $n = 215$ pieces of chopped meat.

   Of course, each of these curves is not observed in a continuous way but on a finite grid (composed of 100 equispaced wavelengths in the example depicted in Fig. 32.1) and there is a first and obvious link between functional and nonparametric statistics which consists in smoothing each of these curves. This is not really a difficult problem since it consists just in successive univariate smoothing processes (exactly 215 univariate smoothings in the example depicted in Fig. 32.1), and this can be done easily (and at low computational costs) by many one-dimensional nonparametric techniques (for instance, Splines have been used in Fig. 32.1).

   The real question is to find models and methods for extracting informations from such kind of data in order to solve some statistical problems: clustering, discrimination, prediction of some extra explanatory variable ... As a matter of example, one practical problem associated with the data of Fig. 32.1 was to predict the quantity of some noxious substance, fatness, moisture, ..., in each piece of meat. In a natural way each among these methodological problems has been firstly investigated through linear point of view, and it was not (maybe it is still not!) easy to convince people of the possible using of nonparametric ideas in this area. At the end of the 1990s, when presenting ideas on this topic, it was not rare to receive very naive negative comments like *One knows that nonparametric does not work for high dimensional data! How can you hope that it works for infinite dimensional ones?*

   However the reason why it can work is very simple. Looking at what happens in multivariate settings, it looks like the problem lies in the sparseness of the data.

While this is true that for standard distances (associated to some functional norm $||.||$) the number of points falling into some fixed ball is exponentially decreasing with the dimensionality, this phenomenon can be easily controlled by using other notion of neighborhood. Said with more mathematical words, by changing the topological structure on the functional space of the variables χ_i one may allow to reduce these sparseness effects. For instance, in terms of local weighting (and many nonparametric techniques are based on local weighting), the key idea is to use neighbourhood based on some general pseudo-distance d

$$\mathcal{V}_d(\chi_0, \epsilon) = \{x, d(\chi, \chi_0) \leq \epsilon\},$$

rather than those based on standard norms

$$\mathcal{V}_{||.||}(\chi_0, \epsilon) = \{x, ||\chi - \chi_0|| \leq \epsilon\}.$$

A "good" choice of the pseudo-distance d may insure more concentration of the functional data, and therefore less sparseness effects. In a natural way, a key parameter for controlling the behaviour of nonparametric functional method is the

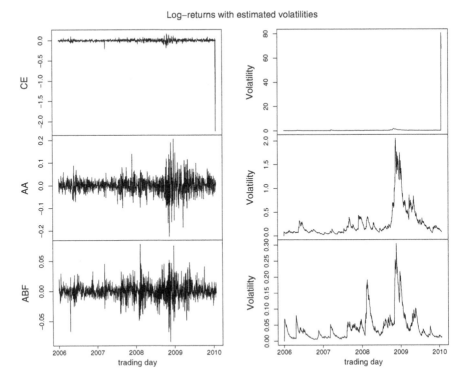

Fig. 32.1 The 215 spectrometric curves

notion of small ball probability function (also called concentration function) defined as the measure (respect to the probability distribution of the variable χ) of the ball $\mathcal{V}_d(\chi_0, \epsilon)$. Once again this is not necessarily the same as the measure of the ball $\mathcal{V}_{\|.\|}(\chi_0, \epsilon)$. In some sense, the using of pseudo-metric is a way for reducing the dimensionality of the problem since the concentration function based on the $\mathcal{V}_d(\chi_0, \epsilon)$ is expected to be much grater than the one based on the standard notion of neighbourhood $\mathcal{V}_{\|.\|}(\chi_0, \epsilon)$.

Said with other words, by a "good" choice of d the door is open for applying to functional data any nonparametric technique based on local weighted average procedures. Of course, things are not so easy, since the pseudo-metric has to increase concentration of the data but without losing too much relevant information on them. Once again this is typical for dimensionality reduction model which has to reduce the sparseness of the data but still keeping their most informative part. For instance, in the data presented in Fig. 32.1, the pseudo-metric based on second derivatives of the curves

$$d(\chi, \eta) = \sqrt{\int (\chi''(t) - \eta''(t))^2 \, dt},$$

is a good compromise which concentrates the curves without losing their predictive power (see [26] for a complete study of these data).[1]

While the precursor papers combining nonparametric approach with functional data (see [21, 22]) were not warmly received, this idea starts to be commonly shared in the statistical community. Thanks to the organizers of the 2002s meeting [5], which allowed to present these ideas to the whole nonparametric community, the general contribution [23] has been much more positively received. From a methodological point of view this field of nonparametric functional data analysis has been popularized by the book [24], while its feasibility and its good behaviour on finite samples has been highlighted by various on-line packages (see [14, 37]). Now, these precursor works have been followed by many further works in this area (the most recent bibliographical surveys can be found in [12, 20, 27]), and also by many specialized sessions in most of international conferences mixing nonparametric and functional data ideas as by various special issues in statistical journals. Even if there are still stubborn people who do not focus on statistical modelling aspects and who still may argue something like *This methodology has no interest since it cannot apply when using any standard (Gaussian for instance) process in "normed space"!*, our guess is that nonparametric functional statistics will still take more and more importance in the next future. We hope that this contribution will help in this sense by convincing the few last sceptics that overpassing the narrow scope of

---

[1]Of course, this specific choice is not the best one for any functional data problem, but it is out of the scope of this short contribution to discuss the crucial question of pseudo-metric choice (see [24] for large discussion). The only purpose here is to convince people that, by this simple idea, many nonparametric techniques can be efficiently adapted to functional data analysis.

normed spaces is a natural and efficient way for doing dimensionality reduction in functional data analysis.

Of course, saying that nonparametric ideas may be used in any problem involving functional data does not mean that it will always outperform other approaches. Even more, our guess is that combinations of the methods (linear and nonparametric for instance) could lead to very interesting results. For instance, some advances in this direction involve functional additive modelling [25, 34, 35], functional single index modelling [1, 2], functional projection pursuit regression [9, 17] and functional partially linear modelling [3, 31]. This field, that one could name functional semiparametric statistics, has undoubtedly a great future ahead.

32.4 Nonparametric Variable Selection Procedure

For the same technological reasons as for functional data, there is now more and more situations in which high dimensional data have been collected and need the attention of the statistician. Basically such data can be expressed as p-dimensional vectors $X_1, \ldots, X_n$, and it is not unusual to have to face situations in which the number of variables p is much greater than the number of statistical units n. An usual and natural way to process is to suppress all the variables being not informative. This field of Statistics has been very active during the last past years (and it still does so). Once again, and for obvious simplicity reasons, this question has been almost exclusively studied under linear modelling assumptions. The most popular contribution is the LASSO procedure introduced by Tibshirani [49], but various other methods (many being extensions of the Tibshirani's LASSO) have been proposed. A selected list of important contributions in this sense involves the SCAD-penalized version of LASSO [16], the Dantzig selector [8], the Least Angle Regression [15], the Elastic Net method [54], the grouped LASSO [52], the adaptive LASSO [53] and the relaxed LASSO [33]. The reader will find more complete list of references in the book [7]. There are a few recent advances on non-linear modelling (for instance, through some additive structural assumptions as in [30, 32, 42]), but the fully nonparametric approach of the problem is still a field being almost virgin. However, as it was the case in Sect. 32.3 before with functional data, the linear modelling is more and more unsatisfactory (and its validity is more and more difficult to assess) when the dimension increases.

So, there is an obvious need for developing fully nonparametric model for this kind of problems. However, saying that, one cannot ignore the difficulty of the task from a computational point of view; basically, the computational costs are higher in nonparametrics than in linear approaches and even more, they are higher and higher as the dimensionality of the problem increases. The precursor papers by Comminges and Dalalyan [11] and Ferraty and Hall [18] are stating the first asymptotic results in this field. In addition, [18] proposes an algorithm for showing the possible implementation and its high interest for finite sample situations of the flexible nonparametric variable selection procedure. Without any doubt nonparametric model selection is a great challenge for the next few years, and we hope that this contribution will help in this sense. Our point of view on this question is that, in the short term, the main task for constructing trustable nonparametric model selection procedures would be to develop artful algorithms combining flexibility of the model and reasonable run time.

32.5 Similarities and Dissimilarities Between Functional and High Dimensional Data

It is worth being noted that functional data and high dimensional problems present not only some similarities but also strong structural differences. As said before, in practice a functional data is always observed on a finite grid. That means that one does not have at hands the whole continuous data $\chi_1, \ldots, \chi_n$ but high-dimensional vectors $X_1, \ldots, X_n$, defined as[2]

$$X_i = (\chi_i(t_1), \ldots, \chi_i(t_p)),$$

where the t_j are the points at which the curves χ_i have been observed. For instance, in the spectrometric example depicted in Fig. 32.1, p is equal to 100 (corresponding to the 100 wavelengths at which the absorbances have been measured). A sample of these discretized data is presented in Fig. 32.2.

A first naive point of view would be to consider such data as usual high dimensional vectors, but in this case most of the commonly used methods (linear ones, for instance) would come against the strong correlation existing between the variables and would fail. Said with other words, functional data analysis can be seen as a special case of high dimensional problem, the specificity coming from these strong correlations. As a matter of consequence, there is a real need for thinking these two fields of modern Statistics as being more complementary than competitive.

To illustrate a first situation where there are obvious interests in thinking both fields in a complementary way, let us go back to the spectrometric data depicted in Fig. 32.1. Chemometricians are quite often interested in two things: firstly in predicting the quantity of some chemical component (for instance, fatness or moisture in the data of Fig. 32.1), but also in knowing which part of the spectrum is of most relevance for predicting this component. While the first question has been widely studied by means of functional regression models (linear or nonparametric) the second one has received much less attention. As proposed in [19] it is possible to solve this problem by combining functional nonparametric ideas described in Sect. 32.3 with the variable selection methods described in Sect. 32.4. Indeed, the

---

[2]For the sake of simplicity we restrict this presentation to balanced situations. In the more general setting, when the curves are not observed on the same grids, there is a preliminary stage which consists in smoothing each among the observed un-balanced curves $X_i = (\chi_i(t_{i,1}), \ldots, \chi_i(t_{i,p_i}))$, and then in constructing a new balanced dataset from these smoothed curves.

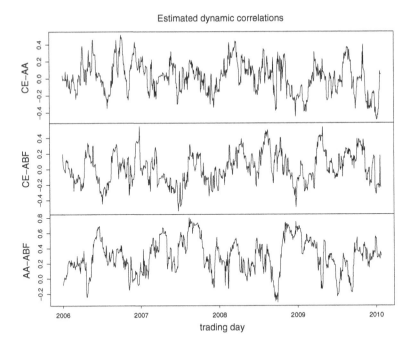

Fig. 32.2 The discretized spectrometric curves, and the four most informative points (*vertical lines*)

most relevant pointwise part of the continuous spectrum χ can be obtained by implementing a nonparametric variable selection procedure acting on the observed variables $\chi_i(t_1), \ldots, \chi_i(t_p)$. For the data in Fig. 32.1, this leads to select as most informative points for predicting fatness the four ones identified by the vertical lines in Fig. 32.2.

An other situation of interest is when one has at hand different kinds of data: some being functional and some other ones being high dimensional vectors. In this case one has to develop models having one infinite dimensional component (to capture the structure of the functional variable) and also being able to reduce the dimensionality of the vector valued variables. The semi-functional model presented in [4] and the variable selection procedure going with it are advances in this direction.

These were just two examples of situations where functional data analysis and variable selection procedures can be very efficiently mixed. Our guess (and our hope) is that this combination could be the source of many other interests. In any case, this is also a great challenge for the nonparametricians.

32.6 Concludings

As a matter of conclusion, one can say that nonparametric statistics had all along its history to face with dimensionality questions. Large world conferences on Nonparametric Statistics like those in Spetzes in 1990 (see [43]), in Creta in 2012 (see [5]) and in Chalkidiki in 2012 (from which this book is issued), have been major moments for making possible strong methodological advances and for overcoming dimensionality obstacles, and we would thank very much all the organizers of these three meetings. At the moment, functional data analysis and variable selection procedures are important fields for which the nonparametric ideas could help in finding interesting structure and/or information in complex datasets, and for which the nonparametricians have to convince the community of the interest (and the feasibility) of these ideas. While people start to be convinced by Nonparametric Functional Data Analysis (and the Creta's meeting in 2012 was a crucial moment for that), this is not so much the case for Nonparametric Variable Selection. We hope that this contribution will help in motivating further advances on both fields and particularly on their possible combinations.

References

1. Ait-Saïdi, A., Ferraty, F., Kassa, R., Vieu, P.: Cross-validated estimations in the single-functional index model. Statistics **42**, 474–494 (2008)
2. Amato, U., Antoniadis, A., De Feis, I.: Dimension reduction in functional regression with application. Comput. Stat. Data Anal. **50**, 2422–2446 (2006)
3. Aneiros-Pérez, G., Vieu, P.: Nonparametric time series prediction: a semi-functional partial linear modeling. J. Multivar. Anal. **99**(5), 834–857 (2008)
4. Aneiros-Pérez, G., Ferraty, F., Vieu, P.: Variable selection in semi-functional regression models. In: Recent Advances in Functional Data Analysis and Related Topics. Contributions to Statistics, pp. 17–22. Physica-Verlag/Springer, Heidelberg (2011)
5. Akritas, M., Politis, D. (eds.): Recent Advances and Trends in Nonparametric Statistics. Elsevier, Amsterdam (2003)
6. Bosq, D.: Linear Processes in Functional Spaces, Theory and Applications. Lecture Notes in Statistics, vol. 149. Springer, New York (2000)
7. Bühlmann, P., van de Geer, S.: Statistics for High-Dimensional Data: Methods, Theory and Applications. Springer, Berlin (2011)
8. Candès, E., Tao, T.: The Dantzig selector: statistical estimation when p is much larger than n. Ann. Stat. **35**, 2313–2351 (2007)
9. Chen, D., Hall, P., Müller, H.-G.: Single and multiple index functional regression models with nonparametric link. Ann. Stat. **39**(3), 1720–1747 (2011)
10. Collomb, G.: Nonparametric regression: an up-to-date bibliography. Statistics **16**, 309–324 (1985)
11. Comminges, L., Dalalyan, A.: Tight conditions for consistency of variable selection in the context oh high dimensionality. Ann. Stat. **40**(5), 2667–2696 (2012)
12. Delsol, L., Ferraty, F., Martinez Calvo, A.: Functional Data Analysis: An Interdisciplinary Statistical Topic. In Statistical Learning and Data Sciences. Chapman & Hall, London (2011)
13. Devroye, L.: A Course in Density Estimation. Progress in Probability and Statistics, vol. 14. Birkhäuser, Boston (1987)

14. fda.usc: R-package on-line (M. Febrero Bande). http://cran.r-project.org/web/packages/fda.usc/index.html
15. Efron, B., Hastie, T., Johnstone, I., Tibshirani, R.: Least angle regression. Ann. Stat. **32**, 407–499 (2004)
16. Fan, J., Li, R.: Variable selection via nonconcave penalized likelihood and its oracle properties. J. Am. Stat. Assoc. **96**, 1348–1360 (2001)
17. Ferraty, F., Goia, A., Salinelli, E., Vieu, P.: Functional Projection Pursuit Regression. Technical report, Test. Quadreni SEMeQ, Univ. Piemonte Orientale, Novara, **12**(2), 293–320 (2013)
18. Ferraty, F., Hall, P.: An algorithm for nonlinear, nonparametric model choice and prediction. Preprint
19. Ferraty, F., Hall, P., Vieu, P.: Most-predictive design points for functional data predictors. With supplementary material available online. Biometrika **97**, 807–824 (2010)
20. Ferraty, F., Romain, Y. (eds.): Oxford Handbook of Functional Data Analysis. Oxford University Press, Oxford (2011)
21. Ferraty, F., Vieu, P.: Dimension fractale et estimation de la régression dans des espaces vectoriels semi-normés. C. R. Acad. Sci. Paris **336**, 403–406 (2000)
22. Ferraty, F., Vieu, P.: The functional nonparametric model and application to spectrometric data. Comput. Stat. **17**, 545–564 (2002)
23. Ferraty, F., Vieu, P.: Functional nonparametric statistics: a double infinite dimensional framework. In: Akritas, M., Politis, D. (eds.) Recent Advances and Trends in Nonparametric Statistics, pp. 61–78. Elsevier, Amsterdam (2003)
24. Ferraty, F., Vieu, P.: Nonparametric Functional Data Analysis. Springer, New York (2006)
25. Ferraty, F., Vieu, P.: Additive prediction and boosting for functional data. Comput. Stat. Data Anal. **53**, 1400–1413 (2009)
26. Ferraty, F., Vieu, P.: Richesse et complexité des donnés fonctionnelles. Rev. de Modulad **43**, 2–43 (2011)
27. Gonzalez Manteiga, W., Vieu, P.: Methodological Richness of Functional Data Analysis. In Statistical Learning and Data Sciences. Chapman & Hall, London (2011)
28. Härdle, W.: Applied Nonparametric Regression. Econometric Society Monographs, vol. 19. Cambridge University Press, Cambridge (1990)
29. Hastie, T., Tibshirani, R.: Generalized Additive Models. Monographs on Statistics and Applied Probability, vol. 43. Chapman and Hall, London (1990)
30. Huang, J., Horowitz, J., Wei, F.: Variable selection in nonparametric additive models. Ann. Stat. **38**, 2282–2313 (2010)
31. Lian, H.: Functional partial linear model. J. Nonparametr. Stat. **23**(1), 11–128 (2011)
32. Meier, L., van de Geer, S., Bühlmann, P.: High-dimensional additive modeling. Ann. Stat. **37**, 3779–3821 (2009)
33. Meinshausen, N.: Relaxed lasso. Comput. Stat. Data Anal. **52**, 374–393 (2007)
34. Müller, H.-G., Yao, F.: Functional additive models. J. Am. Stat. Assoc. **103**, 1534–1544 (2008)
35. Müller, H.-G., Yao, F.: Additive modelling of functional gradients. Biometrika **97**, 791–805 (2010)
36. Nadaraya, E.: On estimating regression. Theory Probab. Appl. **10**, 186–196 (1964)
37. npfda: R-package on-line (F. Ferraty and P. Vieu). http://www.math.univ-toulouse.fr/staph/npfda/
38. Parzen, E.: On estimation of a probability density function and mode. Ann. Math. Stat. **33**, 1065–1076 (1962)
39. Ramsay, J., Silverman, B.: Functional Data Analysis. Springer, New York (1997)
40. Ramsay, J., Silverman, B.: Applied Functional Data Analysis; Methods and Case Studies. Springer, New York (2002)
41. Ramsay, J., Silverman, B.: Functional Data Analysis, 2nd edn. Springer, New York (2005)
42. Ravikumar, P., Lafferty, J., Liu, H., Wasserman, L.: Sparse additive models. J. R. Stat. Soc. B **71**, 1009–1030 (2009)
43. Roussas, G. (ed.): Nonparametric Functional Estimation and Related Topics. Nato ASI Series, vol. 335. Kluwer, Dordecht (1991)

44. Schimek, M. (ed.): Smoothing and Regression. Approaches, Computation and Application. Wiley Series in Probability and Statistics. Wiley, New York (2000)
45. Sperlich, S., Härdle, W., Aydinh, G.: The Art of Semiparametrics. Contribution to Statistics. Physica-Verlag, Heidelberg (2006)
46. Silverman, B.: Density Estimation for Statistics and Data Analysis. Monographs on Statistics and Applied Probability. Chapman & Hall, London (1986)
47. Stone, C.: Optimal global rates of convergences for nonparametric estimators. Ann. Stat. **10**, 1040–1053 (1982)
48. Stone, C.: Additive regression and other nonparametric models. Ann. Stat. **13**, 689–705 (1985)
49. Tibshirani, R.: Regression analysis and selection via the lasso. J. R. Stat. Soc. B **58**, 267–288 (1996)
50. Tuckey, J.: Non-parametric estimation ii: statistically equivalent blocks and tolerance regions - the continuous case. Ann. Math. Stat. **18**, 529–539 (1947)
51. Watson, G.: Smooth regression analysis. Sankhya Ser. A **26**, 359–372 (1964)
52. Yuan, M., Lin, Y.: Model selection and estimation in regression with grouped variables. J. R. Stat. Soc. B **68**, 49–67 (2006)
53. Zou, H.: The adaptive lasso and its oracle properties. J. Am. Stat. Assoc. **101** 1418–1429 (2006)
54. Zou, H., Hastie, T.: Regularization and variable selection via the elastic net. J. R. Stat. Soc. B **67**, 301–320 (2005)

Printed in the United States
by Baker & Taylor Publisher Services

Printed in the United States
By Bookmasters